Cable Construction and Shielding Termination Techniques for Control of EM Interference

Other related titles:

You may also like

- SBEW570 | Felix Vega, Nicolas Mora, Edl Schamiloglu, and Chaouki Kasmi | Recent Advances in High-Power Electromagnetics | 2024
- SBEW571 | Ian B. Darney and Chris M. Hewitt | Darney's Circuit Theory and Modelling: Updated and extended for EMC/EMI | 2023
- SBEW552 | Sourajeet Roy | Uncertainty Quantification of Electromagnetic Devices, Circuits, and Systems | 2021
- SBEW544 | Guillaume Andrieu | Electromagnetic Reverberation Chambers: Recent advances and innovative applications | 2020
- SBEW519 | Christophe Bourlier and Hongkun Li | Shadowing Function from Randomly Rough Surfaces: Derivation and applications | 2019

We also publish a wide range of books on the following topics:
Computing and Networks
Control, Robotics and Sensors
Electrical Regulations
Electromagnetics and Radar
Energy Engineering
Healthcare Technologies
History and Management of Technology
IET Codes and Guidance
Materials, Circuits and Devices
Model Forms
Nanomaterials and Nanotechnologies
Optics, Photonics and Lasers
Production, Design and Manufacturing
Security
Telecommunications
Transportation

All books are available in print via https://shop.theiet.org or as eBooks via our Digital Library https://digital-library.theiet.org.

IET ELECTROMAGNETIC WAVES 547

Cable Construction and Shielding Termination Techniques for Control of EM Interference

Robert (Bob) Scully

The Institution of Engineering and Technology

About the IET

This book is published by the Institution of Engineering and Technology (The IET).

We inspire, inform and influence the global engineering community to engineer a better world. As a diverse home across engineering and technology, we share knowledge that helps make better sense of the world, to accelerate innovation and solve the global challenges that matter.

The IET is a not-for-profit organisation. The surplus we make from our books is used to support activities and products for the engineering community and promote the positive role of science, engineering and technology in the world. This includes education resources and outreach, scholarships and awards, events and courses, publications, professional development and mentoring, and advocacy to governments.

To discover more about the IET, please visit https://www.theiet.org/.

About IET books

The IET publishes books across many engineering and technology disciplines. Our authors and editors offer fresh perspectives from universities and industry. Within our subject areas, we have several book series steered by editorial boards made up of leading subject experts.

We peer review each book at the proposal stage to ensure the quality and relevance of our publications.

Get involved

If you are interested in becoming an author, editor, series advisor, or peer reviewer, please visit https://www.theiet.org/publishing/publishing-with-iet-books/ or contact author_support@theiet.org.

Discovering our electronic content

All of our books are available online via the IET's Digital Library. Our Digital Library is the home of technical documents, eBooks, conference publications, real-life case studies and journal articles. To find out more, please visit https://digital-library.theiet.org.

In collaboration with the United Nations and the International Publishers Association, the IET is a Signatory member of the SDG Publishers Compact. The Compact aims to accelerate progress to achieve the Sustainable Development Goals (SDGs) by 2030. Signatories aspire to develop sustainable practices and act as champions of the SDGs during the Decade of Action (2020–30), publishing books and journals that will help inform, develop, and inspire action in that direction.

In line with our sustainable goals, our UK printing partner has FSC accreditation, which is reducing our environmental impact to the planet. We use a print-on-demand model to further reduce our carbon footprint.

Published by SciTech Publishing an imprint of The Institution of Engineering and Technology, London, United Kingdom

The Institution of Engineering and Technology (the "**Publisher**") is registered as a Charity in England & Wales (no. 211014) and Scotland (no. SC038698).

First published 2025

The Institution of Engineering and Technology
Futures Place,
Kings Way, Stevenage,
Herts, SG1 2UA,
United Kingdom

www.theiet.org

British Library Cataloguing in Publication Data
A catalogue record for this product is available from the British Library

ISBN 978-1-78561-935-9 (hardback)

ISBN 978-1-78561-936-6 (PDF)

Typeset in India by MPS Limited
Cover image: *Derrrek/DigitalVision Vectors via Getty Images*

Contents

Preface

This book aims to describe in a handbook/technical reference fashion, the underlying principles of how and why electrical cables function, why they might or might not require shielding, what sort of shielding might be required, and how that shielding should be terminated. The content is directed towards applications in the areas of hardware and system design, integration, and operation, including computer hardware, communications, industrial instrumentation, and automotive and aerospace industries. Although not intended to serve as a dense, academic treatise full of rigorous solutions, basic electromagnetic theory and some specific extensions thereof underpin the entire body of work. Sufficient references are provided to enable the interested reader/researcher to pursue further, more in-depth information as desired. The primary thrust here is towards the practitioner who needs a solid reference/guide to identify whether they have a cable problem, to capture the essence of their cable design problem if they indeed have one, and to provide something tangible beyond a simple cookbook approach to how to resolve the problem.

This book would not have been possible without the assistance of my friend and colleague, John McCloskey; the gracious patience of my primary contacts at IET, Olivia Wilkins and Brittany Insull; and the always incredible support of my lifelong partner and spouse, Beth Scully. I wish to thank them all for their encouragement and assistance throughout this journey.

Bob Scully
Salida, Colorado
2025

About the author

Robert (Bob) Scully is an EMC consultant to the NASA Engineering and Safety Center, USA. Bob has a PhD in electrical engineering, is an IEEE fellow, is a HAM radio operator, and was also president of the IEEE EMC Society. Bob was Lead EMC Engineer for 20 years at the Johnson Space Center, supporting several major NASA programs. He also worked briefly at the NASA Jet Propulsion Laboratory as a principal EMC engineer.

Chapter 1
Introduction

Contemporary electronic hardware designers suffer relentless pressure to reduce product size, mass, and time to production. This pressure in general drives electronic designs to their simplest forms, sufficient to their intended task, but with few additional design features such as cable shielding to protect from the intrusion or generation of electromagnetic interference (EMI). Many electrical design engineers who have graduated in recent decades with a bachelor's degree have never been exposed to basic electromagnetic theory except perhaps through a one-semester physics class. Engineers who fall into this category tend to lack fundamental and practical understanding of how time-varying or transient currents on externally connected cables can cause radiated emissions, or how time-varying electromagnetic fields can couple harmful or undesirable energy onto externally connected cabling.

For those engineers who develop consumer-grade electronic hardware, or who have the great fortune to develop hardware that operates in a benign and electromagnetically clean environment, the need for such an in-depth understanding may be unnecessary. The addition of protective measures such as cable shielding may well be of little importance to this sort of hardware. But for those working in many other areas of hardware and system design, integration, and operation, including computer hardware, communications, industrial instrumentation, and automotive and aerospace industries, a plethora of EMI requirements exist and are routinely levied on such hardware. A fundamental understanding of the physics of how a time-varying or transient current produces radiated emissions, or the physics of field-to-wire coupling leading to undesirable field-to-wire coupling, can make the difference between a successful and robust design and one that just barely passes its requirements, if at all.

For this latter and perhaps larger class of hardware, proper cable construction and installation are critical aspects of design for a system to be successfully deployed and operated over the expected lifetime of the product. These aspects all too often form the weakest link in the overall system, precisely because many electrical and system design engineers have been driven to minimize their designs and because they lack understanding or appreciation of how the cable designs themselves need to function electromagnetically.

Wiring used for equipment interconnections is typically not well understood by many practising electrical or electronics engineers, who often consider it simply as a convenient means of transferring electrical signals from one point to another within a circuit or between components in a subsystem, exhibiting minimal impedance and near infinite bandwidth. Indeed, the "characteristics of wire" as a topic is not generally taught as part of a standard university-level electrical engineering curriculum. Students learn about topics such as a single wire above an infinite ground plane, infinite two-wire transmission lines with conductor separation large enough to dispense with any concerns over proximity effects or altered dielectric constants, and basic coaxial cable design. These are of course fundamental and form part of the overall knowledge base needed to be successful after graduation. But very few courses deal with the subtleties and design constraints associated with non-shielded or shielded twisted-pair wiring that is ubiquitous in the industry. Radio frequency (RF) engineers working with higher frequencies for wireless networks and printed circuit board designers faced with increasingly faster clock and circuit switching speeds are now being more exposed to advanced transmission line theory, but these engineers are more often dealing with microwave link coverage and margin concerns or circuitry design issues, as opposed to interconnecting wires that join everything together at the subsystem and system levels. Computer equipment and network engineers are faced with steadily increasing speeds over copper interconnects, and the industry seems rife with misunderstanding regarding how the interconnecting wire works or the role it plays in the control and mitigation of EMI. Added to this background, the concept and application of cable shielding often serves to simply complicate the conversation, with the result that a given cable design may well be inadequate or worse, contribute to EMI problems when contrasted against the needs of the installation.

Many cable design problems can be traced to a failure to understand how the current divides and flows along the wiring and the shielding. The most common error seems to be a lack of understanding of the impact of skin depth and proximity effects, with many engineers thinking that the current flows inside the conductor rather than along the surface of the conductor. In fact, the diffusion of electrons through copper is a relatively very slow process. For example, at 50 Hz, the penetration velocity of surface electrons is on the order of 8 m/s [1,2]. The reality is the current flows along the surface of the conductor, supported by displacement currents in the surrounding dielectric that indeed flow at a velocity near that of the speed of light. Stated explicitly, all current changes move into conductors from the outside surface and diffuse into no more than one half of the thickness of the conductor. If not for displacement currents in the surrounding dielectric, energy transmission with any degree of efficiency other than for relatively steady direct currents would be next to impossible to achieve. Hence, there is a need for not only a solid understanding of the electromagnetic characteristics of the conductors themselves but an equally solid understanding for the electromagnetic characteristics of the dielectric materials typically available for conductors.

In a typical shielded twisted-pair cable, we have a pair of conductors with a dielectric insulation covering them. We then have a layer of shielding of some sort, which itself may or may not have a jacket covering. An alternating current flowing along the conductors inside the shielding will radiate because of electron acceleration. The radiated fields from these currents will interact with the layer of shielding, creating induced current and voltage. Via aperture and diffusion coupling, these currents and voltages will appear on the outside of the shield, hopefully attenuated to the point that no interference with other circuits takes place. The topology of multiple shielded twisted pairs all combined and encapsulated by a single larger overshield will then create a complex field structure inside the overshield that will not only appear as potential interference to other shielded twisted pairs that share the same volume but can and will penetrate via the same aperture and diffusion processes and appear on the outside of the overshield. The part of the design wherein most engineers encounter difficulty is how all these individual shields should be terminated in such manner as to mitigate interference from propagating to other circuits. Stated differently, the goal is to manage the various currents flowing in the cable to successfully prevent undesirable interference from either leaking out of the overshield or penetrating the electronics into which the cable passes. Hopefully, this text will aid in the understanding of this last consideration.

Chapter 1 is the introduction and provides historical context and basic motivation for the ensuing material to follow. Chapter 2 discusses the physical characteristics and suitability of various types of materials commonly used for wire and cable to interconnect modern electrical and electronic systems. Topics include discussion of pure, alloyed, and plated or coated materials used as conductors, as well as discussion of several different types of dielectric materials commonly used as wire insulation and cable jacketing. Chapter 3 provides basic electromagnetic theory necessary to understand the material presented throughout the remainder of the text. Topics include Maxwell's equations, development of the wave equation, boundary conditions, and wave propagation. Chapter 4 addresses the electromagnetics of wire structures. Topics include the development of relationships in cylindrical coordinates and the behaviour of wire structures as a function of frequency, including a discussion of the distributed parameters of resistance, inductance, conductance, and capacitance. Transmission line behaviour and characteristics, including crosstalk, the skin effect, and the proximity effect, are discussed. Chapter 5 provides fundamental theory for wire antennas, with emphasis on electric and magnetic dipole configurations. Definitions of the near and far fields of electric and magnetic dipole structures are included, as these are critical to understanding how wire structures work. Chapter 6 discusses basic shielding theory, building on material from previous chapters and addressing wave incidence on and general interaction with planar surfaces, with references to Schelkunoff's theory and the circuit theory of shielding. Chapter 7 builds on information provided in the previous chapters and includes discussions of design characteristics of braided and solid shields, shield transfer impedance and its measurement, the relationship of transfer impedance to shielding effectivity, and the impact of the use of various types of materials in shielding applications. This chapter ties the rest together and addresses various aspects of cable shielding regarding circuit types, functions, and configurations. The chapter includes detailed discussions of why, how, and when inner and outer shields should be connected to common reference planes and/or circuit commons, and the beneficial and detrimental aspects of the various options.

1.1 A bit of history

Metal formed into wire has been used since biblical times, according to references contained in the Bible, Exodus XXXIX. Originally, wire was most likely formed by beating out flat, narrow strips of metal and then shaping them into a cylindrical shape, perhaps by filing the edges. In some cases, flat strips of noble metals might be cut into very fine wires, or threads, used for weaving. The process of forming wire by drawing metal through successively smaller circular holes in a hardened metal plate, as is utilized today, is believed to have originated in the fourteenth century with Rudolph of Nurnberg. In 1570, Anthony Fournier introduced several refinements to the art of forming wire through the drawing process and was able to produce exceedingly thin wire.

In 1592, Frederick Hegelsheimer of Nurnberg was granted a patent from the Nurnberg magistrates for the production of copper wire coated with gold or silver [3]. Sheffield Plated wire, made in England, was originally formed by the cohesion of thin sheets of silver to previously drawn copper wire using an application of heat and burnishing tools. In 1768, a plater in Birmingham named George Whateley filed two patents for making plated drawn wire. Whateley's patents addressed silver plating over copper wire, gold plating over silver wire, and gold plating over silver-plated copper [4].

1.1.1 Forerunners of modern wiring

The earliest use of a recognizable electrical cable was probably early commercialized telegraph lines strung on wooden poles between transmission and reception stations. These lines were configured as single wires made of iron or steel, with the earth forming the return side of the circuit, a technique referred to as single grounded. Copper was known to be a superior conductor and was indeed tried in several early lines, but the tensile strength of copper led to multiple failures, and the engineers fell back on iron and steel. Problems still haunted the construction of these lines related to the manufacturing of the iron and steel conductors. Using available techniques, the metals intended for wire applications were drawn through successively smaller dies. The dies themselves were subject to wear and tear from the drawing process, leading to wire breakage and insufficient reduction of the metal, necessitating frequent replacement and stoppage of the production process during die replacement. The process was eventually improved by lubricating the iron surface of the wires with a thin coating of copper sulphate. Iron conductors formed using this process performed much better than those lacking the coating, and it soon became evident the superior conductive property of the copper was responsible. However, manufacturing wires from copper alone continued to elude the engineers because of its low tensile strength.

Single-grounded telegraph circuits were inherently susceptible to any electrical interference in near proximity to any portion of the lines, but the intentional patterns of code systems, such as Morse code, were easily discernible in the presence of random interference. Like the earlier telegraph lines, telephone lines that began appearing in the late 1800s were located overhead on wooden poles or attached to racks on rooftops. Unlike telegraph systems, telephone systems did not benefit from the use of patterns. Electrical interference could easily introduce distortion and undesirable levels of static and random noise, making voice communications difficult and at times impossible. Phosphor bronze wires and steel wires with a copper coating were made in attempts to decrease the noise in the lines, to no avail. Solid copper conductors lacked the necessary tensile strength for overhead line applications. Then in 1877, Thomas Doolittle developed a process in which soft, annealed copper wire was drawn through a series of dies to harden the wire and increase its tensile strength, thus facilitating its use in overhead communications lines. Although copper wire with demonstrated greater tensile strength enabled the replacement of iron and steel wiring, the use of these stronger copper conductors did not eliminate inherent electrical noise problems.

The metallic, or two-wire, circuit was developed and patented by Alexander Graham Bell in 1881 [5]. This configuration eliminated the return circuit via the earth and introduced twisting of the wires to mitigate the effects of inductance along the length of the line. Various other techniques were employed to eliminate electrical interference on the lines. Conducting wires were wrapped in lead or covered with tin foil. Despite these efforts, poor voice transmission persisted over existing telephone cables.

A technical specification for a standard twisted-pair telephone cable was finally released by Western Electric in 1888. The specification combined the content of several patents issued to William R. Patterson between 1884 and 1888 that stipulated the use of 18 Brown & Sharpe (B&S) gauge copper conducting wire covered with a minimum of two layers of cotton and sheathed in a 97% lead, 3% tin alloy pipe. Cables fabricated in accordance with this specification enabled a two-inch diameter cable to contain as many as 52 circuit pairs. The second improvement, again traceable to work by William R. Patterson, was the development of paper-insulated dry-core cable in which lead–tin alloy sheathing provided adequate water proofing. By 1891, paper-insulated dry-core cable was the standard. The size of the conductor and the electrostatic capacity requirement were further reduced and greatly decreased the noise in the lines.

As the industry was closing in on a suitable means of producing functional and reliable metal conductors and cable designs for telephone communications cables, the advent of wireless radio communications introduced previously unexperienced electromagnetic interference into the existing network of telephone wires. This interference was random but could be quite irritating. Telephone systems began incorporating filtering to eliminate the noise. Still, in certain installations, the filtering was simply not sufficient. Shielding of the cable was determined to be the more efficient approach rather than adding more filtering. Wrapping the cable with grounding wires, also known as a served shield, worked for fixed installations. However, if the cable was subjected to flexing, the wrapped wires in the shield tended to separate, and interference could return. Telephone engineers discovered that tinsel wire could be wrapped or braided around the conductors in telephone earpiece cords, forming a very flexible cable that reduced radio interference. When the cable was not subjected to the high flexing conditions of a telephone cord, tinsel wire could be replaced by simple strands of wire.

1.1.2 Modern wiring

The technology of wire manufacturing has greatly advanced over the decades since the late 1800s. The problems faced by the early telephone system engineers have not changed, however. Longitudinal balance to offset capacitive and inductive loading remains an important aspect of long-distance communications cables. The use of twisted pairs has become ubiquitous, notably in the design and manufacture of hardware used in the computer, communications, industrial instrumentation, and automotive and aerospace industries. Shielded cables are not as commonplace, but their use is widespread in applications where control of radiated emissions and protection from strong radiated fields are paramount.

References

[1] J. Edwards and T.K. Saha, "Diffusion of Current into Conductors," Australasian Universities Power Engineering Conference (AUPEC), Perth, Australia, Sept 23–26, 2001, pp. 401–406 [Online]. https://espace.library.uq.edu.au/view/UQ:9792 (accessed April 19, 2023).

[2] J. Edwards and T.K. Saha, "Establishment of Current in Conductors," Australasian Universities Power Engineering Conference (AUPEC), Christchurch, New Zealand, Sept 28–Oct 1, 2003, pp. 385–390 [Online]. https://espace.library.uq.edu.au/view/UQ:9820 (accessed April 19, 2023).

[3] The Saturday Magazine, Volume 18, January to June, 1841, published by John William Parker, West Strand, London, pp. 142–144.

[4] H.N. Veitch, *Sheffield Plate, Its History, Manufacture, and Art*, George Bell & Sons, London, 1908, pp. 64–71.

[5] A. Graham Bell *Telephone-Circuit*. US Patent No. 244,426. Patented July 19, 1881.

Chapter 2

Materials used for electrical wire and cable

A wide variety of materials have been and still are used for electrical wire and cable, which can be categorized in accordance with their purpose and usage. Categories include conductive materials, insulating materials, protective layer materials, shielding materials, filling materials, and others.

2.1 Conductors

In general, materials used in wires and cables as conductors are metals that include iron, steel, stainless steel, copper, brass, bronze, tin, aluminium, silver, gold, platinum, phosphor bronze, nichrome, constantan, manganin, chromel, and alumel. Insulating materials include polyvinyl chloride (PVC), low-density and high density polyethylene (LDPE and HDPE), polytetrafluoroethylene (PTFE) and fluorinated ethylene propylene (FEP) (both sold as Teflon by Chemours), ethylene tetrafluoroethylene (ETFE, also known as Tefzel), polyimide film (also known as Kapton), polyurethane (PUR), various rubbers, enamel coatings, paper products, and gutta percha. Some of these materials are no longer in use, such as gutta percha, having been replaced by newer, better-performing materials. Others are used for specific applications, such as chromel and alumel, employed in thermocouple applications. Protective layer materials include many of the same materials as are used for insulation. Shielding materials are usually metals but may also include metallized Mylar films. Filling materials range from nylon cord to rigid plastics used to maintain separation between conductors and paper products.

Iron, steel, and most stainless steels are poor electrical conductors compared to silver and copper and are seldom used today for applications as primary electrical conductors. However, the permeability of iron and steel makes these materials highly useful for magnetic applications, such as cores for motors, transformers, generators, actuators, and various types of magnets. Powdered iron is used in recording media. Ferrites made from mixes of iron and various metallic oxides are extensively used for chokes and filter components, both for cables and for printed circuit board applications. Steel is widely used in the electrical power industry as either the core or an overwrapping of aluminium cross-country high-tension lines to provide high tensile strength for long suspended runs and to mitigate potential damage from high wind loading. Steel is also used for catenary wires for lightning protection and in structural applications for antennas. Several types of stainless steel are magnetic as well, including ferritic, martensitic, and so-called duplex stainless steels, and these find uses similar to those enumerated for iron and steel. Austenitic stainless steels in the 304-alloy class are typically non-magnetic but can be made partially magnetic by the application of mechanical stress, also known as cold working, or by thermal stress.

In the early nineteenth century, when the electrical telegraph was invented, copper wire was initially selected for the lines but the copper then available proved to be mechanically too weak for the application. Iron wire was used to repair existing lines after winter storms [1], and within a short time was the material of choice because of its tensile strength. Steel wire became available soon thereafter and was also used for a brief time for telegraph lines. A method for producing hard-drawn copper wire with tensile strength superior to the earlier form of copper was discovered in 1877 [1]. This stronger copper, having much greater conductivity, rapidly replaced iron and steel as primary conductors, and has been used for general-purpose electrical wiring and cabling since. Nearly pure copper and the more than 100 different copper alloys, including brasses and bronzes, constitute one of the major groups of commercial metals that are industrially produced today. They are all relatively good conductors of electricity and heat and are used for these purposes more often than any other metal. Because alloying invariably decreases electrical conductivity and to a lesser extent thermal conductivity, nearly pure copper, such as oxygen-free copper and electrolytic tough pitch copper, and alloys with a high percentage of copper are used extensively for cables and wires, electrical contacts, and a wide variety of other parts that are required to pass electrical current.

Bare, virgin copper is relatively free of corrosion effects, tending to form a combined thin layer of copper oxides and copper sulphides that act to retard further corrosion. Unfortunately, the oxides are semiconductors, and the sulphides are insulative, and if allowed to penetrate interconnections or wick into stranded conductors, they can measurably impact electrical and thermal performance. The presence of oxides or sulphides on the surface of exposed copper conductors will increase the impedance of the conductors in direct proportion to the thickness of the oxide or sulphide layer. If the conductors are inside an insulation layer and water can infiltrate inside the insulation, oxidation or sulphidation may proceed. This can lead to a slow reduction in the velocity of propagation along the conductor and the loss of conductive metal, and if left unchecked, it can effectively negate the radio frequency (RF) performance of the wiring. The formation of copper oxides and sulphides is accelerated by temperature and can occur as part of the ageing process of wire insulation as the materials outgas over time. Sulphidation of bare copper wire insulated with vulcanized rubber was a major problem for older house wiring installations and prompted the use of tin plating as a means of control. Exposure to a saltwater environment can exacerbate galvanic corrosion between bare copper and a variety of other materials, such as nickel, steel, and aluminium, all common engineering materials found in electrical, electronic, and aerospace applications.

Copper and copper alloys exhibit good mechanical ductility and flexibility, are readily able to be formed into bars or straps, and are well suited to the drawing process in the fabrication of electrical wiring. They exhibit outstanding resistance to corrosion, good fatigue resistance, and are generally non-magnetic or diamagnetic. They can be readily soldered and brazed, and many can be welded by various gas, arc, and resistance methods [2].

Tin is seldom used by itself. An exception to this is the use of tin wire in electrical fuses. It is paramagnetic, has a hardness slightly greater than lead, and has an electrical conductivity roughly 14% of that of sliver. It readily alloys with almost all metals, and is generally used as a coating or plating, or a constituent component of an alloy product [3]. Tin/lead electroplates find excellent usage for the plating of printed circuit boards and various electronic components. Eutectic grade solder is composed of 63% lead and 37% tin and exhibits the lowest melting point of soft solders used in electrical and electronic assembly and repair activities.

A hard, brittle oxide may form on the surface of exposed tin. This oxide is stable and helps to prevent further corrosion of the base metal. This oxide is an electrical insulator and must be disrupted to provide good electrical continuity through an electrical interface between mating tin surfaces, between tin surfaces, and the surfaces of other materials. With sufficient force normal to the mating surfaces, the oxide film can be broken through, which may allow for good electrical contact. This is related to the fact the underlying tin is much softer and more ductile than the tin oxide on the surface and will tend to deform slightly under pressure, facilitating the oxide breakthrough [4].

In addition to the oxide formation on the exposed tin surface, tin readily forms intermetallic compounds with underlying metals such as copper when used as a plating material [5]. Given sufficient time, the intermetallic formation process will completely convert the layer of tin plating to a solid layer of copper-tin intermetallic, resulting in a surface that is hard, brittle, easily oxidized, and very difficult to solder. Copper-tin intermetallic compounds cannot be easily removed without mechanical removal of the surface of the wire.

Another consideration for the use of tin is the possible formation of tin whiskers. Tin whiskers are electrically conductive, crystalline structures of tin that sometimes grow from surfaces where tin (especially electroplated tin) is used as a final finish. Tin whiskers have been observed to grow to lengths of several mm and in rare instances to lengths more than 10 mm. Numerous electronic system failures have been attributed to short circuits caused by tin whiskers that bridge closely spaced circuit elements maintained at different electrical potentials [6]. No consensus opinion has been reached regarding the formation of tin whiskers, although the most commonly held opinion indicates that conditions of high mechanical stress are conducive to whisker formation and growth [7].

Although aluminium is less conductive than copper, it has been used in many electrical applications, including house wiring and power bus systems in commercial airliners and cross-country high-tension power line applications wherein mass and cost are of paramount concern. Aluminium is abundantly available and is a less expensive alternative to copper for use as electrical conductors. In comparing conductors of the same size, one of copper and one of aluminium, the aluminium conductor exhibits roughly 60% of the conductivity of the copper conductor. Thus, an aluminium conductor requires a larger cross-sectional area than a copper conductor to support the same current magnitude. The larger size conductor requires a proportionately greater amount of insulation material compared to the copper conductor. This in turn can result in a greater volume requirement in certain applications. But the aluminium conductor has roughly one third of the mass by volume of the copper conductor, a characteristic that can make the aluminium conductor the preferred solution for mass-sensitive applications despite the volume increase. Copper is more thermally conductive than aluminium, but when conductor sizes are factored in, the difference between the two materials is for the most part mitigated. Aluminium has less tensile strength than copper, but this is only a concern if the wiring needs to traverse long unsupported lengths such as in cross-country power distribution systems. In these applications, it is common to find a steel core or steel strapping tape wound over a core comprising several individual aluminium conductors. Copper-plated aluminium conductors have found application in coaxial cable manufacture. Pure aluminium is paramagnetic. Metallic aluminium is very reactive with atmospheric oxygen, and a thin passivation layer of aluminium oxide (≤ 4 nm thickness) forms on any exposed aluminium surface in a matter of hundreds of picoseconds [8]. This layer protects the metal from further oxidation.

Aluminium was first used for utility distribution applications in the early 1900s. Use of aluminium expanded greatly after World War II and quickly replaced copper as the conductor of choice in utility grids because of its conductivity-to-mass ratio and significant cost and availability advantages. Use in homes became widespread in the 1960s through the late 1970s for similar reasons, but wiring devices such as circuit breakers and convenience outlets were not designed to accommodate the use of the softer and more ductile aluminium, resulting in improper installation, faulty interconnections, and dissimilar metal issues, including different electrical resistances and coefficients of thermal expansion.

Silver has long been valued as a precious metal and has important and broad application in technological and electronic applications. Silver is found in cell phones, computers, water-purification systems, and spacecraft solar radiation tiles. Silver is strongly diamagnetic, weakly paramagnetic, and roughly 6% more conductive than an amount of copper of the same dimensions. Silver is significantly more rare and thus comparatively more expensive than copper. Solid silver wire is generally reserved for sensitive, high-reliability systems and speciality electronics where high conductivity over a small distance is prioritized. Silver coating on wires and cables offers flexibility, improved solderability, and a possible enhancement of electrical conductivity at frequencies greater than 100 MHz. Silver is subject to corrode in areas of high moisture or where various sulphides are present. Detailed discussion of silver corrosion behaviour is provided in Section 2.4.2.

Gold is slightly less conductive than copper, and platinum is less conductive than iron. Gold is strongly diamagnetic, weakly paramagnetic, and roughly 78% as conductive as an amount of copper of the same dimensions. Because of cost considerations, the use of gold as electrical conductors is limited primarily to precision engineering and scientific instrumentation. Gold is commonly used as a coating over less expensive metal. Gold coating on wires and cables offers durability and flexibility, making this metal a key choice in electrical wires and cables where high reliability is a paramount concern. Gold is readily solderable and does not chemically react with the environment even at high temperatures. Gold exhibits excellent corrosion resistance, and when alloyed with small amounts of nickel or cobalt, it exhibits good wear resistance. Gold can be fabricated as extremely fine wire for use in precision scientific equipment and in small semiconductors with complex design structures. Gold has been used in home electronic wiring because of its longevity and efficiency.

Platinum is paramagnetic and is valuable as wiring material in high-temperature environments and complex configurations. Platinum's conductivity is much lower, being roughly 16% as conductive as an amount of copper of the same dimensions. Like gold, platinum wire can be fabricated as extremely fine wire. It is highly corrosion and oxidation resistant and withstands most acids. Platinum's excellent malleability and ductility make it ideal for complex or precise wire fabrications. Aerospace and military manufacturers use platinum wiring for monitoring equipment and instrumentation and thermocouples. Thermocouples made with platinum are preferred for measuring temperatures in environments with temperatures as high as 1650°C. Other uses for platinum in aerospace applications include aircraft engine wiring harnesses, heat exchangers, and infrared detectors.

Phosphor bronzes, also sometimes referred to as tin bronzes, are alloys containing Cu, Sn, and P. Phosphor bronze is non-magnetic and was first produced by Kunzel and Montefiore between 1868 and 1873 [9]. In 1897 Montefiore and Manne conducted numerous experiments with a view to replacing iron and steel overhead wires, which had many disadvantages, with copper wires. The discovery of phosphor-bronze wire and its refinement helped greatly to solve the iron and steel overhead wire problem [10].

Phosphor bronzes contain 0.5% to 11% Sn and 0.01% to 0.35% P, with traces of Fe, Zn, and Pb with the remainder being Cu. The most widely used phosphor bronze alloy is UNS C51000, also known as CDA510. The introduction of Sn produces a significant increase in the strength and corrosion resistance, while the P increases stiffness and wear resistance. The combination of fair electrical conductivity and low thermal conductivity offered by phosphor bronze allows the making of electrical interconnections for devices at ultra-low temperatures without conducting heat that can otherwise create problems for sensors and devices designed to operate at cryogenic temperatures.

Over the span of several years, the alloy colloquially known as Nichrome was co-invented by Albert Leroy Marsh and William Hoskins in 1905 [11] to create a metallic alloy from which a high-resistance wire could be made that could be used as a durable and safe heating element. When perfected, the new alloy was 300 times stronger than other types of resistance wire available at that time. The material was awarded US Patent # US811859A in Feb 1906 under the title of Electric Resistance Element, marketed at present as nichrome. For this invention, Marsh was acclaimed as the "father of the electrical heating industry".

Nickel–chromium alloys [12] are prized for their high corrosion resistance, high-temperature strength, and high electrical resistance. For example, the alloy

NiCr 70/30, also designated as Ni70Cr30, Nikrothal 70, Resistohm 70, and X30H70, has a melting point of 1380°C and an electrical resistivity of 1.18 μΩ-m. Heating elements in toasters and other electrical resistance heaters most frequently employ wires fabricated from nickel–chromium alloys. When produced in wire form, they are known as Nichrome wire.

Constantan is a diamagnetic Cu–Ni alloy discovered by Edward Weston in the late nineteenth century [13]. Constantan typically comprises 55% Cu and 45% Ni. Constantan is widely used for strain gauges, electrical resistance heating, and thermocouple applications. It exhibits a very small temperature coefficient of resistance and a nearly constant low thermal variation of resistivity over a wide range of temperatures up to about 65°C (149°F). Above this temperature, constantan exhibits a continuous drift in resistivity that can negatively impact applications requiring stability over an extended period of time. Constantan is characterized by good fatigue life and relatively high elongation capability and is used to form thermocouples with wires made of Fe, Cu, or chromel.

Manganin is the trademark name for a diamagnetic alloy typically comprising 84.2% Cu, 12.1% Mn, and 3.7% Ni. It was first developed by Edward Weston in the late nineteenth century as the product of additional research after his discovery of constantan [13]. Manganin foil and wire are used in the manufacture of resistors because of manganin's extremely low temperature coefficient of resistance and long-term stability. This same characteristic enables manganin wire to be used as an electrical conductor in cryogenic systems, minimizing heat transfer between sensitive electronic hardware.

Chromel is a diamagnetic alloy made of approximately 90% Ni and 10% Cr by weight that is used to make the positive conductors of ANSI Type E and K thermocouples. It can be used at temperatures up to 1100°C (2010°F) in oxidizing atmospheres. Two other alloys of chromel are both used in various heating applications. Chromel A is an alloy containing approximately 80% Ni and 20% Cr, with additional alloying additives of 1% Si and 0.5% Fe. It is used for its excellent resistance to high-temperature corrosion and oxidation. This alloy is more commonly referred to as Nichrome 80-20 and is used for electric heating elements. Chromel C is an alloy containing 60% Ni, 16% Cr, and 24% Fe. This alloy is more commonly referred to as Nichrome 60 and like Nichrome 80-20 is used for heating elements, resistance windings, and hot wire cutters. Chromel is a registered trademark of Concept Alloys, Inc.

Alumel is a paramagnetic alloy consisting of approximately 95% Ni, 2% Al, 2% Mn, and 1% Si. Alumel is used to make the negative conductors of ANSI Type K thermocouples and thermocouple extension wire. Alumel is a registered trademark of Concept Alloys, Inc.

Table 2.1 illustrates the relative resistivities and conductivities of selected metals used for electrical wires and cables. As might be surmised by the astute reader, a great deal of information is available for the broader scope of wire and cable as a general topic. This text will, as indicated in previous chapters, narrow its focus on wire and cable designs used in areas of hardware and system design, integration, and operation that are dominant within industries including computer

Table 2.1 Relative electrical resistivities and conductivities of selected metals at 293.15 K (20°C)

Material	Resistivity (Ω-m)	Conductivity (S/m)
Silver, pure (Ag)	1.59×10^{-8}	6.29×10^{7}
Copper, pure (Cu)	1.68×10^{-8}	5.95×10^{7}
Copper, annealed (IACS)	1.7241×10^{-8}	5.80×10^{7}
Gold, pure (Au)	2.21×10^{-8}	4.52×10^{7}
Aluminium, pure (Al)	2.65×10^{-8}	3.77×10^{7}
Iron, pure (Fe)	9.61×10^{-8}	1.04×10^{7}
Platinum, pure (Pt)	1.06×10^{-7}	9.43×10^{6}
Palladium, pure (Pd)	1.05×10^{-7}	9.26×10^{6}
Tin, pure (Sn)	1.10×10^{-7}	9.09×10^{6}
Phosphor Bronze (CDA510)	1.15×10^{-7}	8.70×10^{6}
Chromium (Cr)	1.25×10^{-7}	7.74×10^{6}
Steel (cast)	1.61×10^{-7}	6.21×10^{6}
Alumel	2.94×10^{-7}	3.40×10^{6}
Manganin	4.55×10^{-7}	2.20×10^{6}
Constantan	4.94×10^{-7}	2.02×10^{6}
Stainless Steel, 301	6.94×10^{-7}	1.45×10^{6}
Chromel	7.06×10^{-7}	1.42×10^{6}
Nichrome	1.08×10^{-6}	0.93×10^{6}

hardware, communications, industrial instrumentation, and automotive and aerospace industries. Almost without exception, in the industries cited, plated or coated copper wire is employed for many reasons. The next sections discuss these reasons in more detail.

2.2 General considerations for plating of electrical conductors

While it is true that many applications exist wherein "pure" metals are used as electrical conductors, it is very common to find the preferred configuration is a "pure" conductor that has been plated by one or more layers of different metals. Electrical conductors have been plated to prevent insulation from adhering to the primary conductor, improve solderability, reduce corrosion, and improve electrical properties at high frequencies. Aside from these benefits, porosity of the plating itself has always been a serious problem, leading to the increase of attenuation caused by rough surface finish, as well as penetration of the surface by water and other chemicals such as fuels, lubricants, and cleaning agents that can accelerate the corrosion process. Pores in the plating will exacerbate electrolytic action at the interface between the conductor metal and the plated metal surface, leading to rapid acceleration of corrosion. In general, contacts between metals at the extremes of the electrochemical series are undesirable because of the large potential differences set up between them.

To achieve satisfactory protection from corrosion in harsh environments, stringent specifications must be set up for the electroplating process. Nickel provides high corrosion resistance when carefully prepared. Chromium is tarnish-resistant under ordinary conditions but will fail rapidly in the presence of chlorides. Because chromium plating is also highly porous, it is generally necessary to protect the base metal with nickel. Gold is highly corrosion-resistant but the cost of deposits of suitable thickness prohibits its extensive use. In general, it is difficult to electroplate thick deposits of platinum, and cost is a factor in limiting its use.

Because of their high electrical conductivity, silver and copper are the most desirable plating materials, particularly from the microwave standpoint. Unfortunately, these two metals will corrode or tarnish under exposure to harsh environmental conditions and may profitably be protected with thin coats of low-loss varnish or a thin "flash" plating of a non-corroding metal like palladium. If copper or brass substrate material corrosion products combine with silver sulphide (Ag_2S) tarnish film, conductor resistivity can become elevated, leading to increased conductor losses.

Copper has about the same electrical conductivity as silver but tends to corrode very rapidly under harsh environmental conditions. Success has been achieved in obtaining non-porous silver-plated deposits on copper conductors. These results have been achieved, however, under ideal laboratory conditions. For example, ordinary decorative silver plating requires 15 or more individual steps to produce a smooth, adherent surface. To minimize porosity and maximize electrical conductivity, several additional and very exacting controls must be imposed upon the usual plating process. Not only is it necessary to exercise a very precise and elaborate control over the actual plating process, but it is also essential to select and carefully prepare the metallic surfaces that are to be plated. Heterogeneity in chemical composition and roughness of the base-metal surface are the major factors that contribute to low-quality electroplated deposits.

2.3 A brief history of electroplating

The earliest known examples of the application of thin sheets of gold have been found in burial plots in the cemetery of Varna, dating from between 4550 and 4450 BCE. Early examples of the application of gold foil, mercury and lead gilding, silver, and imitations of silver made from high-tin alloys have been found in Egypt, the Iberian Peninsula, and the Mediterranean [14]. Additional finds dated from the second millennium BCE appear to have employed metal leaf instead of foil to decorate statues. Evidence that ancient Greeks applied metal foil and wire around entire objects has been found dating between the twelfth and fifth centuries BCE. The Romans appear to be the first to use displacement plating to coat objects. In the Early Roman period between the first and fourth centuries CE, artefacts show evidence of electrochemical plating achieved using the natural potential difference between different types of metals. Mercury gilding, invented in Central Asia, was developed around the same time. Pliny the Elder was the first to describe the process in the first century CE. In this process, also called fire gilding, small pieces of gold were mixed with mercury in a one-to-eight ratio, creating a viscous amalgam that could be brushed onto the surface of a substrate. Once complete, the object was then heated until the mercury vaporized, leaving behind the gold plating. This technique was also called the Lost Apprentice Technique because the process created toxic mercury fumes, which, when combined with poor ventilation, resulted in mercury poisoning that destroyed the sanity of and then killed gilders after about four coatings. The mediaeval period between the fifth and fifteenth centuries saw an expansion of techniques developed during the Roman period. In the ninth century, Europeans developed displacement plating techniques for iron armour, which plated the surface of the armour with copper to prepare it for mercury gilding. Damascene was also developed during this period, named after Damascus, Syria. This inlay technique involved cutting a design into a substrate metal, then cutting and hammering a decorative metal into the substrate. False Damascene was also developed, which uses a similar process but involves scoring the substrate metal instead of cutting into it. The Renaissance period from the fifteenth to the seventeenth century saw further development of displacement plating [15].

While previous techniques used displacement plating and set the groundwork for the development of electroplating, the true origin of electroplating came during the Industrial Revolution, which took place between the late eighteenth and early nineteenth centuries. Alessandro Volta invented what was functionally the first electric battery in 1800 [16]. Named after Volta, the voltaic pile stacked galvanic cells to produce an electric current. Volta's invention was lauded by French Emperor Napoleon Bonaparte, who made Volta a count for his work. This key invention was the catalyst needed to take displacement plating to the next level, which occurred shortly after Volta's invention. After Volta invented and published his findings on his electrochemical batteries, many people began to experiment with the technique to discover new applications. It was in the subsequent decades that electroplating was born.

In 1801, British scientist William Cruickshank first reported on electroplating. Cruickshank described his experiment of depositing dendritic metallic lead and copper onto a surface using Volta piles [17,18]. Cruickshank achieved this by attaching silver wires to the top and bottom layers of the Volta pile and placing the ends in an acetate solution, then a copper solution in subsequent experiments. As a result, fine needles of metallic material deposited on the bottom of the Volta pile. However, while Cruickshank's experiment was the first reported incident of electroplating, he is not credited with the invention of electroplating since the technique was not fully explored for another couple of years.

Luigi Brugnatelli, an Italian chemist and personal friend of A. Volta, is most widely credited for inventing electroplating. In 1803 Professor Luigi Brugnatelli, a close friend and colleague of Volta's at the University of Pavia, wrote to his correspondent Professor Van Mons of Brussels describing how he had "recently gilt in a perfect manner two large silver medals, by bringing them into communication, by means of a steel wire, with the negative pole of a Voltaic pile, and keeping them, one after the other, immersed in ammoniuret of gold newly made and well saturated". Brugnatelli later gave details of his electrolyte: "To one part of the saturated solution of gold in nitro-muriatic acid, add six parts of solution of ammonia, by which the solution is decomposed and oxide of gold is precipitated, and a portion is set free, forming ammoniuret of gold" [19]. Van Mons published Brugnatelli's letter in 1803 [20].

Unfortunately, while Napoleon had celebrated Volta's work, the French Emperor disapproved of Brugnatelli's work. In Napoleon's view, this discovery would make gold plating accessible to the lower class, which was unacceptable in his opinion. Napoleon's disapproval subsequently barred Brugnatelli's work from the French Academy of Sciences, preventing it from reaching the wider scientific community [15].

In the 1820s to 1830s, observations made by Michael Faraday led to the establishment of the laws of modern electroplating. Faraday published his results in 1834 [21]. These tenets bear Faraday's name and are still widely used today.

Around 1838, multiple individuals claimed to have invented a process known as electrotyping, using techniques quite similar to those employed by Brugnatelli [22]. Around 1839, British inventor John Wright realized he could use potassium cyanide as an electrolyte for electroplating [23], and in 1840, Wright began working with the Elkington cousins, George and Henry, who bought Wright's patent and received several more for similar electroplating processes using gold and silver. The Elkingtons found commercial success plating silverware and manufacturing decorative products, including the flatware aboard the RMS Titanic when it sank.

Electroplating became enormously popular after it was patented by the Elkingtons, and by the 1850s, Elkington & Co. had experienced widespread success due to the popularity of plating various surfaces with gold, silver, copper, and other metals. With the Industrial Revolution came a new wealthy class, made so by the expansion of factories and the abundance of work opportunities in cities. It became fashionable to show off this wealth by any means necessary, and the ornate styles of the recent Baroque and Rococo periods seemed fitting. Products reminiscent of these stylistic periods could be made more cheaply using manufacturing techniques and electroplating, making them more widely affordable and accessible. While Western Europe's masses clamoured to get plated products, Russia experienced a similar rush for gold electroplating among its religious and economic elite. The Eastern Orthodox Church invested heavily in gilded icon cases, statues, and other decorations for their churches, while the Romanov royal family showed off their enormous wealth with gold-plated items throughout their properties [15].

In 1871, Belgian inventor Zénobe Gramme created a direct current generator that used rotary motion. The Gramme dynamo allowed electroplating to become even more affordable and accessible, with an increasing number of electroplating businesses cropping up across Europe [24]. The late nineteenth century saw a huge expansion of the British Empire across the world, and electroplating quickly followed wherever the Empire went. The spread of British culture, inventions, and fashions also spread the popularity of electroplated items to populations that otherwise would not have used the technology.

While these factors allowed electroplating to flourish very quickly in the nineteenth century, several problems arose that hindered its growth. Most notable among these was the fact that electroplating was considered a trade, not a science. Plating formulas and processes were considered trade secrets and were closely guarded by various industry leaders, preventing knowledge from freely passing between producers. As a result, electroplating processes became increasingly unreliable and irreproducible, leading to a decline of quality and innovation in the electroplating industry. By the beginning of the twentieth century, the electroplating industry had effectively come to a standstill.

In the 1940s, because of its excellent conductivity and corrosion resistance, gold electroplating in electric circuits played a key role in reviving the static electroplating industry. Safer and more effective electroplating methods were created, replacing hazardous cyanide baths with safer acid baths. With these advances, the electroplating industry has grown alongside the continued development of technology, resulting in electroplating opportunities Brugnatelli could never have imagined in 1803.

The production of integrated circuits represented a major turning point in the electroplating industry when IBM started using damascene techniques and electroplating in the production of computer chips in the 1970s. Electroplating is now used widely throughout the electronics industry for protective coatings, contacts, and other applications. Plating onto non-metal substrates has gained a great deal of popularity due to its ability to impart glass, plastic, ceramic, Kevlar, and other materials with the physical and mechanical properties of metals, a key capability in the provision of electromagnetic shielding for cell phones, tablets, laptops, and automotive and aviation electronics to achieve lower mass without sacrificing finish quality.

One major development resulted from research conducted by Abner Brenner and G.E. Riddell, who essentially rediscovered electroless plating in 1944 while researching how to plate the inside of machine gun barrels. Brenner and Riddell experimented with different levels of current, substrates, additives, and other variable factors, eventually discovering nickel could be plated without electricity by using sodium hypophosphite as a source of electrons. Brenner and Riddell presented their findings at the 34th annual American Electroplaters Society (AES) conference in 1946. A second paper was presented a year later, and the work eventually led to the award of two separate patents. During the presentation, the term "electroless" was suggested and is still used today [25]. Electroless plating has since been used to deposit nickel, cobalt, copper, silver, gold, or similar metals onto various substrates for a wide range of applications.

2.4 Plating of copper wire

Modern copper wire used in electrical and electronic applications is plated for many reasons, including the enhancement of solderability, to prevent the adhesion of insulative materials to the metal, and to prevent accelerated oxidation of the copper during high-temperature service, thereby extending the operating temperature range.

However, plating copper wire introduces new complexity with respect to electromagnetic performance. Cylindrical copper conductors, routinely plated with tin, silver, or nickel, are found throughout the electrical, electronic, automotive, and aerospace industries. While each of these plating metals offers unique advantages over bare copper wire, they all interact with copper in a metallurgic sense as well as an electromagnetic sense, resulting in intermetallic growth, metal migration, and potential galvanic corrosion issues. Electromagnetic impacts are discussed at length in Chapter 4.

The plating of copper wire used for electrical and electronic purposes using pure tin, tin–lead alloys, silver, or nickel is accomplished today with precision, in adherence with various specifications [26–29]. Copper wire can be plated by hot dipping in a molten metal bath, electroplating, or cladding. With the advent of continuous processes, electroplating has become the dominant process, especially because it can be completed easily and in sequence, either prior to or following the wire-drawing operation. These processes can be controlled to produce relatively uniform plating thicknesses ranging from 1 micron to more than 50 microns, depending on the application. Note that many other forms of electrical conductors having some form of plating are available and in common use [30–35].

2.4.1 Tin plating of copper wire

Copper wire plated with pure tin, or a tin alloy, is the type of plated wire most commonly found throughout the electrical, electronic, and aerospace industries. Tin plating provides sacrificial protection for copper and arrests corrosion caused by sulphur-bearing reagents. It enhances solderability in newly produced wire and increases the operating temperature range for copper up to 150°C. The use of tin plating does present some drawbacks.

Because tin is between 6 and 7 times less conductive than copper, tin-plated conductors should be limited to low-frequency applications to limit signal attenuation related to skin depth effects. The formation of a tin–copper intermetallic layer over time, as previously discussed, can dramatically increase signal attenuation in aged wiring at moderate to high frequencies. Intermetallic formation can also introduce high resistance in joints found in older house wiring, motivating the replacement of older house wiring during remodelling.

Tin-plated wire cannot be used with common insulation materials that require high-temperature processing, such as PTFE, FEP, ETFE, and polyimide film, more commonly known by the DuPont brand name Kapton. Note all these materials are commonly used as wire jacketing material by the military and commercial aerospace industries. The non-aerospace industry, particularly in computer networking applications, still relies on lower-temperature and less expensive jacketing materials such as PVC. Additional information on insulative materials and dielectrics is provided in Section 1.4.

2.4.2 Silver plating of copper wire

Depending on the applied thickness, silver-plated copper wire may exhibit an enhanced electrical conductivity at frequencies greater than 100 MHz, making it an attractive option for RF applications where skin effect plays a significant role, possibly allowing the use of smaller diameter substrate conductors. It offers good immunity to corrosion under exposure to high processing and operating temperatures up to 200°C. Unlike tin-plated wire, no intermetallics form between the plating and the underlying copper, and so silver-plated wire has excellent solderability, whether new off the shelf or after many years of exposure to uncontrolled environments. Like tin, the use of silver electroplating also presents some drawbacks.

Silver conductors or silver-plated surfaces exposed to the environment are subject to oxidation and corrosion in the form of tarnish films. Like copper oxide, silver oxide (Ag_2O) forms very rapidly on contact with the environment. The oxide is semi-conductive and has lesser conductivity than that of the substrate. It forms a very thin layer on the surface, so it is essentially transparent [36]. Sulphur and chlorine in the environment react readily with the Ag_2O to form Ag_2S and silver chloride (AgCl). Ag_2S tarnish films are semi-conductive at ambient temperatures and are inherently soft but are insoluble in most solvents. AgCl is harder, less prevalent, and electrically resistive. The higher the level of the harder insulative AgCl in the tarnish film relative to the level of the softer semi-conductive Ag_2S, the more insulative is the tarnish. Ag_2S is the major product of silver corrosion in environments containing various sulphuric gases such as hydrogen sulphide (H_2S), carbonyl sulphide (OCS), carbon disulphide (CS_2), and sulphur dioxide (SO_2). Silver reacts with extremely low levels of H_2S in parts per billion and even in parts per trillion. Silver and H_2S react to form dark brown or black Ag_2S.

When there is a thick enough layer of Ag_2S and a sufficiently high temperature and humidity, the process of growing thin filaments (whiskers) begins. Silver whiskers usually grow in areas exposed to H_2S. They grow practically everywhere but more intensely in the areas with higher temperatures, such as bus joints and sliding contacts, outside edges, and corners of the contacts. Temperature gradients in these areas may encourage this phenomenon. The corrosion process creates an increasingly high resistance that produces more heat associated with current flow, which in turn stimulates further tarnishing and whisker growth [37–39]. There is no precise definition of the minimum concentration of H_2S required for tarnishing to be observed on silver – however, the silver is exposed. Sulphuric corrosion for electrical and electronic apparatus may arise from the outgassing of H_2S or other sulphuric compounds from seals, gaskets, and other polymers. The amount of tarnishing is a function of the relative humidity, ambient temperature, gas concentration, and time of exposure. Even minor concentration levels of H_2S will produce Ag_2S on parts exposed to the environment.

Storage or use of silver-plated copper in a moist or high-humidity environment is conducive to corrosion through galvanic action at pinholes or breaks in the silver plating. The presence of moisture in the region of exposed copper produces an electrolytic cell formed between the copper and the silver. Galvanic copper corrosion will progress in the presence of oxygen, producing cuprous oxide and possibly black cupric oxide corrosion products. If not stopped, the corrosion process will continue until all the copper is destroyed, leaving only a shell formed of the silver plating and a core of copper oxide. The cuprous oxide crystals, which have a characteristic red appearance, precipitate out onto the silver plating [40,41]. This type of corrosion is known as red plague and is readily identifiable by the presence of a brown-red deposit on the surface.

Corrosion of silver-plated conductors insulated with PTFE, perfluoroalkoxy (PFA, aka perfluoroalkoxy alkane), and several other similar fluoropolymers has been observed multiple times. During the manufacture of fluoropolymer-insulated silver-coated copper electrical wires and cables, the extrusion of fluorocarbon resin generally occurs at sufficiently high temperatures to enable oxidative degradation in the presence of air and water to produce carbonyl fluoride end groups such as carbonyl difluoride (COF_2), an extremely reactive compound. These end groups remain in the polymer and can later hydrolyse in the presence of moisture to produce carbon dioxide (CO_2) and hydrofluoric acid HF (aq), a colourless and highly corrosive reagent, capable of etching glass and corroding plated surfaces of connectors and other components. The rate of production of HF (aq) as a part of this reaction is limited by the diffusion rate of water into and through the fluoropolymer material and may occur over an extended time duration. When HF (aq) penetrates through the plating of a silver-plated copper conductor, it will react with the copper core and form copper fluoride (CuF_2). As the amount of CuF_2 increases on the copper surface, it will spread and eventually break through the silver plating, exposing the copper, thereby dramatically increasing the risk of the onset of red plague. The presence of a powdery white finish comprising white to bluish-white crystals on a conductor surface that makes the surface non-solderable, now referred to as fluorine attack, or white plague, indicates the presence of CuF_2, and strongly suggests this process is underway.

In wire insulation, this process is sometimes referred to as fluoride off-gassing or fluoride outgassing. While fluorine outgassing is a concern for all fluoropolymer insulations, ETFE and cross-linked ETFE (XL-ETFE) have been reported to have a higher evolution rate, possibly due to the blending and extrusion processes typically used for this polymer.

Antimony oxide (Sb_2O_3) is known for its ability to scavenge HF, protecting wires from the corrosive attack by the HF (aq). Because of this capability, wire manufacturers heavily dope fluoropolymer insulation with Sb_2O_3 to reduce the amount of outgassing. Sb_2O_3 in the insulation will reduce HF outgassing and the resultant corrosion in the short term, but eventually the Sb_2O_3-scavenging capacity will be exhausted, and outgassing and corrosion may proceed unabated. Published incidents of white plague have to date been confined to the satellite manufacturing industry, specifically in instances where wire harnesses or assemblies are stored in sealed bags for months or years before use [42,43].

When silver is plated over copper with some areas of bare copper exposed, the copper may corrode under the plating (underplating corrosion) [44]. Silver-plated surfaces may react with substances such as glycol to produce a potential flammability hazard [45]. Larger gauge wires such as AWG 0 cannot be used with silver plating, as cold welding at points of contact between the wire strands may increase wire rigidity [40].

2.4.3 *Nickel plating of copper wire*

Nickel-plated copper wire is more expensive than tin-plated copper wire and is less expensive than silver-plated copper wire. It offers superior corrosion control and strength when compared to either tin or silver plating. It extends the useful operating temperature range of plated copper wire up to 260°C while still exhibiting good flexibility after long-term ageing at 200°C. An excellent discussion of the merits of nickel plating on copper conductors is available from Sarkuysan Electrolytic Copper Company [46].

Nickel-plated wire is very difficult to solder with non-activated fluxes. Active fluxes typically contain strong acids or other active ingredients that help remove oxidation from metal surfaces. Common active ingredients include rosin, zinc chloride, or other organic acids. In comparison, non-active fluxes generally contain milder agents, such as rosin or synthetic materials, without strong acids. They may include inert substances that do not react with the metal. Even with the use of activated fluxes, nickel-plated wire has a variable solderability, dependent on the type and thickness of the nickel plating. Moreover, the flux may wick under the insulation and promote corrosion long after the soldering process is complete. Otherwise, nickel-plated wire is insusceptible to corrosion over a wide range of applications. Crimpability is acceptable, but ohmic resistance varies with the cleanliness of the surface, which is paramount to long-term success and reliability. Depending on the plating layer thickness, nickel-plated copper conductors may exhibit anywhere from 4% to 30% higher resistivity than IACS copper wire, with a correspondingly higher insertion loss in a given circuit application. Nickel-plated wire may exhibit a magnetic moment that can be detected by an external observer or that can introduce unsuitable perturbations into magnetically sensitive spacecraft experiments. A very interesting application of nickel-plated copper wire that capitalized on the magnetic properties of nickel plating was developed and refined to a high degree of success during the late 1950s and throughout the 1960s [47–51]. Moderate lengths of copper wire plated with a nickel-iron ferrite material were arrayed in a mechanical matrix. These arrays were referred to as plated wire memory cells. Arrays of this ilk were used successfully in a plethora of high-speed memory applications, including the Remington Rand UNIVAC 1110 computer, all UNIVAC 9000 series computers, the Minuteman III missile, the Viking Mars Lander, the LANDSAT-D satellite, and the Hubble Space telescope.

Related to nickel plating of copper wire is the use of ferrite sheathing in the insulation of copper wire to produce a distributed filtering action. In the so-called FilterLine wire [52], controlled by specifications MIL-C-85485 and SAE AS85485, the electrical energy associated with high-speed transients and high-frequency noise components above 100 MHz is trapped in the ferrite layer by an overlying metallic shield and is therein dissipated. Similar commercial applications, made by EMC Eupen [53,54], ARC Technologies [55], and others, also incorporate ferrite material into the wire sheathing, but do not rely on the presence of a metallic shield for the associated attenuation.

2.5 Insulators and dielectrics

Electrical insulation prevents human contact with conductors that may be carrying hazardous levels of electrical current or voltage and prevents contact between conductors that share a cable assembly or between conductors and electrically conductive external structures. It also provides protection from environmental conditions that prevail in the intended application such as temperature extremes, atmospheric pressure variations, or water immersion over an extended period. The term "insulation" is usually reserved to refer to the material in direct contact with a conductor, whereas the terms "jacket" or "sheath" refer to the outermost layer of a cable whose primary function is to protect the insulation and conductor core from external physical forces and chemical deterioration. In some applications, insulative materials may be used in dielectric applications for RF cables, particularly coaxial and triaxial cables.

The type of insulation or dielectric material used on cables of any sort is equal in importance to the selection of a conductor. An insulator is a material that has no "free" electrons and as a result exhibits a high resistivity to the flow of electrons. Usually the term "dielectric" is reserved for an insulative material in which the polarizability, or polarization, of the insulative material dominates the material's behaviour or response to electric fields. In such a material polarization refers to a displacement of otherwise bound charges that is a function of frequency. Because this displacement of charge does in fact constitute a movement of charge, it can be called a displacement or polarization current, at least during the time the charges are in motion. This polarization or displacement current was originally conceived by James Clerk Maxwell, who famously added this current to Ampere's Circuital Law in his paper "On Physical Lines of Force", published in the years 1861 and 1862 in four parts. [56–60]. Additional information on displacement current is provided in Section 3.2.2 of Chapter 3.

Insulation intended for use in electrical wire and cable applications has many physical requirements, including but not limited to insulation resistance, dielectric strength, thermal shock characteristics, bend testing, flammability, resistance to water or electric "treeing", and abrasion resistance. In this text we will be focused on electrical performance characteristics known by terms such as insulation resistance, dielectric strength, dielectric loss, and dielectric absorption. Additional specific dielectric characteristics related to performance over frequency are of paramount interest for RF applications including dielectric constant variation with frequency and velocity of propagation.

2.5.1 Types of insulation materials

Insulation may be fabricated from a variety of materials. Major types of insulation include thermoplastics, thermosets, fibrous coverings, and enamels. Thermoplastics are a class of polymers that are solid at room temperature but can be softened and melted by the application of heat because they are composed of linear molecular chains that separate once heat is applied. Thermoplastics used for wire insulation are typically formed into their final shape by injection or extrusion moulding.

Thermoplastics tend to resist shrinking while offering good elasticity and strength. Thermoplastics can soften, deform, and lose some of their physical properties if exposed to high temperatures. Thermoplastics are resistant to chemical exposure, exhibit high impact resistance and good electrical insulation properties, are easy to colour, tend to work well in automated stripping processes, and tend to be less expensive than thermosets. Examples of thermoplastics used for wire insulation include PVC, polyethylene (PE), chlorinated polyethylene (CPE), fluoropolymers, polypropylene (PP), and PUR.

Thermosets are a class of polymers that are liquid at room temperature and harden, or cure, upon application of heat or the addition of certain chemicals. Thermoset materials are moulded once and will burn if heat is continually applied. During the curing process, a thermoset's polymer chains cross-link with other molecules to become a natural or synthetic rubber material. Covalent bond formation occurs during the cross-linking reaction that occurs during the curing process. Once cured, a thermosetting plastic cannot be reshaped and will tend to decompose upon application of high heat. Thermosets are most often used for cable jackets and like thermoplastics are typically formed into final shape by injection or extrusion moulding. Thermosets exhibit a comparatively high strength to weight ratio, can be used in elevated heat applications, are water resistant, offer good elasticity and flexibility at cold temperatures, exhibit superior electrical insulation properties, and tend to be more expensive than thermoplastics. Examples of thermoset plastics used for wire insulation include cross-linked polyethylene (XLPE), cross-linked chlorinated polyethylene (XL-CPE), chloroprene rubber (CR, aka polychloroprene or neoprene), chlorosulphonated polyethylene (CSPE), silicone rubbers, and ethylene propylene rubbers (EPRs).

Other materials used for wire and cable insulation, mentioned but not discussed herein, include enamel coatings and fibrous coverings such as oil-impregnated paper and textile braids fabricated from various blends of aramids, phenolic fibres, polyimides, nylon, and fibreglass.

Thermoplastic Materials: Note unless otherwise indicated the following information in this and the following section on thermosets is compiled from these multiple sources: [61–67].

PVC is among some of the oldest known synthetic materials in the plastics industry. As with many other examples, the discovery of PVC was by happenstance in 1838 when French chemist Henri Victor Regnault was experimenting with ethylene chloride, also known as Dutch oil. In his attempt to decompose the oil by heating it with an alcoholic mixture of potassium hydroxide, he unknowingly created vinyl chloride in the form of a gas. After completing his experiment, he left the flask where it was exposed to sunlight, and later found a hard white substance in the flask. Thirty-four years later, a German chemist named Eugen Baumann performed experiments very similar to those of Regnault, and like Regnault later found a hard white substance in a flask left exposed to sunlight. Neither Regnault nor Baumann was able to identify any practical use for the hard white substance they found, and it remained obscure. In 1913, another German chemist named Friedrich Klatte realized the potential for the hard white substance discovered earlier by Regnault and

Baumann and patented the material. Unfortunately, Klatte was, like his two predecessors, unable to identify any practical commercial use for the material, which was brittle and easily fractured. Finally, in 1926, Waldo Lonsbury Semon was hired by B.F. Goodrich to identify and produce a rubber that would bond to metal. While experimenting with a dehydrohalogenation process involving PVC very similar to that patented by Klatte, Semon produced what is now known as plasticized PVC. This newly found material was inexpensive, elastic, flexible, and more durable than natural rubber and could be poured into moulds of complex shapes. It could also be coloured as pellets, making it easy to transport and easy to combine into mixtures to produce every colour of the rainbow. Semon had the brilliant idea of using the new material as a water-resistant coating for fabrics. When World War II broke out, PVC was adopted by the US Navy as an insulation for wiring harnesses. Semon was awarded two US patents for his work with PVC: U.S. Patent 1929453, Synthetic Rubber-Like Composition and Method of Making Same, awarded Oct. 10, 1933, and U.S. Patent 2188396, Method of Preparing Polyvinyl Halide Products, awarded Jan. 30, 1940 [68–72].

Today, PVC has many formulations of composition that find application in diverse applications. PVC is relatively inexpensive, although the available variations differ in pliability and electrical properties with corresponding variation in cost. PVC is odourless, tasteless, and is non-toxic, making it safe for medical and food-related purposes. As already mentioned, colour may easily be added to PVC insulation during the manufacturing process. PVC exhibits a wide range of dielectric constant values between 3.5 and 6.5 and is commonly used as wire insulating material for low-voltage power and control cables. Some PVC formulations exhibit a wide temperature range, while others are more restricted. Certain formulations may have a –55°C to 105°C rating, while others may exhibit a –20°C to 60°C rating. PVC is inherently flame retardant, but it is known to produce HCl fumes that may pose a health hazard when exposed to flame. Fortunately, the Cl in the fumes serves to scavenge free radicals and acts to support the material's fire-retardant capability. PVC generally provides resistance to petroleum-based products and solvents, various acids and alkalis, exposure to sunlight, heat, weathering, bacteria, moisture, and abrasion. This range of properties makes PVC a suitable outer covering or as a jacket over and/or under metal armour for cable used in direct burial underground feeders and for externally exposed power and control lines.

Polyethylene, also known as polythene in the UK, was first discovered at the end of the nineteenth century by German chemist Hans von Pechmann, who observed the formation of a white precipitate during the autodecomposition of diazomethane in ether. Two of his colleagues subsequently identified the material as polymethylene, a polymer essentially identical to PE. Then in 1933, British chemists Eric Fawcett and Reginald Gibson, working for Imperial Chemical Industries (ICI), obtained a waxy solid material during experiments reacting ethylene with benzaldehyde under high pressure. They found the material could easily be melted and drawn into threads. When they attempted to repeat the experiment, it resulted in an explosion and fire. ICI management decided to abandon further pursuit of research into the material until safer equipment and facilities could be established. Fawcett was able to obtain permission to present the results of his and Gibson's work at a scientific conference in September of 1935. Several leading German chemists were in attendance, but they did not believe the conclusions presented by Fawcett. In December of the same year, Sir Michael Perrin, working with a new team at ICI, was able to develop safe and reliable means of production, and a patent application was filed by ICI in 1936 and was awarded on Sept. 6, 1937, as British patent 471,590. Another chemist with ICI, Bernard Habgood, realized that PE could readily replace gutta percha (refer to Section 2.6.1 of this chapter), and with this realization ICI went into full production on Sept. 1, 1939, the same day that Germany invaded Poland. As it turned out, the LDPE produced in the ICI processes proved to be a superior wire insulation and was used widely for radar installations on land and in airborne applications, providing a significant technological advantage for British forces during World War II [73–76].

All formulations of PE are less dense than water and exhibit excellent dielectric characteristics, including high dielectric strength, typically between 17 and 100 MV/m, low dielectric constant, between 2.28 and 2.35, and low dissipation factor, between 0.0002 and 0.0005 across a wide range of frequencies. Note that foamed PE (FMPE) exhibits a much lower range of dielectric constant between 1.2 and 1.55, and chlorinated PE exhibits a somewhat higher dielectric constant between 5.1 and 6.1 and a much worse dissipation factor between 0.01 and 0.1. These characteristics make most PE products an ideal wire insulation for multiple different cable types including controlled impedance high-speed data transmission, RF signal cables, telephone, electrical and electronic system control cables, and power cables for low-, medium-, and high-voltage applications. PE tends to be waxy, translucent, soft, tough, and non-toxic. PE exhibits excellent weather resistance. It is flammable and will continue to burn once it is ignited. PE melts and drips while burning, and combustion products exude the smell of burning paraffin. Flame-retardant formulations are available, but they tend to have poorer electrical properties. PE tends to be comparatively stiff and very hard, depending on molecular weight and density, but it can be cross-linked to produce high resistance to cracking, cut-through, soldering, and solvents, and generally can be used in temperatures ranging anywhere from −65°C to 80°C. When ethylene is polymerized to form PE, relatively straight polymer chains form, with more or fewer branches off the primary chains. The various types of PE can be separated based on the amount of branching into one of several general sub-categories: LDPE, linear low-density polyethylene (LLDPE), medium-density polyethylene (MDPE), HDPE, and high molecular weight polyethylene (HMWPE). FMPE is essentially a niche product used for dielectric applications wherein a low dielectric constant and a high velocity of propagation are important.

LDPE is the least dense of the PE series of thermoplastics, with a range between 0.91 and 0.925 g/cm^3. LDPE exhibits good flexibility and elongation at low temperatures and is an excellent electrical insulating material. LDPE exhibits low shrinkage percentage, good abrasion resistance and ductility, high impact strength, and water resistance. LDPE exhibits good resistance to alkalis, salt solutions, and organic and inorganic acids but has poor resistance to hydrocarbons. LDPE does exhibit some less-desirable characteristics, including poor tensile strength and low heat resistance with a melting point near 105°C, and is likely to crack and fracture under stress.

LLDPE has a density comparable to that of LDPE, with a range between 0.91 and 0.94 g/cm^3. LLDPE exhibits improved performance in several characteristics as compared to LDPE. It has higher tensile strength and elongation, good resistance to frictional loads, better impact resistance, and greater capability to recover from flexure, and is less likely to crack and fracture under stress. LLDPE is resistant to UV exposure and has a melting point near 122°C.

MDPE was created to solve room temperature cracking observed in LDPE. MDPE has a density comparable to that of both LDPE and LLDPE, with a range between 0.926 and 0.94 g/cm^3. MDPE tends to be very hard and consequently exhibits excellent impact and fracture resistance and a very high resistance to abrasion. The hardness of MDPE cable sheathing protects cables from sharp objects dropped or loaded onto the cable, and because of this characteristic, it is often used for cables located in environments with high ambient temperatures such as are found in tropical and subtropical countries. MDPE has superior oxidation resistance, and various additives may be included in MDPE compounds to improve fire retardant properties, UV and weathering resistance, and chemical degradation. MDPE has a melting point near 120°C.

HDPE is thought to be the toughest, most rugged form of cable insulation, with a density range between 0.94 and 0.97 g/cm^3. It is typically characterized by high rigidity, wear, and chemical resistance. HDPE has excellent abrasion resistance, higher tensile and shear strength, and better shrink percentage performance but lacks the flexibility of less dense PE formulations. HDPE insulation is non-corrosive and very resistant to UV rays, making it ideal for otherwise unprotected outdoor applications. HDPE has a melting point near 130°C.

HMWPE is very tough and exhibits the highest impact compared to other PE formulations, with a density range between 0.928 and 0.941 g/cm^3. It is resistant to crushing, abrasion, and disfigurement caused by large amounts of weight and pressure and is frequently employed to provide insulation for direct burial cables used to provide cathodic protection for applications in areas of high moisture or corrosive chemicals. HMWPE presents the same characteristics as HDPE with enhanced resistance to acids, alkalis, and organic solvents. HMWPE has a melting point near 130°C.

An additional and very important addition to the foregoing is FMPE. FMPE is most often used as the dielectric inside high-performance RF coaxial and triaxial cables. FMPE exhibits a low dielectric constant and an excellent velocity of propagation, both highly desirable features for critical RF applications. FMPE may be formulated either chemically or by injecting an inert gas such as nitrogen into molten PE during the extrusion process. Typical foam content with the gas injection process ranges between 35% and 40%, and use of gas injection eliminates any chemical foaming agent residue. The dielectric constant of chemically foamed PE foam can usually only be reduced to about 1.55. Use of the gas injection process allows the dielectric constant of FMPE to be reduced to as low as 1.20, nearly the same as that of air insulation.

Chlorinated polyethylene (CPE) was developed by DuPont in the late 1950s by a simple chlorination of PE in a solution process using a solvent [77]. In 1971, Dow Chemical Company introduced a CPE polymer using a more efficient and cost-effective aqueous suspension chlorination known as the slurry process. CPE can be regarded as a random copolymer of ethylene, polyethylene, and 1.2-dichloroethylene. It has full molecular chains and a random distribution of polar chlorine atoms. The ASTM D1418 designation for chlorinated polyethylene is CPE, where C denotes chlorine and M denotes a saturated chain of polymethylene in the chlorinated polymer. CPE is produced by reacting LDPE with Cl to introduce Cl groups along the polymer chains. Because the main polymer chains are saturated, CPE is highly resistant to oxidation, exposure to ozone, weather, and solar radiation including UV, and exhibits excellent abrasion resistance. CPE is resistant to most strong acids and bases and many solvents except for chlorinated organics, making it well-suited for chemical plant use in both above- and below-ground applications. Appropriately compounded CPE is capable of withstanding prolonged immersion in water. CPE can be conveniently coloured over a wide range and will maintain colour upon ageing. The chlorine content provides resistance to swelling by hydrocarbon oils and improves chemical and flame resistance, exhibiting a melting point near 150°C. CPE maintains its flexibility at –18°C (0°F) and does not become brittle unless temperatures are below –40°C (–40°F). Its low-temperature impact resistance is excellent. CPE jackets are suited to 105°C (221°F) and intermittently to higher temperatures and will maintain flexibility after repeated ageing at elevated temperatures. CPE does not support the growth of mould, mildew, or fungus.

CPE exhibits a dielectric strength on the order of 12 MV/m. It also exhibits a relative dielectric constant between 2.70 and 5.5 and a dissipation factor < 0.025, both measured at 1 MHz. CPE will not support combustion, but under the right conditions of excessive heat, oxygen supply, and flame source, it will burn slowly. Removal of the ignition source will extinguish the flame. CPE-jacketed cables pass the IEEE 383, UL, CSA, and ICEA flame tests. Like PVC, CPE is known to produce HCl fumes that may pose a health hazard when exposed to flame. Fortunately, the Cl in the fumes serves to scavenge free radicals and acts to support the material's fire-retardant capability.

Current commercial grades of CPE contain between 25% and 42% chlorine by weight. Higher chlorine content CPE is self-extinguishing and exhibits low residual crystallinity and is generally targeted for rubber-like applications, whereas lower chlorine content CPE exhibits high residual crystallinity and is generally targeted for thermoplastic applications. Thermoplastic CPE is most often used in blends with other polymers, particularly PVC. CPE may also be cross-linked to obtain a thermoset material. The thermoplastic material is more readily available than the thermoset material and tends to be less expensive. The presence of polar chlorine atoms in CPE restricts its use to low-voltage applications. CPE is typically used as the outer jacket for power and control cables for industrial power plant applications, mining applications, appliance wire, coil and transformer leads, and motor lead wire. CPE often serves as a lower-cost and more environmentally friendly alternative to CSPE.

Fluoropolymers are defined as polymers made up of carbon and fluorine. In April of 1938, Dr. Roy J. Plunkett, a relative newcomer to the DuPont Jackson Laboratory in 1936, discovered what was later known as PTFE resin while working with tetrafluoroethylene gas to develop refrigerants. During his experiments, after checking a frozen, compressed sample of tetrafluoroethylene, he and his associates discovered the gas sample had polymerized spontaneously into a white, waxy solid, with the iron surface of the cylinder acting as a catalyst [78]. Subsequent tests showed the substance was extremely heat resistant, non-adhesive, and proved to be inert to exposure to virtually all chemicals including most acids, bases, and solvents. US patent 2230654, tetrafluoroethylene polymers, was awarded to Plunkett for his discovery on Feb. 4, 1941. PTFE was introduced to the market in 1946, followed by polyvinylidene fluoride (PVDF) in 1948, FEP in 1960, ETFE and ethylene-chlorotrifluoroethylene (ECTFE, aka Halar) in 1970, and PFA in 1972. Fluoropolymers used for wire insulation typically have a specific gravity typically between 1.7 and 2.2 and exhibit excellent dielectric characteristics, including high dielectric strength (measured on 10 mil (0.254 mm) samples) ranging between 40 and 80 MV/m and a low dielectric constant between 2.04 and 2.6 (except for PVDF that ranges between 7.2 and 8.5) and a low dissipation factor between 0.0001 and 0.005 (except for PVDF, which ranges between 0.03 and 0.09).

Several fluoropolymers exhibit excellent thermal performance that enables their use in high-temperature applications and for high-magnitude electrical currents. For example, Ag and Ni wires coated with PTFE are rated 200°C and 260°C maximum, respectively. In comparison, FEP has a maximum service temperature equal to 205°C; PFA is rated up to 250°C. Conversely, ETFE and ECTFE both are rated up to 150°C; PVDF has the lowest maximum temperature rating of 135°C. Fluoropolymers generally exhibit excellent chemical resistance, extremely high insulation resistance, a very low dielectric constant, exceptionally low water absorption, and are non-flammable. Fluoropolymers are generally more expensive per pound than PVC or PE compounds.

Each of the fluoropolymers has notable variations in capability. PTFE has the lowest coefficient of friction, approximately equal to 0.1, and was once listed in the Guinness Book of World Records as the world's most slippery substance. FEP has a glossy surface, may be easily coloured, yields the smallest diameter of any high-temperature wire, and is often used as non-flammable jacketing material for multiconductor cables. PFA exhibits excellent high-temperature creep resistance, low-temperature toughness, and flame resistance. ETFE is commonly used in wire-wrap applications such as computer backplanes and offers the highest abrasion and cut-through resistance of any fluoropolymer. ECTFE has properties very similar to those of ETFE and is also used in wire-wrap applications, but because it is less expensive, it is often used as insulation on multipair plenum telephone cables. PVDF is one of the least expensive fluoropolymers and, like ECTFE, is frequently used as a jacketing material on plenum cables. Because of its comparatively high dielectric constant, PVDF tends to be a poor insulator for controlled impedance or RF applications.

PTFE is extruded as a paste using a hydraulic ram process onto wire plated with either Ag or Ni and then must be sintered in situ. Lengths for PTFE extrusion are limited due to the amount of material in the ram, thickness of the insulation, and preform size. As an alternative approach, solid PTFE may be wrapped as a tape around a conductor. Fluoropolymers such as PVDF, FEP, ETFE, ECTFE, and PFA may be extruded in continuous lengths.

PP is typically harder than PE, making it suitable for thin-wall applications, and exhibits electrical properties very similar to them. This material is popular for wire and cable insulation and is often used as an alternative to PVC, particularly in low-ventilation environments, because it emits less smoke and no toxic halogens, which may lead to the production of acid in high-temperature conditions. The normal temperature range for PP is between −30°C and 80°C. Polypropylene has a specific gravity between 0.89 and 0.91, the smallest of commonly used thermoplastics in use at present. It exhibits excellent mechanical strengths, has the highest softening temperature among thermoplastic resins, and exhibits good resistance to low-temperature exposure and ageing. The dielectric constant is typically 2.25 for solid and 1.55 for foam designs.

PUR has poor electrical properties, with a typical dielectric constant between 5 and 7.5 and a typical dielectric strength on the order of 20 MV/m, but despite this has been used for cable sheathing since the 1960s and is a thermoplastic material with superior physical properties to that of even the toughest thermosetting rubber. It is extremely wear-resistant and mechanically tough, being very difficult to cut or tear, with excellent flexibility and flex life. It also has excellent ageing resistance against environmental humidity, ozone, UV radiation, and microbes. When specially formulated, a PUR jacket can exhibit good flame resistance. PUR offers excellent water resistance, exhibiting superior performance in marine applications. PUR remains flexible even in low temperatures and exhibits a wide application temperature range from −62°C to 125°C. This material works well in retractile cord applications and can be a good option for salt-spray and low-temperature military purposes.

Thermoset Materials: Note unless otherwise indicated the following information in this and the following section on thermosets is compiled from these multiple sources: [79–86].

Examples of thermoset plastics used for wire insulation include XLPE, XL-CPE, CR, CSPE, silicone rubbers, and EPRs.

PE material itself has excellent dielectric strength, high insulation resistance, and a low dissipation factor at all frequencies, making it an ideal insulator; however, it is limited in its temperature range. Cross-linking the PE to become XLPE increases the temperature range of the insulation while maintaining the electrical properties. Almost all XLPE compounds for wire and cable applications are based on LDPE. Al Gilbert and Frank Precopio invented XLPE in March 1963 in the GE Research Laboratory located in Niskayuna, New York. XLPE is a thermoset material produced by the compounding of LDPE with a crosslinking agent. In this process, the long-chain PE molecules "cross-link" during a curing, or vulcanization, process to form a material that has electrical characteristics similar to those of thermoplastic PE, but with better mechanical properties, particularly at high temperatures. XLPE-insulated cables have a rated maximum conductor temperature of 90°C and an emergency rating of up to 140°C, with a short-circuit rating of 250°C. XLPE insulation is strong, durable, and temperature conducting and is intended for lighting, power, and general-purpose applications. XLPE is suitable for voltage ranges from low to extra high voltage, surpassing other insulation materials such as PVC, EPR, and silicone rubbers. XLPE has excellent dielectric properties, making it useful for medium voltage (<69 kV AC), and high-voltage cables (≤380 kV AC), and several hundred kV DC. The mechanical properties of the XLPE are superior to many other insulations, offering greater tensile strength, elongation, and impact resistance. The addition of carbon black can be used to further enhance hot deformation and cut-through resistance. The XLPE insulation will not melt or drip, even at the temperatures of soldering irons, and it has increased flow resistance and improved ageing characteristics. Also, XLPE insulation can be modified to limit the amount of by-product gases generated during the cross-linking process. This is particularly useful for high-voltage cable and extra-high-voltage cable applications, where degassing requirements can significantly lengthen cable manufacturing time. The cross-linking process also enhances the chemical and oil resistance of XLPE at elevated temperatures and makes it suitable for use as a Low Smoke Zero Halogen material. XLPE exhibits electrical properties similar to those of PE, with a relative dielectric constant on the order of 2.3, a dissipation factor on the order of < 0.05, and a dielectric strength between 35 and 50 kV/mm.

As indicated previously, CPE may also be cross-linked to obtain a thermoset material, although the thermoplastic material is more readily available than the thermoset material and tends to be less expensive. In a process similar to vulcanization, cross-linking of the molecules in CPE can be effected through the chlorine or chlorosulphonyl groups, making the material into a rubbery solid. The cross-linked variety exhibits very similar properties and characteristics as the non-cross-linked variety and offers superior high-temperature capability in comparison.

CR, a vulcanized synthetic rubber, also known as chlorobutadiene rubber, polychloroprene, and neoprene, is an important diene-based elastomer. The name Neoprene® was a registered trademark of DuPont Performance Elastomers. CR was first prepared in 1931 by Arnold Collins, a chemist in Wallace Hume Carothers' research group at DuPont, while he was investigating by-products of divinylacetylene. CR has been used as a jacketing material since the 1950s for large power cables.

CR primarily comprises cross-linked carbon, hydrogen, and chlorine polymers. The chlorine in the polymer reduces its reactivity to many oxidizing agents and is primarily responsible for its resistance to cracking caused by exposure to UV and atmospheric ozone and heat ageing. The cross-linking of molecules that occurs during the vulcanization process produces sulphur bridges that unite individual chains of chloroprene to create a larger molecule. The number of sulphur links affects the overall properties of the batch of CR. Depending on the vulcanization process and the number of sulphur bonds produced, CR can express an array of traits in varying degrees without changing its basic structure. CR is often compounded with other ingredients to enhance desired properties including resistance to many chlorofluorocarbons, aliphatic hydrocarbons, mineral oils, and greases, as well as improved low-temperature characteristics, improved physical strength, and better moisture resistance. CR jackets are tough, strong, and resilient, and generally exhibit excellent resistance to abrasive wear, impact, crushing, and chipping. They are also resistant to damage from mildew, fungus, and other biological agents. In contrast, CR can be variously susceptible to acids, solvents, and fuels and is known to degrade in the presence of some common chemicals including hydrochloric acid, acetone, xylene, acetic acid, and hydrogen peroxide.

Compounds made from CR are generally characterized by poorer electrical properties than natural rubber, styrene-butadiene rubber (SBR), and butyl compounds, typically relegating their use to cable jacketing or non-critical applications.

Depending on the compound, most CR jackets remain pliable at low temperatures and may be flexed without damage in ambient temperatures as low as –40°C (–40°F). CR jackets also perform well at elevated temperatures and resist degradation for prolonged periods at temperatures as high as 121°C (250°F). Satisfactory performance at even higher temperatures is possible if exposures are brief or intermittent.

CR will burn slowly in the presence of a flame, but as soon as the flame is removed, the CR will self-extinguish.

CR typically exhibits a dielectric strength on the order of 20 kV/mm, a dielectric constant in the range of 3.2 to 8.0, and a loss tangent in the range of 0.008 to 0.05.

In 1951, DuPont began marketing a thermosetting, cross-linked CSPE under the name Hypalon. CSPE is an excellent material for wiring applications where the heat associated with overloaded circuits might otherwise damage plastic insulation or jacketing materials. CSPE exhibits a high degree of UV and ozone stability and can withstand extended exposure to direct sunlight with limited degradation of its electrical and mechanical properties. Aside from its UV and ozone stability, CSPE exhibits many excellent physical and electrical properties. CSPE is resistant to compression set resulting from clamping pressures and other forms of external loading and generally exhibits extremely low water absorption. CSPE is known for its rugged nature and resistance to abrasion, exhibiting high resistance to impact, crushing, and chipping, exposure to chemicals, hydrocarbon oils, and fuels, and flexibility. It is especially useful in applications involving contact with oils at elevated temperatures. CSPE works well as low-voltage insulation up to 600 volts or as jacketing for most wiring applications. These characteristics make CSPE an ideal material choice for applications in severe environments including heavy construction and mining applications. It is particularly useful as a cable sheathing in plant processing areas, where airborne chemicals are likely to attack jacketing materials and metal conduit.

CSPE remains flexible at –18°C (0°F) and will not become brittle at –40°C (–40°F). At high temperatures, Hypalon will perform satisfactorily after short-term exposure at temperatures up to 148°C (300°F) – even higher if compounded for maximum heat resistance.

Like CR, CSPE sheathing will not support combustion. CSPE will burn slowly in the presence of a flame, but as soon as the flame is removed, the CSPE will self-extinguish. Hypalon-jacketed constructions pass both the Underwriters Laboratories' vertical flame test and the U.S. Bureau of Mines' flame test for mining cable.

This insulation material can be found in appliance wire, lead wire, coil leads, transformer leads, and motor lead wire.

Unfortunately, Hypalon had specific formulation requirements and is now either legally restricted or avoided as a part of good design practice. CSPE materials would not melt in high-temperature environments, but they will eventually ignite and burn, producing gases that exhibit varying levels of toxicity, including HCl, H2S, and HBr. These gases can combine with water (in the order cited) to form hydrochloric acid, a weak acid that dissociates into HS and H3O with S precipitated out of the mixture over time, and hydrobromic acid. Additionally, some compounds of Hypalon contain additives made from lead or lead compounds that can be released in a fire. Environmental and commercial pressures, high manufacturing costs, and the hazardous nature of manufacturing the product, all probably contributed to DuPont's decision to discontinue the production of their Hypalon products at the end of June 2009.

Alternatives do exist, and CSPE remains available from some sources. Other compounds are now available that perform similarly to CSPE without the drawbacks mentioned above. Manufacturers have generally either (1) switched to other suppliers of CSPE materials that do not use the trade name Hypalon or (2) switched to other materials such as thermoset CPE, which offers performance characteristics very similar to those offered by CSPE.

CSPE typically exhibits a dielectric strength on the order of 20 kV/mm, a dielectric constant in the range of 9.0 to 11.0, and a loss tangent in the range of 0.05 to 0.07.

Silicone rubber is a high-temperature vulcanizing material formulated from siloxanes. Siloxanes were first characterized as macromolecules by the English chemist Frederic Stanley Kipping in 1927. Based on his observations of the siloxane structure, Kipping incorrectly called them silicones, a name that has persisted. In 1943 Eugene George Rochow at the General Electric Company Laboratories in Schenectady, NY, USA, prepared silicones by the hydrolysis of dialkyldimethoxysilane. Rochow patented his process in 1945, and that process remains the basis of modern polymerization methods.

Silicone rubber is extremely heat-resistant and can be used in temperatures ranging between −80°C and +200°C. Silicone rubber exhibits excellent electrical properties including excellent corona and radiation resistance. Silicone rubber is extremely stable and possesses strong resistance to exposures of UV and ozone. It typically has low mechanical strength and abrasion resistance, although it is extremely flexible and exhibits low moisture absorption. Silicone rubber does not support fungal growth, and exhibits good resistance to acids, bases, salts, and minerals found in soil.

Silicone rubber is difficult to ignite. Once ignited, it burns to a non-conductive ash. While burning, silicone does not give off toxic materials.

The dielectric strength of silicone rubber varies between 4 and 30 kV/mm. The dielectric constant ranges between 3.0 and 3.5, and the loss tangent ranges between 0.001 and 0.010.

Two major types of ethylene-propylene polymers with elastic properties are presently available in the market, generally known as EPRs. Both are prepared by dissolving gaseous ethylene and propylene (and liquid diene) in an organic solvent such as hexane and subjecting the solute to the action of Ziegler-Natta catalysts, developed in the 1950s. The first major type is a copolymer known as EPM (ethylene-propylene monomer). EPM is made with ethylene and propylene alone. The ethylene and propylene constituents in EPM are hydrocarbons in which there is only a single carbon–carbon double bond. The second major type is a terpolymer known as EPDM (ethylene-propylene-diene monomer, aka EPT), made with ethylene, propylene, and approximately 5% of a third hydrocarbon with two pairs of carbon atoms joined by a double bond, known as a diene, most often either ethylidene norbornene or 1,4-hexadiene. Both copolymer types contain approximately 60% by weight ethylene.

Both polymers must be vulcanized (cured) to achieve a rubber-like consistency and performance. At the time EPRs were discovered, the only cure process known made use of peroxides. EPM may only be vulcanized with a peroxide-based cure system or can be irradiated. The use of peroxide offers enhanced heat stability, and as expected, peroxide-cured EPR products perform well in service temperatures as high as 150°C. EPM also offers superior compression set resistance. Unfortunately, EPM exhibited a tendency to stick to the mould, was hard to deflash, and was susceptible to scorching. These difficulties prompted the addition of the third monomer, allowing the use of sulphur instead of peroxides. EPDM may be vulcanized with either a sulphur-based or peroxide-based system. The use of sulphur offers improved mechanical properties. Sulphur cured EPDM elastomers are easier to process and generally cost less than peroxide-cured materials. Sulphur-cured EPR products perform best in service temperatures that do not exceed 100°C and are often preferred where resistance to weather is important.

Both EPM and EPDM are non-polar elastomers and thus offer good performance in many polar fluids, including Methyl Ethyl Ketone (MEK), acetone, most phosphate-ester-based hydraulic fluids, silicone oils and greases, glycol- and silicon-based brake fluids, and sodium and potassium alkalis. Both EPM and EPDM are strongly swollen and weakened in non-polar fluids such as hydrocarbon oils, lubricants, and greases. For applications involving exposure to acetic acid, EPDM typically exhibits better performance than EPM.

EPM and EPDM both exhibit excellent resistance to heat, oxidation, and ozone, are abrasion resistant, and remain flexible at temperatures as low as –60°C. Flame-retardant versions are often referred to as "FREP" or "flame-retardant EP". Peroxide-cured grades of EPM and EPDM can reach a maximum temperature of approximately +150°C in hot water, alcohols, and organic and inorganic acids and bases. Sulphur-cured grades of EPDM offer a typical temperature range of –50°C to +120°C and optimal mechanical properties.

Because EPM and EPDM exhibit very similar characteristics, they are frequently used as insulation in 600 V through 69 kV power cables, as an integral insulation/jacket on welding cables and as insulation in many cords, portable mining cables, and control/instrumentation cables. EPM and some EPDM formulations are usable in the nuclear industry.

The dielectric strength of EPR varies between 35 and 45 kV/mm. The dielectric constant ranges between 2.8 and 3.2 depending on the specific EPR formulation, and the loss tangent ranges between 0.01 and 0.04.

2.6 A brief history of wire insulation

Wires and cables that interconnect electrical systems and subsystems can be broadly categorized into two types: power and communication. The needs of these two separate classes were quite similar in the early to mid-nineteenth century. Telegraph system wiring and wires and cables used for control of explosives in mining operated at low voltages and low frequencies. As the push for electrifying cities grew, the same design techniques used for telegraph and mining cables were applied to power transmission and distribution cables. Early telegraph system wires were often bare iron supported by glass and sometimes wooden insulators coated with shellac. As time and experience accumulated, it soon became evident some sort of covering to provide protection from physical and environmental damage was needed. The first known use of paper as an insulation was documented in a paper by Don Francisco Salva and published by the Academy of Science in Barcelona in 1795 [87]. Many materials were tried with mixed success, including cotton, cotton impregnated with paraffin, silk, and even asbestos thread, all of which may have been coated with natural resins like shellac and copal or impregnated with melted natural waxes or even asphalt in organic solutions. Unvulcanized natural rubber was tried but exhibited several faults. Legend has it that circa 1820, William Henry once ripped up a silk skirt belonging to his wife to use as insulation for magnet coil windings [88]. Many cables were wrapped in, or covered with, lead or tin foil layered together with additional insulation layers. In many cables, once the lead or tin foil covering was complete, the cable would be filled with paraffin, resin, and sometimes illuminating oil.

2.6.1 Gutta percha, forerunner of modern cable insulation

W.F. Cooke and C. Wheatstone were awarded a UK patent for their five-wire telegraph system in 1837 and a second UK patent for their two-wire telegraph system in 1840, based on an 1832 design by S.F. Ronalds [89,90]. In 1832, Samuel Morse conceived of a single-wire telegraph, and he was awarded a US Patent [91] in 1840. The invention and successful implementation of a telegraph system in the UK and in the US was quickly followed by the epochal creation of the network of underwater telegraph cables that tied the world together. Key to the success of this worldwide communications network relied on the use of the best electrical conductor material available at the time and, perhaps even more importantly, on the discovery and use of a suitable and reliable insulation for the underwater environment. As indicated above, many different types of insulation were in use at the time, but for the undersea cables, all were superseded by gutta percha.

Gutta percha is very similar to natural rubber and is a natural latex material produced from the sap of trees of the Sapotaceae family. It is a hard, dark substance at room temperatures. Chemically, gutta percha is a gum plastic and is a stereoisomer of rubber, sharing a similar chemical composition known as a polyisoprene. Unlike rubber, gutta percha will not bounce when dropped or stretch and return to its original shape and size. It becomes very pliable in boiling water and is easily moulded into intricate shapes, retaining its shape and becoming inelastic once the substance has cooled back to room temperature. It is very dense and watertight, more so than rubber, making it an ideal insulation for a telegraph cable lying on the sea floor. Gutta percha gradually and inevitably replaced many other contemporary insulating materials starting in about 1841. Gutta percha was once widely used for domestic, medical, and industrial purposes but, with the discovery of plastics such as PE, has since become largely forgotten.

In 1656, John Tradescant, an English traveller, collected and returned to England with samples of gutta percha. In 1842 Dr. Jose d'Almeida, convinced of its scientific merits, took some raw and manufactured samples of gutta percha with him to England and submitted a portion of his samples to the Royal Asiatic Society in April of 1843. The previous month, Scottish physician Dr. William Montgomerie sent a communication and three samples of gutta percha to the Medical Board of Calcutta [92]. Montgomerie became acquainted with the material in his encounters with Malay farmers and labourers who had been making handles for tools with the material for many years. Montgomerie was convinced the material could be used for making surgical splints. Montgomerie also submitted a similar communication and four samples of gutta percha to the Society of Arts in April of 1843. Montgomerie was awarded a Society Gold Medal for his efforts in June 1845.

In 1848, Faraday wrote about the uses of gutta percha as an electrical insulator in *The Philosophical Magazine* [93]. Word spread among British telegraphers, who used it to insulate wires for underground telegraph lines and found it to be far superior to anything they had tried before. Engineers found gutta percha would not crumble underwater as the natural India rubbers that were available at the time. In August of 1850, a 25-nautical-mile-long cable with a single copper conductor core and covered in unprotected gutta percha was laid by brothers Jacob and John Watkins Brett across the English Channel between Dover and Cape Gris Nez in France, near Calais. Although signals were received at either end, they were garbled and indecipherable, and the line was soon abandoned. In 1891, Willoughby Smith, with experience between the 1840s and his retirement in the 1880s, wrote a book entitled The Rise and Extension of Submarine Telegraphy. Smith had personal experience with the 1850 cable and indicated his belief the garbled transmissions resulted from "..the retardation caused by induction ..." in his first chapter. Retardation was to become the focal point for success or failure of these undersea cables, as was developed by William Thomson, Oliver Heaviside, and others in the intervening years. In September 1851, a much-improved cable design with four copper conductors and a surrounding armour of iron wires was laid by the brothers Brett between South Foreland in England, near Dover, and Sangatte in France, near Calais, commonly referred to at the time as the Dover to Calais cable. Communication over this line was definitively established on or about 15 October 1851, and the cable remained in service until 1875 [94,95]. The submarine cables laid globally grew from about 15,000 nautical miles in 1866 to over 200,000 nautical miles by 1900. Insulated by a sheath of gutta percha, this global cable network was often referred to in the nineteenth century as the "nervous system" of the British Empire and of world commerce, as it constituted a key means by which modern technology extended its reach around the world. No better material than gutta percha was found for this purpose until PE was discovered in the 1930s [96].

With the advent of the telephone, the needs of power and communication wire and cable systems soon led to a divergence in design techniques. While telegraph, telephony, and power transmission and distribution cables all needed to exhibit low attenuation, telephony cables also needed to exhibit minimal speech distortion and negligible cross talk between adjacent circuits. Prior to the introduction of the metallic circuit, telegraph and telephony cables used a single conductor with the return circuit through the earth. This worked well for the telegraph but led to unacceptable levels of crosstalk in voice circuits. Once the metallic circuit, developed and patented by Alexander Graham Bell [97,98], replaced the single-line network, telephony cables required twice the number of conductors, mitigating crosstalk problems but increasing line capacitance and potentially dielectric losses as well. Because telegraph and telephony cables operated at low voltages and currents, conductors for these cables could be small. The smaller conductor size meant the insulation for each conductor could be correspondingly small as well, but as line lengths grew, line capacitance increased.

As line lengths increased, transmission speeds on telegraph lines decreased and speech distortion on telephony lines increased. The delay and distortion on the unloaded lines was caused by variations in the signal propagation velocity, leading to dispersion across frequency. The problems of telegraph signalling speeds and speech distortion were closely related. These problems, referred to as "retardation", figured prominently in the design of telegraph cables stretching across the Atlantic [99] that not only had to convey intended electrical signals but also needed protection from constant immersion in seawater and the temperatures and pressure levels existing on the sea floor.

In October of 1853, Michael Faraday and several others joined Latimer Clark at the Gutta Percha Company works on Wharf Road in Islington and then later at the offices of the Electric Telegraph Company at Lothbury for the purpose of observing the retardation effects Clark had observed on long sections of telegraph cable immersed in a canal versus a similar length of cable in dry storage. Faraday reported his observations and opinions in a paper presented to the Royal Institution in January of 1854 [100], in which he indicated the retardation effect was entirely consistent with his theories of electromagnetism. Faraday regarded the space around the wire as the real seat of electromagnetic phenomena and asserted the observed behaviour of the immersed cable depended on the nature and arrangement of the surrounding thin layer of gutta percha sandwiched between the central conducting wire and the surrounding iron armour and seawater in the case of a submerged cable.

The retardation effect that Clark and others had observed showed in an unmistakable fashion that the propagation of electrical actions through space required time and was equally dependent on the surrounding dielectric as it was on the conductor itself. The observations made by Clark, Faraday, and others served to emphasize the importance of Faraday's "specific inductive capacity" of different dielectrics, in which electrical phenomena such as the charge stored by a condenser (aka capacitor) are found to depend on the characteristics of both the conductors and the surrounding medium [99].

In October of 1854, Professor George Gabriel Stokes wrote a letter to William Thomson (aka Lord Kelvin) regarding Faraday's observations. In his January 1854 paper, Faraday had demonstrated how the construction of a cable would limit the rate at which messages could be sent. Thomson produced an analysis [101] based on how he believed electrical energy diffused along the length of the cable in a fashion entirely similar to that of heat diffusing along a metal bar and concluded the data rate was proportional to the square of the length of the cable, as shown here:

$$t = \frac{RCl^2}{2} \tag{2.1}$$

where

t = the time between signals
R = the resistance of the line
C = the capacitance of the line
l = the length of the line

Notably, the data rate was also proportional to the capacitance of the cable, a direct function of the type of insulation and the physical relationships of the inner conductors and the outer sheath and armour. Unfortunately, Thomson omitted to account for the inductance of the line. It remained for Oliver Heaviside to work out the mathematics of transmission lines in a complete fashion. He was able to state a simple relationship known as the Heaviside Condition [102], given by

$$\frac{L}{C} = \frac{R}{G} \tag{2.2}$$

where

L = the inductance of the line
C = the capacitance of the line
R = the resistance of the line
G = the conductance of the line

The Heaviside Condition emphasizes the need for balance of line characteristics for distortion-less propagation on the transmission line. The conductance and capacitance of the line are characteristics largely dominated by the characteristics of the insulation used in the cable construction, whereas the resistance of the line is a characteristic of the conductor itself, and the inductance is a characteristic of the conductor and its physical relationship with the return. The selection of suitable insulation materials was limited in the nineteenth century to primarily gutta percha and natural rubber. This, coupled with the conductor size constraints, effectively minimized the ability to introduce control by changing either the conductance or capacitance of the communication or telegraph lines. The resistance of a transmission line is directly proportional to the quality and size of the conductor, and the selection in the nineteenth century was limited to metals such as iron, steel, and copper. This left adding inductance along the length of the line to balance out the capacitive effects and thereby minimize propagation velocity variations. Based on the Heaviside Condition, inserting so-called loading coils at distance intervals small in relation to wavelengths of the expected maximum frequencies enabled relatively distortion-free operation, although it did increase the line impedance. The impedance increase, however, was more than acceptable as a price to pay for distortion-free operations. Additional information concerning transmission lines is contained in Chapter 4.

2.6.2 Gutta percha was not found suitable for power transmission and distribution lines

In contrast, power transmission and distribution cables were operated at much higher voltages with correspondingly larger current flows in their conductors. These cables required large conductors to carry systemic load currents without excessive voltage drop across the line resistance and without overheating to the point of damaging the insulation and the conductors. Initially power transmission and distribution cables were run on elevated poles similar in manner to how communication cables were run, but this soon led to safety and reliability concerns, and many of these cables were eventually buried in the ground. Insulation used on these power cables needed to be able to protect the cables from moisture intrusion and temperature degradation and had to be thick enough with a sufficiently high dielectric strength to withstand high operating voltage gradients or transient stresses. Like communication cables, the capacitance per unit length and dielectric losses needed to be low to limit capacitive charging currents and efficiency loss. This was a more difficult problem for power transmission and distribution cables because of their much larger size.

In 1880, Thomas Edison created what he referred to as “street pipes” or “street tubes” or “Edison tubes”, iron pipes within which he passed two or three copper rods, mutually insulated from each other by wrapping jute fibre around and between the conductors, then filling the void with a mixture of pitch, paraffin wax, beeswax, and linseed oil [103]. Later, in 1881, Edison filed US Patent 251,552 [104], in which he modified his design to remove the jute fibre and replace it with circular slotted discs of insulating material such as paper or “Manila board” spaced periodically along the length of the tube. Edison developed and deployed the first power distribution system in New York City using his “street pipes”. The design was effective but very time-consuming and labour-intensive, not only for fabrication but also for installation and maintenance. Still later, in 1883, Edison adopted the three-wire system introduced by Dr. John Hopkinson in UK Patent No. 3576 of 1882. Edison's new design followed his earlier efforts closely, employing three solid rods of copper of the same circular cross section, mutually wrapped with jute fibre, then bound together by means of a fourth cord. As he had done previously, he then placed the three-conductor arrangement into an iron pipe and filled the void with the same pitch admixture [105]. Glass tubes were employed for some underground cables, but their use again required a great deal of time and was labour intensive. By the 1890s, mass impregnated paper insulation was being used on cables and voltages as high as 10 kV and had become the standard [106]. In 1897, vulcanized rubber-insulated cable was used for 11,000-volt circuits installed for the Niagara Falls power project, which could be stated as the first use of insulated cables for high-voltage circuits. Edison himself later employed rubber insulation for electrical conductors [107]. In 1906, armoured cables were introduced, which had flexible sheathing and two cloth-covered, rubber-insulated conductors.

2.6.3 Modern cable insulation for data and power transmission/distribution

German scientists and engineers started testing PVC as insulation material for cables in the 1930s, and by the end of the Second World War, a wide selection of synthetic rubbers and PEs were available. By the 1950s, PVC had become commercially viable and slowly started replacing rubber cables in many areas, particularly in domestic wiring. XLPE insulation came into play in the early 1970s, largely replacing paper-insulated cables for medium-voltage applications, followed by many significant improvements in polymer compound manufacturing and insulation techniques. In the 1980s use of XLPE was expanding into high-voltage transmission lines between 66 and 240 kV. PVC alternatives safer in fires were being developed in response to several tragic public fires, which demonstrated the dangers of smoke and toxic gases from various materials, including PVCs. By the 1990s, polymers were being extended to use in extra-high-voltage applications [106,108,109].

References

[1] G. Calhoun, "Why Wire Mattered: Building U.S. Networked Infrastructures, 1845–1910," *Technology and Culture*, vol. 62, no. 1, pp. 156–184, 2021.

[2] ASM Handbook, *Vol. 2: Properties and Selection: Nonferrous Alloys and Special-Purpose Materials, Introduction to Copper and Copper Alloys*, ASM Handbook Committee, ASM International. DOI: 10.31399/asm.hb.v02.a0001065.

[3] Goodfellow Materials Hub, "Tin" [Online]. https://www.materialshub.com/material/tin/ (accessed Nov. 20, 2022).

[4] "Materion Brush Performance Alloys", *Technical Tidbits Newsletter*, no. 32, Aug 2011. [Online]. https://materion.com/-/media/files/alloy/newsletters/technical-tidbits/issue-no-32—tin-as-a-coating-material.pdf (accessed Nov. 20, 2022).

[5] Copper Development Association Inc., "Copper-Tin Intermetallic Compounds" [Online]. https://www.copper.org/applications/industrial/DesignGuide/performance/coppertin03.html (accessed Nov. 30, 2022).
[6] NASA, "Tin Whisker (and Other Metal Whisker) Homepage" [Online]. https://nepp.nasa.gov/whisker/background/index.htm (accessed Nov. 30, 2022).
[7] L. Wishart, "Detailed Discussion of MIL-STD-1353 (Part 1): Tin Plating" [Online]. https://lectromec.com/tin_plating_and_connectors/ (accessed Nov. 30, 2022).
[8] T. Campbell, R.K. Kalia, R. Nakano, P. Vashishta, O. Shuji, and S. Rodgers, "Dynamics of Oxidation of Aluminum Nanoclusters using Variable Charge Molecular-Dynamics Simulations on Parallel Computers," *Physical Review Letters*, vol. 82, no. 4866, 1999.
[9] A. Dick, "Phosphor-Bronze and Its Applications," *Journal of the Society of Arts, Royal Society for Arts, Manufactures, and Commerce*, vol. 25, no. 1276, pp. 551–566, 1877.
[10] Obituary Notice, "Jacques Manne," *Journal of the Institution of Electrical Engineers*, vol. 57, no. 285, pp. 620–621, DOI:10.1049/jiee-1.1919.0097.
[11] Thomas A Xometry Company, "Types of Nickel Alloys" [Online]. https://www.thomasnet.com/articles/metals-metal-products/types-of-nickel-alloys/ (accessed Nov. 20, 2022).
[12] "Albert Marsh, Inventor, Scientist" [Online]. https://web.archive.org/web/20060822130023/http://www.toaster.org/marsh.html [From Pana News-Palladium, June 5, 1997], (accessed Nov. 20, 2022).
[13] D.O. Woodbury, *A Measure of Greatness: A Short Biography of Edward Weston*, McGraw-Hill, NY, 1949, pages 168.
[14] A. Giumlia-Mair "Plating and Surface Treatments on Ancient Metalwork," *Advances in Archaeomaterials*, vol. 1, no. 1, pp. 1–26, 2020, DOI:10.1016/j.aia.2020.10.001.
[15] Sharretts Plating Company, "History of Electroplating" [Online]. https://www.sharrettsplating.com/blog/history-of-electroplating/ (accessed Nov. 23, 2022).
[16] A. Volta, "On the Electricity Excited by the Mere Contact of Conducting Substances of Different Kinds," *Philosophical Transactions of the Royal Society of London (in French)*, vol 90, pp. 403–431, DOI:10.1098/rstl.1800.0018.
[17] W. Cruickshank, "Some Experiments and Observations on Galvanic Electricity," *Journal of Natural Philosophy, Chemistry, and the Arts (AKA Nicholson's Journal)*, vol. 4, pp. 187–191, 1800.
[18] W. Cruickshank, "Additional Remarks on Galvanic Electricity," *Journal of Natural Philosophy, Chemistry, and the Arts (AKA Nicholson's Journal)*, vol. 4, pp. 254–264, 1800.
[19] L.B. Hunt, "The Early History of Gold Plating," *Gold Bulletin*, vol. 6, no. 1, pp 16–27, 1973.
[20] L.V. Brugnatelli, "Extrait d'une Lettre de Brugnatelli, I. Galvanisme," *Journal de chimie et de physique, ou recueil périodique des découvertes*

dans les sciences chimiques et physiques, tant en France que chez l'etranger, vol. 5, pp. 357–358, 1803 (English translation printed in *The Philosophical Magazine*, vol. 21, p. 187, 1805).

[21] M. Faraday, "VI. Experimental Researches in Electricity-Seventh Series," *Philosophical Transactions of the Royal Society of London*, vol. 124, pp. 77–122, DOI:10.1098/rstl.1834.0008.

[22] H. Heinrich, "The Discovery of Galvanoplasty and Electrotyping," *Journal of Chemical Education*, vol. 15, no. 12, pp. 565–574, 1938, DOI:10.1021/ed015p565.

[23] G. Williams, "The Cyanides of Gold," *Gold Bulletin*, vol. 11, no. 2, pp. 56–59, 1978.

[24] "Zénobe Theophile Gramme Direct-Current Dynamo" [Online]. https://www.invent.org/inductees/zenobe-theophile-gramme, (accessed Nov. 24, 2022).

[25] C.R. Shipley, Jr., "Historical Highlights of Electroless Plating," *Plating and Surface Finishing*, vol. 71, no. 6, pp. 24–27, 1984 (republished under the same title in NASF Surface Technology White Papers, vol. 82, no. 9, pp. 13–24, June 2018.

[26] *Standard Specification for Tinned Soft or Annealed Copper Wire for Electrical Purposes*, American Society for Testing and Materials (ASTM) B33.

[27] *Standard Specification for Silver-Coated* Soft or Annealed Copper Wire, American Society for Testing and Materials (ASTM) B298.

[28] *Standard Specification for Silver Coated Copper and Copper Alloy Stranded Conductors for Electronic Space Application* (ASTM) B961.

[29] *Standard Specification for Nickel-Coated Soft* or Annealed Copper Wire, American Society for Testing and Materials (ASTM) B355.

[30] *Standard Specification for Silver-Coated, Copper-Clad Steel Wire for Electronic Application*, American Society for Testing and Materials (ASTM) B501.

[31] *Wire, Steel, for Electronic Applications, Tin-Coated, Copper Clad*, American Society for Testing and Materials (ASTM) B520.

[32] *Standard Specification for Nickel-Coated, Copper-Clad Steel Wire for Electronic Application*, American Society for Testing and Materials (ASTM) B559.

[33] *Standard Specification for High-Strength, High-Conductivity Copper Alloy Wire for Electronic Application*, American Society for Testing and Materials (ASTM) B624.

[34] *Standard Specification for Rope-Lay-Stranded Copper Conductors Having Concentric-Stranded Members, for Electrical Conductors*, American Society for Testing and Materials (ASTM) B173.

[35] *Standard Specification for Copper Conductors for Use in Hookup Wire for Electronic Equipment*, American Society for Testing and Materials (ASTM) B286.

[36] G.T. Burstein, "The Tarnishing Conundrum of Silver," Keynote speech, day2, The Goldsmiths' Company Jewelry Materials Congress, 8/9 Jul

2019, Goldsmiths' Hall, Gutter Lane, London [Online]. https://www.assayofficelondon.co.uk/media/2566/the-tarnishing-conundrum-of-silver-g-tim-burstein.pdf.

[37] B.H. Chudnovsky, D. Swindler, and J.R. Thompson, "Silver Corrosion and Whiskers Growth on Power Contacts in Industrial Atmosphere of Pulp and Paper Plants," *Proceedings of the IEEE/IAS Pulp & Paper Industry Technical Paper Conference*, pp. 155–162, June 2001.

[38] B.H. Chudnovsky, D. Swindler, and J.R. Thompson, "A Touch of Gray," *IEEE Industry Application Magazine*, vol. 8, no. 5, pp. 45–52, 2002.

[39] B.H. Chudnovsky, "Degradation of Power Contacts in Industrial Atmosphere: Silver Corrosion and Whiskers," *Proceedings of the 48th IEEE Holm Conference on Electrical Contacts*, pp. 140–147, 2002 [Online]. https://nepp.nasa.gov/whisker/reference/tech_papers/chudnovsky2002-paper-silver-corrosion-whiskers.pdf (accessed Nov. 30, 2022).

[40] B.D. Dunn, A. de Rooij, and D.S. Collins, "Corrosion of Silver-Plated Conductors," *European Space Agency Journal*, vol 8, pp. 307–335, 1984 [Online]. https://escies.org/download/webDocumentFile?id=62170 (accessed Nov. 30, 2022).

[41] *Investigation of Silver-Plated Conductor Corrosion (Red Plague)*, SAE Standard AIR4487.

[42] Guidance for the Development and Implementation of a White Plague Control Plan (WPCP), IPC-WP-114 Rev A, IPC International (IPC), Oct 2021.

[43] Fluoride Offgassing in Fluoropolymer Insulations, SAE Standard ARP6229

[44] "Under Coating Corrosion of Silver-Plated Copper in a Circuit Breaker," Corrosion Testing Laboratories [Online]. http://www.corrosionlab.com/Failure-Analysis-Studies/20034.under-coating.silver-plated-copper.htm (accessed Dec. 2. 2022).

[45] M. Traskos, "Glycol and Silver-Plated Conductors" [Online]. https://lectromec.com/glycol-and-silver-plated-conductors/ (accessed Nov. 30, 2022).

[46] S. Harput, "Fields of Application of Nickel-Plated Copper Conductor" [Online]. http://www.sarkuysan.com/Upload/Document/document_3456096f90eb4f61a619f1376fdee351.pdf (accessed Nov. 20, 2022).

[47] T.R. Long, "Electrodeposited Memory Elements for a Nondestructive Memory," *Journal of Applied Physics*, Supp to vol 31, no. 5, pp. 123S–124S, 1960.

[48] H.K. Lotsch, "Magnetic-Field Design Considerations for a Plated-Wire memory," *IEEE Transactions on Computers*, vol. C-18, no. 10, pp. 894–899, 1969.

[49] J.S. Mathias and G.A. Fedde, "Plated-Wire technology: A Critical Review," *IEEE Transactions on Magnetics*, vol. MAG-5, no. 4, pp. 728–751, 1969.

[50] W.A. England, "Applications of Plated Wire to the Military and Space Environments," *IEEE Transactions on Magnetics*, vol. MAG-6, no. 3, pp. 528–534, 1970.

[51] L.D. Bolton, C.D. Crosby, and J.A. Howe. "Plated Wire Memory Usage on the UNIVAC Minutemen Weapon System Computer" [Online]. http://www.vipclubmn.org/Articles/PlatedWire.pdf (accessed Nov. 20, 2022).

[52] R.M. Rowe, "Filter Line Wiring Designs in Aircraft," in SAE Aerospace Technology Conference and Exposition, Sep. 1990, pp. 1653–1658.

[53] *EMC Eupen EMI Cable Solutions* [Online]. https://emceupen.com/ (accessed Nov. 20, 2022).

[54] M. Kirschvink,L. Halbach, and M. Louges, "Interference Suppression on Motor Control Cables for Frequency Converters," in *Applied Power Electronics Conference and Exposition Annual IEEE Conference – APEC*, 1996, pp. 593–600.

[55] Wave-X™ Z [Online]. Available: https://www.hexcel.com/Products/WaveXZ (accessed Nov. 20, 2022).

[56] J.C. Maxwell, "XXV. On Physical Lines of Force Part I. The Theory of Molecular Vortices Applied to Magnetic Phenomena," *The London, Edinburgh, and Dublin Philosophical Magazine and Journal of Science, 4th Series*, vol. 21, no. 139, pp. 161–175, 1861.

[57] J.C. Maxwell, "XXLIV. On Physical Lines of Force Part II. The Theory of Molecular Vortices Applied to Magnetic Phenomena," *The London, Edinburgh, and Dublin Philosophical Magazine and Journal of Science, 4th Series*, vol. 21, no. 140, pp. 281–291, 1861.

[58] J.C. Maxwell, "LI. On Physical Lines of Force Part II. The Theory of Molecular Vortices Applied to Magnetic Phenomena (concluded from p. 291)," *The London, Edinburgh, and Dublin Philosophical Magazine and Journal of Science, 4th Series*, vol. 21, no. 141, pp. 338–347, 1861.

[59] J.C. Maxwell, "III. On Physical Lines of Force Part III. The Theory of Molecular Vortices Applied to Statical Electricity," *The London, Edinburgh, and Dublin Philosophical Magazine and Journal of Science, 4th Series*, vol. 23, no. 151, pp. 12–24, 1862.

[60] J.C. Maxwell, "XIV. On Physical Lines of Force Part IV. The Theory of Molecular Vortices Applied to the Action of Magnetism on Polarized Light," *The London, Edinburgh, and Dublin Philosophical Magazine and Journal of Science, 4th Series*, vol. 23, no. 152, pp. 85–95, 1862.

[61] "Wire Insulation – A Material Guide Types and Applications" [Online]. https://www.electroprep.com/blog/wire-insulation-a-material-guide-types-and-applications/ (accessed Jan. 7, 2023).

[62] "Wire Wisdom Cable Jacket Types 101" [Online]. https://www.anixter.com/content/dam/anixter/resources/wire-wisdom/anixter-cable-jacket-types-101-wire-wisdom-en.pdf (accessed Jan. 7, 2023).

[63] "Cable Insulation" [Online]. https://www.picwire.com/resources/cable-gram/cable-insulation/ (accessed Jan. 7, 2023).

[64] D. Hill. "Types of Wire and Cable Insulation" [Online]. https://precmfgco.com/blog/wire-cable-insulation-types/ (accessed Jan. 7, 2023).

[65] J.L. Mead, Z. Tao, and H.S. Liu, "Insulation Materials for Wire and Cable Applications," *Rubber Chemistry and Technology*, vol. 75, no. 4, pp. 701–712, 2002, DOI:10.5254/1.3544996

[66] Anixter, *Technical Information Handbook Wire and Cable* [Online]. https://www.anixter.com/en_us/resources/literature/technical-references/wire-and-cable-technical-information-handbook.html (accessed Jan. 7, 2023).

[67] M.P. Stevens, F. Rodriguez, and J. Preston, "Major Industrial Polymers," Brittanica [Online]. https://www.britannica.com/topic/industrial-polymers-468698 (accessed Mar. 26, 2023).

[68] "Who Invented Vinyl? The History of PVC" [Online]. https://frshcustoms.com/vinyl-pvc-invention-history/ (accessed Jan. 7, 2023).

[69] "History of PVC in the Window" [Online]. https://www.museoagostini.com/en/pvc-history/ (accessed Jan. 7, 2023).

[70] "A Brief History of PVC" [Online]. https://www.piper-plastics.com/2017/03/27/a-brief-history-of-pvc/ (accessed Jan. 7, 2023).

[71] "Waldo Semon American Chemist," [Online]. https://www.britannica.com/biography/Waldo-Semon (accessed Jan. 7, 2023).

[72] "Waldo L. Semon Polyvinyl Chloride" https://www.invent.org/inductees/waldo-l-semon (accessed Jan. 7, 2023).

[73] A. Jagger, "Polyethylene: Discovered by Accident 75 Years Ago" [Online]. https://www.icis.com/explore/resources/news/2008/05/12/9122447/polyethylene-discovered-by-accident-75-years-ago/ (accessed Jan. 14, 2023).

[74] M. Lauzon, "PE: The resin that helped win World War II" [Online]. https://www.plasticsnews.com/article/20070806/NEWS/308069961/pe-the-resin-that-helped-win-world-war-ii (accessed Jan. 14, 2023).

[75] S. Cotton, "Polyethylene" [Online]. https://www.chemistryworld.com/podcasts/polythene/3005911.article (accessed Jan. 14, 2023).

[76] "Gibson and Fawcett" [Online]. https://plastiquarian.com/?page_id=14255 (accessed Jan. 14, 2023).

[77] "CPE: A Unique Polymer for Rubber and Plastics Extrusion" [Online]. https://www.thefreelibrary.com/CPE%3A+A+unique+polymer+for+rubber+and+plastics+extrusion.-a0665228936 (accessed Mar. 19, 2023).

[78] "This Month in Physics History April 6, 1938: Discovery of Teflon" APS News, series II, vol. 30, no. 4, pp. 2 & 7, Apr 2021 [Online]. https://www.aps.org/publications/apsnews/202104/upload/April-2021.pdf (accessed Jan. 22, 2023).

[79] N. Hampton, R. Hartlein, H. Lennartsson, H. Orton, and R. Ramachandran, "Long Life XLPE Insulated Power Cable" [Online]. https://www.neetrac.gatech.edu/publications/jicable07_C_5_1_5.pdf (accessed Mar. 19, 2023).

[80] G.B. Kauffman, G.P. Bierwagen, M.P. Stevens, "Major Industrial Polymers" [Online]. https://www.britannica.com/topic/industrial-polymers-468698/Polychloroprene-chloroprene-rubber-CR (accessed Apr. 9, 2023).

[81] Anixter, "Technical Information Handbook Wire and Cable" [Online]. https://www.anixter.com/content/dam/Anixter/Guide/7H0011X0_W%26C_Tech_Handbook_Sec_03.pdf (accessed Apr. 9, 2023).

[82] "CR-Polychloroprene (Neoprene)" [Online]. https://polymerdatabase.com/Elastomers/Chloroprene.html (accessed Apr. 9, 2023).

[83] C.C. Landinger, "Sheaths, Jackets, and Armors," in *Electrical Power Cable Engineering*, W.A. Thue, Ed., 2nd ed., Marcel Dekker, Inc., New York, NY, USA, 2003, p. 125.

[84] "All About Neoprene – Properties, Uses and Applications" [Online]. https://www.thomasnet.com/articles/plastics-rubber/traits-applications-neoprene/ (accessed Apr. 9, 2023).
[85] "Types, Properties, and Uses of Hypalon" [Online]. http://www.inforubber.com/product/CSM/Hypalon%2020.pdf (accessed Apr. 9, 2023).
[86] "Hypalon: Hype About Discontinuing" [Online]. https://www.anixter.com/en_us/resources/literature/wire-wisdom/hypalon-hype-about-discontinuing.html (accessed Apr. 9, 2023).
[87] F. Salva, "Memoria Sobre la Electricidad Aplicada a la Telegrafia," Memorias de la Real Academia de Ciencias Naturales y Artes de Barcelona (Segunda Epoca), Tomo 1, pp. 2–12, 16 Diciembre 1795.
[88] K.N. Mathes, "A Brief History of Development in Electrical Insulation," presented at the 20th Electrical and Electronics Insulation Conf., Boston, MS, USA, Oct. 7–10, 1991, DOI: 10.1109/EEIC.1991.162590.
[89] W.F. Cooke and C. Wheatstone, "Improvements in Giving Signals and Sounding Alarums in Distant Places by Means of Electric Currents Transmitted Through Metallic Circuits," England & Wales, A.D. 1840, No. 7390.
[90] W.F. Cooke and C. Wheatstone, "Improvements in Giving Signals and Sounding Alarums in Distant Places by Means of Electric Currents Transmitted Through Metallic Circuits," England & Wales, A.D. 1837, No. 8345.
[91] S.F.B. Morse, "Improvement in the Mode of Communicating Information by Signals by the Application of Electro-Magnetism," U.S. Patent 1647, Jun. 20, 1840.
[92] F.J. Mouat, "Correspondence Relative to a Remarkable Variety of Caoutchouc, Termed Gutta Percha or Gutta Tuban, etc.," *The Journal of the Agricultural & Horticultural Society of India*, vol. 3, no. 5, pp. 101–106, 1843.
[93] M. Faraday, "XXV. On the Use of Gutta Percha in Electrical Insulation," *Philosphical Magazine, Series III*, vol. 32, no. 214, pp. 165–167, 1848. DOI:10.1080/14786444808645958
[94] "Specimen of 'Crampton's Dover and Calais Telegraph Cable, Laid in 1851 and Remained in Use Until 1875, Made by R S Newall and Company, Millwall, London, England, 1851" [Online]. https://collection.sciencemuseumgroup.org.uk/objects/co32953/sample-of-1851-dover-calais-telegraph-cable-1851-cable (accessed Jan. 7, 2023).
[95] "Specimen of Dover to Calais Telegraph Cable Showing Corrosion, Made by R S Newall and Company, Millwall, London, England, 1851" [Online]. https://collection.sciencemuseumgroup.org.uk/objects/co33775/specimen-of-dover-calais-telegraph-cable-1851-cable (accessed Jan. 7, 2023).
[96] B.J. Hunt, "Insulation for an Empire: Gutta-Percha and the Development of Electrical Measurement in Victorian Britain," in Semaphores to Shortwaves: Proceedings of a Conference on the Technology and Impact of Early Telecommunications, Held by the Royal Society for the Encouragement of Arts, Manufactures and Commerce, London, UK, July 29, 1996, Organized

by The British Society for the History of Science, The Newcomen Society and the RSA, Published by Royal Society for the Encouragement of Arts, Manufactures & Commerce, London, UK, 1998.

[97] A.G. Bell, "Improvement in Telephone Circuits," U.S. Patent 220 791, Oct. 21, 1879.

[98] A.G. Bell, "Telephone-Circuit," U.S. Patent 244 426, Jul. 19, 1881.

[99] B.J. Hunt, ""An Ill-Understood Effect of Induction" Telegraphy and Field Theory in Victorian Britain," in *Imperial Science Cable Telegraphy and Electrical Physics in the Victorian British Empire*, Cambridge Univ. Press, Cambridge, United Kingdom, 2021, ch. 1, pp. 3–36.

[100] M. Faraday, "On Electric Induction-Associated Cases of Current and Static Effects," *Notices of the Proc. at the Mtgs of the Members of the Roy. Inst. (1851–1854)*, vol. 1, pp. 345–355, 1854.

[101] W. Thomson, "On the Theory of the Electric Telegraph," *Proceedings of the Royal Society of London (1854–1855)*, vol. 7, pp. 382–399, 1855, DOI:10.1098/rspl.1854.0093.

[102] C. Donaghy-Spargo, "On Heaviside's Contributions to Transmission Lines Theory: Waves, Diffusion and Energy Flux," *Philosophical Transactions of the Royal Society A*, vol. 376, no. 2134, art. 20170457, 2018, DOI:10.1098/rsta.2017.0457.

[103] "Historic Electric Cable arrives in Tamworth" [Online]. https://tamworth-powerstationmuseum.com.au/historic-electric-cable-arrives-in-tamworth/ (accessed Jan. 7, 2023).

[104] T.A. Edison, "Underground Conductor," U.S. Patent 251 552, Dec. 27, 1881.

[105] "History of Electrical Systems and Cables" [Online]. https://www.myin-sulators.com/acw/bookref/histsyscable (accessed Jan. 7, 2023).

[106] "What is the History of Electrical Cables?" [Online]. https://www.elandcables.com/the-cable-lab/faqs/faq-what-is-the-history-of-electrical-cables (accessed Jan. 7, 2023).

[107] T.A. Edison, "Electric Conductor," U.S. Patent 470 924 426, Mar. 15, 1892.

[108] "A Blog on the History of Electrical Wire" [Online]. https://nicab.co.uk/a-blog-on-the-history-of-electrical-wire/ (accessed Jan. 7, 2023).

[109] "History of Cable Manufacturing-Major Milestones in Development" [Online]. https://cablecommunity.com/history-of-electric-cables-major-mile-stones-in-development/ (accessed Jan. 7, 2023).

Chapter 3
Basic electromagnetic theory

This chapter provides background information and a brief review of vector analysis and fundamental electromagnetic theory. Unless otherwise indicated, the reader is referred to references [1–14] or similar texts for more detail.

3.1 Basic review of vector analysis

3.1.1 Fundamental concepts of vector algebra

A vector is a mathematical object that has both magnitude and direction. The various elements of Maxwell's equations in their present form use vector quantities that describe fields in three-dimensional space. In this text, vectors are denoted by bold letters and are graphically represented as a directed line segment. Using mathematical notation, the magnitude or length of a vector is denoted as $|\mathbf{A}| = A$. In graphical form, a vector's magnitude can be represented by the relative length of one vector in comparison to one or more other vectors.

Mathematical operations using vectors fall into two separate categories: (1) vector algebra, wherein algebraic operations on vectors including addition, subtraction, and multiplication are defined, and (2) vector calculus, wherein vectors may be differentiated or integrated. All these mathematical operations require that a coordinate system be selected. Many different types of orthogonal coordinate systems exist, and the final choice of a particular system for a particular problem is typically based on the geometry of the problem itself from considerations of symmetry and simplicity in representation. In general, there are three orthogonal coordinate systems that play a major role in the study of electromagnetics because many problems in electromagnetics are most easily represented using one of these three systems: the Cartesian system, the cylindrical system, and the spherical system.

Each of these three systems comprises three families of surfaces that are mutually perpendicular. In the Cartesian system, we have three axes usually denoted as x, y, and z. At any given value of x, y, and z, a set of three mutually perpendicular, or orthogonal, planes may be constructed. A cylindrical system may be superimposed on the Cartesian system such that a cylindrical surface with radius ρ may be constructed surrounding a central axis, often chosen as z. For any point on the cylindrical surface, we have the intersection with that surface an infinite half-plane surface parallel to and rotated about the central axis through an angle ϕ, often measured from the x-axis of the underlying Cartesian system to the point of interest, with positive values measured in the anticlockwise direction. In similar fashion, a spherical system may be superimposed on the Cartesian system such that a spherical surface with radius r may be constructed surrounding a central axis, often chosen as z. For any point on the spherical surface, we have the intersection with that surface of an infinite half-plane surface parallel to and rotated about the central axis through an angle ϕ, often between the x-axis of the underlying Cartesian system and the point of interest, with positive values measured in the anticlockwise direction, and the intersection of the spherical surface with the surface of a right-circular cone, the axis of which coincides with the chosen central axis of the underlying Cartesian system, whose apex is at the centre of the underlying Cartesian system, and whose half-angle, θ, is measured from the central axis to the point of interest, with positive values measured in the clockwise direction. Figure 3.1 illustrates these three orthogonal coordinate systems.

In a more general sense, we can assign variables a_1, a_2, and a_3 to represent the three mutually perpendicular curvilinear surfaces in these orthogonal coordinate

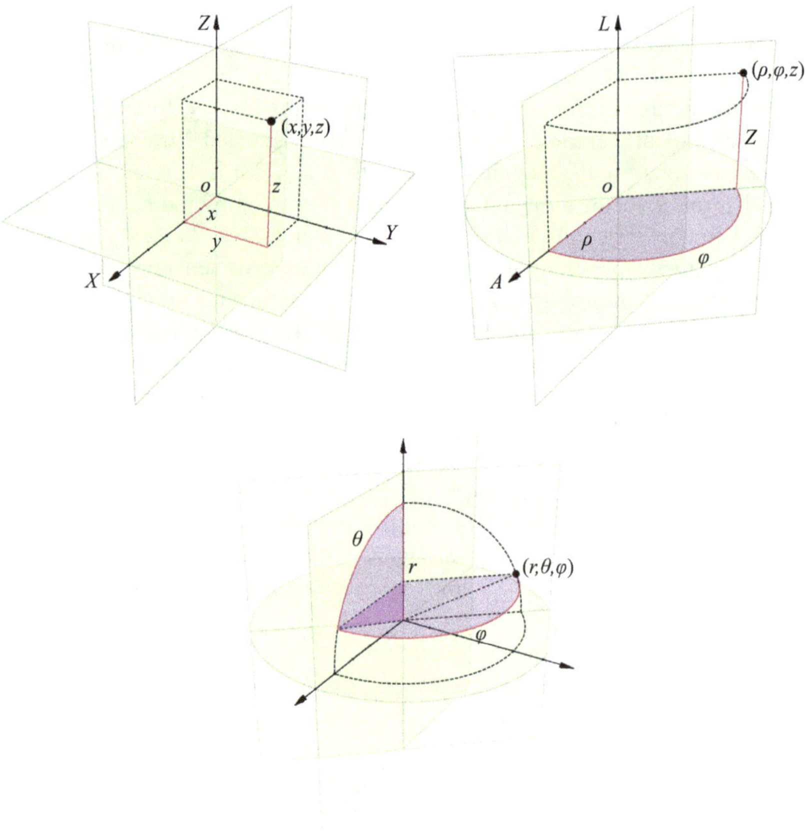

Figure 3.1 Illustration of coordinate systems: (a) Cartesian, (b) cylindrical, and (c) spherical

systems. The intersections of surfaces in these orthogonal coordinate system form three families of lines. At any point of intersection of these surfaces, a set of base, or unit, vectors may be constructed that are tangent to the lines of intersection of the surfaces. These unit vectors enable the unambiguous construction of electromagnetic field vector quantities in any of the three orthogonal coordinate systems by providing direction along each of the orthogonal surfaces. With the foregoing in mind, we can now proceed to make a series of statements about vectors that hold true regardless of the chosen system.

A vector, **A**, in an orthogonal coordinate system can be described by the sum of the products of the components and the corresponding unit vectors in the three orthogonal directions:

$$\boldsymbol{A} = A_1\boldsymbol{a}_1 + A_2\boldsymbol{a}_2 + A_3\boldsymbol{a}_3 \tag{3.1}$$

In this expression, the unit vectors denoted by $\boldsymbol{a}_1$, $\boldsymbol{a}_2$, and $\boldsymbol{a}_3$ provide the direction information, and the components denoted by A_1, A_2, and A_3 provide the magnitude information. The magnitudes of the components may vary as a function of the location and direction of **A**. The magnitude of **A** is found by taking the root of the sum of the squares of each component:

$$A = |\boldsymbol{A}| = \sqrt{\left(A_1{}^2 + A_2{}^2 + A_3{}^2\right)} \tag{3.2}$$

Vectors may be added by adding their respective components as

$$\boldsymbol{A} + \boldsymbol{B} = (A_1 + B_1)\boldsymbol{a}_1 + (A_2 + B_2)\boldsymbol{a}_2 + (A_3 + B_3)\boldsymbol{a}_3 \tag{3.3}$$

The length of a vector may be altered by simply multiplying each component of the vector by a scalar factor k:

$$k\boldsymbol{A} = kA_1\boldsymbol{a}_1 + kA_2\boldsymbol{a}_2 + kA_3\boldsymbol{a}_3 \tag{3.4}$$

Two vectors may be subtracted one from the other by assigning the value of -1 to the scalar k.

$$\boldsymbol{A} + k\boldsymbol{B} = \boldsymbol{A} + (-\boldsymbol{B}) = (A_1 - B_1)\boldsymbol{a}_1 + (A_2 - B_2)\boldsymbol{a}_2 + (A_3 - B_3)\boldsymbol{a}_3 \tag{3.5}$$

Two vectors may be multiplied together in one of two ways. These are the scalar or dot product, sometimes referred to as the inner product, and the vector or cross product, sometimes referred to as the outer product.

The dot product is defined as the scalar product of the magnitude of each of the two vectors multiplied by the cosine of the lesser of the two angles between the two vectors:

$$\boldsymbol{A} \cdot \boldsymbol{B} \equiv |\boldsymbol{A}||\boldsymbol{B}|\cos\theta_{AB} \tag{3.6}$$

Equation (3.6) may also be written as

$$\boldsymbol{A}\cdot\boldsymbol{B} = (A_1B_1) + (A_2B_2) + (A_3B_3) \tag{3.7}$$

The dot product is less than or equal to the product of their respective magnitudes, may be positive or negative, depending on whether θ_{AB} is less than or greater than $\pi/2$ radians, and is equal to 0 if the two vectors are perpendicular to one another. A vector dotted with itself is equal to the magnitude of the vector squared.

The dot product is commutative and distributive:

$$\boldsymbol{A} \cdot \boldsymbol{B} = \boldsymbol{B} \cdot \boldsymbol{A} \tag{3.8}$$

$$\boldsymbol{A} \cdot (\boldsymbol{B} + \boldsymbol{C}) = \boldsymbol{A} \cdot \boldsymbol{B} + \boldsymbol{A} \cdot \boldsymbol{C} \tag{3.9}$$

The dot product is not associative, as no more than two vectors may multiplied in this fashion.

The cross product is defined as a vector product, the magnitude of which is equal to the scalar product of the magnitude of each of the two vectors multiplied by the sine of the lesser of the two angles between the two vectors.

$$\boldsymbol{A} \times \boldsymbol{B} \equiv |\boldsymbol{A}||\boldsymbol{B}| \sin \theta_{AB} \boldsymbol{u}_{\text{normal}} \tag{3.10}$$

Equation (3.10) may also be written as

$$\boldsymbol{A} \times \boldsymbol{B} = (A_2B_3 - A_3B_2)\boldsymbol{a}_1 + (A_3B_1 - A_1B_3)\boldsymbol{a}_2 + (A_1B_2 - A_2B_1)\boldsymbol{a}_3 \tag{3.11}$$

or in matrix form as

$$\boldsymbol{A} \times \boldsymbol{B} = \begin{bmatrix} \boldsymbol{a}_1 & \boldsymbol{a}_2 & \boldsymbol{a}_3 \\ A_1 & A_2 & A_3 \\ B_1 & B_2 & B_3 \end{bmatrix} \tag{3.12}$$

The direction of the vector resulting from the cross product, represented by the unit vector, $\boldsymbol{u}_{\text{normal}}$, is normal to the plane formed by the two vectors. The sense of the resulting vector is that of advance of a right-hand screw rotated from the first vector to the second through the angle θ_{AB} between them, where θ_{AB} is less than π radians, as illustrated in Figure 3.2.

Note that order matters in the cross product, such that

$$\boldsymbol{A} \times \boldsymbol{B} = -\boldsymbol{B} \times \boldsymbol{A} \tag{3.13}$$

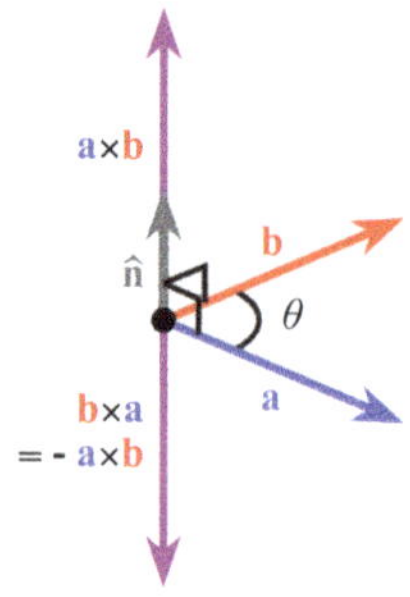

Figure 3.2 Illustration of cross product

It is evident the cross product is not commutative. The cross product is however distributive:

$$\boldsymbol{A} \times (\boldsymbol{B} + \boldsymbol{C}) = \boldsymbol{A} \times \boldsymbol{B} + \boldsymbol{A} \times \boldsymbol{C} \tag{3.14}$$

Three vectors may be multiplied together in one of three ways. The first of these is expressed as

$$\boldsymbol{A}(\boldsymbol{B} \cdot \boldsymbol{C}) \tag{3.15}$$

which is the same mathematically as (3.4) in which a scalar multiplies a vector.

The scalar triple product is the second of these three methods. This product may be expressed as

$$\boldsymbol{A} \cdot (\boldsymbol{B} \times \boldsymbol{C}) = \boldsymbol{B} \cdot (\boldsymbol{C} \times \boldsymbol{A}) = \boldsymbol{C} \cdot (\boldsymbol{A} \times \boldsymbol{B}) \tag{3.16}$$

or equivalently as

$$\boldsymbol{A} \cdot (\boldsymbol{B} \times \boldsymbol{C}) = -\boldsymbol{A} \cdot (\boldsymbol{C} \times \boldsymbol{B}) = -\boldsymbol{B} \cdot (\boldsymbol{A} \times \boldsymbol{C}) = -\boldsymbol{C} \cdot (\boldsymbol{B} \times \boldsymbol{A}) \tag{3.17}$$

The third of these three methods is known as the vector triple product. This product may be expressed as

$$\boldsymbol{A} \times (\boldsymbol{B} \times \boldsymbol{C}) = \boldsymbol{B}(\boldsymbol{A} \cdot \boldsymbol{C}) - \boldsymbol{C}(\boldsymbol{A} \cdot \boldsymbol{B}) \tag{3.18}$$

Like the dot product, the vector triple product is not associative:

$$\boldsymbol{A} \times (\boldsymbol{B} \times \boldsymbol{C}) \neq (\boldsymbol{A} \times \boldsymbol{B}) \times \boldsymbol{C} \tag{3.19}$$

Division by a vector is not defined.

3.1.2 Scalar versus vector fields

The concept of a field is central to the understanding of electromagnetic theory. The term field can be used to describe some quantity existing within some volume or region of space, such as temperature or fluid flow, and this quantity can vary continuously throughout the volume or region as a function of spatial position and perhaps of time as well. Two types of fields exist that will have interest in the study of electromagnetics, those being a scalar field and a vector field.

A scalar field has a variation of magnitude in space and time but has no directionality. A classic example of a scalar field is the temperature distribution throughout a room. At any given location within the room, the temperature will have some magnitude. This same magnitude will exist at many different locations, and a curvilinear surface that extends within the volume of the room can be constructed upon which this same magnitude exists. Many other similar surfaces upon which other temperatures exist can also be constructed throughout the room. Each surface so constructed is unique with respect to the others, in that on each surface only a single temperature can be found. A second example is the sphere of light surrounding a single light bulb. At any given distance from the light bulb, the light intensity will have some magnitude that may be expressed in lumens. This light intensity will vary as a function of the distance from the light bulb. A series of curvilinear surfaces, each having a single value of lumens, can be constructed surrounding the light bulb, entirely similar to the contours of temperature in the room example.

A vector field exhibits similar characteristics to those of a scalar field with the very important difference that the quantity of interest will have not only a variation of magnitude in space and time but also a variation in a specific direction or angle with respect to a reference, such as the x-axis in the Cartesian coordinate system. A classic example of a vector field is the wind blowing over a geographic area. This wind has a magnitude, or speed, that varies as a function of location but also has a specific direction or angle. The velocity of the wind can be depicted by vectors of length proportional to the speed of the wind at a particular location together with the angle of the vectors with respect to some coordinate system depicting the direction away from or towards which the wind is blowing.

3.1.3 Fundamental concepts of vector calculus

As stated previously, Maxwell's equations in their present form are expressed in terms of vector quantities. Maxwell's equations are described and manipulated mathematically by several useful vector calculus operations. It is thus important to gain understanding of these operations.

The first of these is the gradient, a mathematical concept that describes the space rate of change of a scalar field. The gradient is a vector that points in the direction of the greatest change in a function value and is zero-valued at a local maxima or minima of that function because in either case, no single direction of greatest change can be defined. From the discussion in the previous section, a scalar field consists of a set of single scalar-valued surfaces with some distance of separation between each surface. In a generalized orthogonal coordinate system, each surface whose value is constant can be expressed as a function of the three variables a_1, a_2, and a_3 that represent the three mutually perpendicular curvilinear surfaces in the system:

$$f(a_1, a_2, a_3) = c \tag{3.20}$$

If we have two such surfaces separated by an infinitesimal distance dl, it is possible to define the second surface as $f = c + df$. The total differential change can then be expressed as

$$df = \frac{\partial f}{\partial a_1} da_1 + \frac{\partial f}{\partial a_2} da_2 + \frac{\partial f}{\partial a_3} da_3 \tag{3.21}$$

$$= \frac{\partial f}{\partial l_1} dl_1 + \frac{\partial f}{\partial l_2} dl_2 + \frac{\partial f}{\partial l_3} dl_3 \tag{3.22}$$

The vector distance between these two surfaces is given as

$$\boldsymbol{dl} = dl_1 \boldsymbol{a}_1 + dl_2 \boldsymbol{a}_2 + dl_3 \boldsymbol{a}_3 \tag{3.23}$$

We can then write (3.22) as the dot product of two vectors; thus

$$df = \left(\frac{\partial f}{\partial l_1} \boldsymbol{a}_1 + \frac{\partial f}{\partial l_2} \boldsymbol{a}_2 + \frac{\partial f}{\partial l_3} \boldsymbol{a}_3 \right) \cdot \boldsymbol{dl} \tag{3.24}$$

In (3.24), the quantity inside the parentheses is known as the gradient of *f*. This quantity may then be written as

$$\mathbf{grad}\, f = \left(\frac{\partial f}{\partial l_1} \boldsymbol{a}_1 + \frac{\partial f}{\partial l_2} \boldsymbol{a}_2 + \frac{\partial f}{\partial l_3} \boldsymbol{a}_3 \right) = \nabla f \tag{3.25}$$

The symbol "∇" is known as "del". The gradient shown in (3.25) is defined as the vector that represents both the magnitude and direction of the maximum space rate of change of a scalar as the gradient of that scalar. Substituting (3.25) into (3.24) yields

$$df = \nabla f \cdot \boldsymbol{dl} \tag{3.26}$$

which, from (3.6), can be written as

$$df = |\nabla f||\boldsymbol{dl}|\cos\theta = |\nabla f|dl\cos\theta \tag{3.27}$$

where θ is the angle between the direction of ∇f and the direction of the length vector $\boldsymbol{dl}$ tangent to the path of dl.

If $\theta = 0$, the maximum space rate of change will occur and we then have

$$\left.\frac{df}{dl}\right|_{\max} = |\nabla f| \tag{3.28}$$

To enable proper descriptions of differential path lengths, surface areas, and volumes in different coordinate systems wherein not all of these are strictly lengths, a means is needed whereby it is possible to define the relationship between a differential change in position of a variable on a surface to a corresponding differential change in the length along that surface. Scaling factors, aka metric coefficients, are used for this purpose and are denoted simply as h_i. Accordingly, we have

$$dl_i = h_i da_i \tag{3.29}$$

where h_i may itself be a function of the a_i coordinates.

Substituting metric coefficients into (3.25) yields for general orthogonal coordinate systems

$$\nabla f = \left(\frac{1}{h_1} \frac{\partial f}{\partial a_1} \boldsymbol{a}_1 + \frac{1}{h_2} \frac{\partial f}{\partial a_2} \boldsymbol{a}_2 + \frac{1}{h_3} \frac{\partial f}{\partial a_3} \boldsymbol{a}_3 \right) \tag{3.30}$$

In a Cartesian system, all the metric coefficients are equal to 1. Plugging in these metric coefficients into (3.30) yields the expression for the gradient in a Cartesian system as

$$\nabla f = \left(\frac{\partial f}{\partial x} \boldsymbol{a}_x + \frac{\partial f}{\partial y} \boldsymbol{a}_y + \frac{\partial f}{\partial z} \boldsymbol{a}_z \right) \tag{3.31}$$

where $\boldsymbol{a}_x$, $\boldsymbol{a}_y$, and $\boldsymbol{a}_z$ are unit vectors in the corresponding axis directions of the Cartesian coordinate system. From (3.31), we can infer the so-called "del operator" in the Cartesian system as

$$\nabla = \left(\frac{\partial}{\partial x}\boldsymbol{a}_x + \frac{\partial}{\partial y}\boldsymbol{a}_y + \frac{\partial}{\partial z}\boldsymbol{a}_z\right) \tag{3.32}$$

Metric coefficients differ between coordinate systems. In the Cartesian system, we have

$$h_x = h_y = h_z = 1 \tag{3.33}$$

For the cylindrical system, we have

$$h_\rho = 1;\ h_\phi = \rho;\ h_z = 1 \tag{3.34}$$

and for the spherical system, we have

$$h_r = 1;\ h_\theta = r;\ h_\phi = r\sin\theta \tag{3.35}$$

Using the metric coefficients in (3.34) and (3.35) associated with the cylindrical and spherical systems, (3.30) can be written as expressions for the determination of gradients in those coordinate systems as follows:

$$\nabla f(\rho, \phi, z) = \frac{\partial f}{\partial \rho}\boldsymbol{a}_\rho + \frac{1}{\rho}\frac{\partial f}{\partial \phi}\boldsymbol{a}_\phi + \frac{\partial f}{\partial z}\boldsymbol{a}_z \tag{3.36}$$

$$\nabla f(r, \theta, \phi) = \frac{\partial f}{\partial r}\boldsymbol{a}_r + \frac{1}{r}\frac{\partial f}{\partial \theta}\boldsymbol{a}_\theta + \frac{1}{r\sin\theta}\frac{\partial f}{\partial \phi}\boldsymbol{a}_\phi \tag{3.37}$$

where $\boldsymbol{a}_\rho$, $\boldsymbol{a}_\phi$,$\boldsymbol{a}_z$,$\boldsymbol{a}_r$,$\boldsymbol{a}_\theta$, and $\boldsymbol{a}_\phi$ are unit vectors in the corresponding axis directions of the two coordinate systems.

Note that although (3.30) will generate correct values for gradients in the Cartesian, cylindrical, and spherical coordinate systems, and for the divergence or the curl of a vector field in the Cartesian system, it must be remembered that for use in the determination of the divergence or the curl of a vector field in other orthogonal coordinate systems, the differentiation of base vectors can, and will, lead to new vectors in different directions. For this reason, the so-called "del operator" is only defined in the Cartesian system, and "del" itself serves a symbolic role otherwise. The foregoing comments will become clearer in the discussion of divergence and curl.

The second vector calculus operation useful for the study of electromagnetics is the line integral of a vector field $\boldsymbol{F}$ along a contour C between two points, $P1$ and $P2$, expressed as

$$\int_{P1}^{P2} \boldsymbol{F}\cdot d\boldsymbol{l} = \int_{P1}^{P2} \boldsymbol{F}\cdot(dl_1\boldsymbol{a}_1 + dl_2\boldsymbol{a}_2 + dl_3\boldsymbol{a}_3) \tag{3.38}$$

Like (3.26) and (3.27), (3.38) may be expressed as

$$\int_{P1}^{P2} \boldsymbol{F}\cdot d\boldsymbol{l} = \int_{P1}^{P2} |\boldsymbol{F}|\cos\theta dl \tag{3.39}$$

Here θ represents the angle between the direction of the applied force, $\boldsymbol{F}$, and the path or contour between points $P1$ and $P2$. dl represents the path length tangential to the path between points $P1$ and $P2$.

If $\boldsymbol{F}$ is a force, this integral represents work accomplished in moving from $P1$ to $P2$ along a specified path or contour. If the force $\boldsymbol{F}$ is replaced by an electric field intensity $\boldsymbol{E}$, the integration evaluates to an electromotive force. If the line integral of $\boldsymbol{F}$ is dependent on the path taken from $P1$ to $P2$, the vector field $\boldsymbol{F}$ is said to be non-conservative. If the line integral of $\boldsymbol{F}$ is independent of the path taken from $P1$ to $P2$, the vector field $\boldsymbol{F}$ is said to be conservative. An example of this is given by

$$\oint \nabla f \cdot d\boldsymbol{l} = 0 \tag{3.40}$$

Equation (3.40) expresses that the line integral of the gradient of a scalar field around a closed loop or contour is identically equal to zero. An example of a conservative vector field is the electric field surrounding a point charge. The gradient from one potential to the next on any two spherical surfaces at any location surrounding the point charge will always have the same value. Integration from one scalar potential to the next and back will always equal zero, regardless of the path or paths taken.

If (3.38) is taken around a closed loop or contour, we have

$$\oint \boldsymbol{F} \cdot d\boldsymbol{l} \tag{3.41}$$

Equation (3.41) may be thought of as representing a circulation, or force, around the contour. Consider a paddlewheel with flat vanes placed in a stream of moving fluid such that the shaft of the paddlewheel is normal to the surface of the fluid. If the fluid is flowing at the same speed on opposite sides of the paddlewheel, the force on the paddlewheel vanes will be the same on either side and the paddlewheel will not rotate. The rotational force on the paddlewheel is balanced. If the fluid is flowing with a variation in speed such that it is greater on one side of the paddlewheel than the opposite side, the paddlewheel will rotate, with the direction of its rotation determined by which side of the paddlewheel is exposed to the greater fluid speed. The rotational force on the paddlewheel is not balanced.

The circulation about a point may be determined by shrinking the contour and the surface area enclosed by the contour. In the limit, we have the definition of circulation density, or circulation per unit area:

$$\lim_{\Delta S \to 0} \frac{\oint \boldsymbol{F} \cdot d\boldsymbol{l}}{\Delta S} \tag{3.42}$$

The circulation density in (3.42) is maximized by suitable orientation of the contour with respect to the vector field $\boldsymbol{F}$. When this is done, and the limit is taken of the product of a unit vector normal to the plane of the contour and the line integral, it yields the curl of $\boldsymbol{F}$. Curl is sometimes referred to as rotation, often abbreviated as "rot".

$$\text{curl}\, \boldsymbol{F} = \lim_{\Delta S \to 0} \frac{1}{\Delta S} \left[(\boldsymbol{u}_{\text{normal}}) \left(\oint \boldsymbol{F} \cdot d\boldsymbol{l} \right) \right]_{\max} = \nabla \times \boldsymbol{F} \tag{3.43}$$

The direction of the unit vector $\boldsymbol{u}_{\text{normal}}$ is normal to the plane of the contour C, and by convention, the positive sense is determined by the direction of the thumb of the right hand when the fingers of the right hand are wrapped around the periphery of the contour C in an anticlockwise direction.

The curl of $\boldsymbol{F}$ may be expressed using a matrix, as shown here:

$$\nabla \times \boldsymbol{F}(a_1, a_2, a_3) = \frac{1}{h_1 h_2 h_3} \begin{bmatrix} h_1 \boldsymbol{a}_1 & h_2 \boldsymbol{a}_2 & h_3 \boldsymbol{a}_3 \\ \dfrac{\partial}{\partial a_1} & \dfrac{\partial}{\partial a_2} & \dfrac{\partial}{\partial a_3} \\ h_1 F_{a_1} & h_2 F_{a_2} & h_3 F_{a_3} \end{bmatrix} \tag{3.44}$$

Using (3.33), (3.34), (3.35), and (3.44), the curl for the Cartesian, cylindrical, and spherical coordinate systems is given by

$$\nabla \times \boldsymbol{F}(x,y,z) = \left(\frac{\partial F_z}{\partial y} - \frac{\partial F_y}{\partial z}\right)\boldsymbol{a}_x + \left(\frac{\partial F_x}{\partial z} - \frac{\partial F_z}{\partial x}\right)\boldsymbol{a}_y + \left(\frac{\partial F_y}{\partial x} - \frac{\partial F_x}{\partial y}\right)\boldsymbol{a}_z \tag{3.45}$$

$$\nabla \times \boldsymbol{F}(r,\phi,z) = \left(\frac{1}{r}\frac{\partial F_z}{\partial \phi} - \frac{\partial F_\phi}{\partial z}\right)\boldsymbol{a}_r + \left(\frac{\partial F_r}{\partial z} - \frac{\partial F_z}{\partial r}\right)\boldsymbol{a}_\phi + \frac{1}{r}\left(\frac{\partial (rF_\phi)}{\partial r} - \frac{\partial F_r}{\partial \phi}\right)\boldsymbol{a}_z \tag{3.46}$$

$$\nabla \times \boldsymbol{F}(r,\theta,\phi) = \frac{1}{r\sin\theta}\left(\frac{\partial (F_\phi \sin\theta)}{\partial \theta} - \frac{\partial F_\theta}{\partial \phi}\right)\boldsymbol{a}_r + \frac{1}{r}\left(\frac{1}{\sin\theta}\frac{\partial F_r}{\partial \phi} - \frac{\partial (rF_\phi)}{\partial r}\right)\boldsymbol{a}_\theta + \frac{1}{r}\left(\frac{\partial (rF_\theta)}{\partial r} - \frac{\partial F_r}{\partial \theta}\right)\boldsymbol{a}_\phi \tag{3.47}$$

For a very small differential areas, $\Delta \boldsymbol{s}$, bounded by the contour C, (3.43) can be stated as

$$(\text{curl } \boldsymbol{F}) \cdot \Delta \boldsymbol{s} = \oint \boldsymbol{F} \cdot d\boldsymbol{l} \tag{3.48}$$

which is a form of Stokes' theorem

$$\int (\text{curl } \boldsymbol{F}) \cdot d\boldsymbol{s} = \oint \boldsymbol{F} \cdot d\boldsymbol{l} \tag{3.49}$$

Stokes' theorem (3.49) equates the net flux of a vector field over an arbitrarily shaped open surface S to the line integral of that same vector field around the contour C that bounds the surface S. Stokes' theorem facilitates the computation of surface integrals or line integrals that would ordinarily be quite difficult by transforming a line integral into a surface integral or vice versa. Stokes' theorem thus connects the concepts of curl and circulation.

Another equally useful integral is that of a vector field $\boldsymbol{F}$ over a surface S, expressed as

$$\int \boldsymbol{F} \cdot d\boldsymbol{s} = \int (\boldsymbol{F} \cdot \boldsymbol{u}_{\text{normal}}) ds = \int |\boldsymbol{F}| \cos \theta ds \tag{3.50}$$

The direction of the unit vector $\boldsymbol{u}_{\text{normal}}$ is normal to the surface S, and by convention, the positive sense is determined by the direction of the thumb of the right hand when the fingers of the right hand are wrapped around the periphery of the surface S in a anticlockwise direction. In (3.50), the angle θ is measured between the vector $\boldsymbol{F}$ and the vector $\boldsymbol{u}_{\text{normal}}$.

If the surface is closed, (3.50) becomes

$$\psi = \oint \boldsymbol{F} \cdot d\boldsymbol{s} \tag{3.51}$$

Equation (3.51) represents the net outward flux through the closed surface S. For a closed surface, the positive sense of the unit normal vector by convention is directed orthogonally outward from the enclosed volume. Note that (3.50) and (3.51) are both scalar quantities. Flux in this context is directly related to the strength of the field passing through a defined surface area, and the sign of ψ indicates whether a source or a sink exists within the enclosed volume. A positive ψ indicates a source is present; a negative ψ indicates a sink is present. If ψ is equal to zero, neither a source nor a sink is present, or both are present with equal strength. If (3.51) is divided by a differential element of the enclosed volume and the volume is allowed to shrink to zero, we have

$$\operatorname{div} \boldsymbol{F} = \lim_{\Delta v \to 0} \frac{\oint \boldsymbol{F} \cdot d\boldsymbol{s}}{\Delta v} = \nabla \cdot \boldsymbol{F} \tag{3.52}$$

Equation (3.52) yields the net outward flux, a scalar quantity, of a vector field per unit volume as the volume shrinks to zero. This is the definition of the divergence of a field.

The divergence can be expressed for general orthogonal coordinate systems as

$$\operatorname{div} \boldsymbol{F} = \nabla \cdot \boldsymbol{F} = \left(\begin{array}{c} \dfrac{1}{h_1 h_2 h_3} \dfrac{\partial(h_2 h_3 F_i)}{\partial a_i} + \dfrac{1}{h_1 h_2 h_3} \dfrac{\partial(h_1 h_3 F_j)}{\partial a_j} \\ + \dfrac{1}{h_1 h_2 h_3} \dfrac{\partial(h_1 h_2 F_k)}{\partial a_k} \end{array} \right) \tag{3.53}$$

Substituting (3.33), (3.34), and (3.35) into (3.53), the divergence for the Cartesian, cylindrical, and spherical coordinate systems is given by

$$\nabla \cdot \boldsymbol{F} = \left(\frac{\partial}{\partial x} F_x + \frac{\partial}{\partial y} F_y + \frac{\partial}{\partial z} F_z \right) \tag{3.54}$$

$$\nabla \cdot \boldsymbol{F}(\rho, \phi, z) = \frac{1}{r} \frac{\partial}{\partial \rho} (\rho F_\rho) + \frac{1}{r} \frac{\partial F_\phi}{\partial \phi} + \frac{\partial F_z}{\partial z} \tag{3.55}$$

$$\nabla \cdot \boldsymbol{F}(r, \theta, \phi) = \frac{1}{r^2} \frac{\partial}{\partial r} (r^2 F_r) + \frac{1}{r \sin \theta} \frac{\partial}{\partial \theta} (F_\theta \sin \theta) + \frac{1}{r \sin \theta} \frac{\partial F_\phi}{\partial \phi} \tag{3.56}$$

If we integrate (3.52) throughout the volume whose boundary is the surface enclosing that volume as that volume shrinks to zero, we obtain a very important relationship known as the divergence theorem, sometimes referred to as Gauss' law:

$$\int \operatorname{div} \boldsymbol{F} dV = \oint \boldsymbol{F} \cdot d\boldsymbol{s} \tag{3.57}$$

Equation (3.57) states the integral of the normal component of any vector field over a closed surface is equal to the integral of the divergence of this vector field throughout the enclosed volume.

Stokes' theorem and the divergence theorem can be used to validate two vector identities we will encounter as we move forward. The first of these is

$$\nabla \times \nabla f = 0 \tag{3.58}$$

which states the curl of the gradient of some scalar field is equal to zero. We can show this using Stokes' theorem

$$\int (\nabla \times \nabla f) \cdot d\boldsymbol{s} = \oint (\nabla f) \cdot d\boldsymbol{l} = 0 \tag{3.59}$$

We showed the line integral around a closed loop is equal to zero in (3.40), and that the field in this case is conservative. Application of Stokes theorem also shows that such a field is irrotational (the curl equals zero).

The second vector identity we can validate is

$$\nabla \cdot (\nabla \times \boldsymbol{F}) = 0 \tag{3.60}$$

which states the divergence of the curl of a vector field is equal to zero. We can show this using the divergence theorem

$$\int \nabla \cdot (\nabla \times \boldsymbol{F}) dV = \oint (\nabla \times \boldsymbol{F}) \cdot d\boldsymbol{s} \tag{3.61}$$

The closed surface integral can be split into two open surfaces that connect one to the other, with contours C_1 and C_2 on either side of the split. This is expressed as

$$\int_{S1} (\nabla \times \boldsymbol{F}) \cdot d\boldsymbol{s} + \int_{S2} (\nabla \times \boldsymbol{F}) \cdot d\boldsymbol{s} = \oint_{C1} \boldsymbol{F} \cdot d\boldsymbol{l} + \oint_{C2} \boldsymbol{F} \cdot d\boldsymbol{l} \tag{3.62}$$

Because the two contours C_1 and C_2 are opposing each other in direction, they have opposite signs and thus cancel out, leaving (3.62) equal to zero, which means that (3.60) and (3.61) must also equal zero. If a vector field has zero divergence, it is said to be divergenceless or solenoidal.

Three additional vector identities that will prove valuable are given by

$$\nabla \times (\nabla \times \boldsymbol{F}), \ \nabla \cdot (\nabla f), \text{ and } \nabla(\nabla \cdot f)$$

These are interrelated in a single expression of a very important identity

$$\nabla^2 \boldsymbol{F} = \nabla(\nabla \cdot f) - \nabla \times (\nabla \times \boldsymbol{F}) \tag{3.63}$$

Equation (3.63) is known as the vector Laplacian. A scalar Laplacian also can be written as

$$\nabla^2 f = \nabla \cdot \nabla f \tag{3.64}$$

Equation (3.63) is used extensively for solution of magnetostatic problems, and (3.64) is used extensively for solution of electrostatic problems involving Poisson's and Laplace's equations.

3.2 Basic review of Maxwell's equations

3.2.1 Early forms of Maxwell's equations

In 1873, James Clerk Maxwell published "A Treatise on Electricity and Magnetism" [15]. In his treatise, Maxwell referred to and expanded on his original 20 equations, shown in Table 3.1, previously published in "A Dynamical Theory of Electromagnetic Fields", describing the relationship between electric and magnetic fields and the associated potentials. At the time of publication, Maxwell's work was not universally accepted. Controversy remained within the physics communities in England and on continental Europe regarding Maxwell's work until Heinrich Hertz successfully demonstrated in experiments performed in 1887 and 1888 that electromagnetic waves existed in accordance with the implications of Maxwell's Treatise.

Table 3.1 James Clerk Maxwell's equations from "A Dynamical Theory of the Electromagnetic Field" [16]

Total currents	$p' = p + \frac{df}{dt}$ $q' = q + \frac{dg}{dt}$ $r' = r + \frac{dh}{dt}$	Electric elasticity	$P = kf$ $Q = kg$ $R = kh$
Magnetic force	$\mu\alpha = \frac{dH}{dy} - \frac{dG}{dz}$ $\mu\beta = \frac{dF}{dz} - \frac{dH}{dx}$ $\mu\gamma = \frac{dG}{dx} - \frac{dF}{dy}$	Electric resistance	$P = -\rho p$ $Q = -\rho q$ $R = -\rho r$
Currents	$\frac{d\gamma}{dy} - \frac{d\beta}{dz} = 4\pi p'$ $\frac{d\alpha}{dz} - \frac{d\gamma}{dx} = 4\pi q'$ $\frac{d\beta}{dx} - \frac{d\alpha}{dy} = 4\pi r'$	Free electricity	$e + \frac{df}{dx} + \frac{dg}{dy} + \frac{dh}{dz} = 0$
Electromotive force	$P = \mu\left(\gamma\frac{dy}{dt} - \beta\frac{dz}{dt}\right) - \frac{dF}{dt} - \frac{d\psi}{dx}$ $Q = \mu\left(\alpha\frac{dz}{dt} - \gamma\frac{dx}{dt}\right) - \frac{dG}{dt} - \frac{d\psi}{dy}$ $R = \mu\left(\beta\frac{dx}{dt} - \alpha\frac{dy}{dt}\right) - \frac{dH}{dt} - \frac{d\psi}{dz}$	Continuity	$\frac{de}{dt} + \frac{dp}{dx} + \frac{dq}{dy} + \frac{dr}{dz} = 0$

Between 1881 and 1885, Oliver Heaviside and Josiah W. Gibbs independently formulated what is today known as vector calculus, as a means of simplifying the use of Sir William Rowan Hamilton's quaternionic calculus [17] used originally by Maxwell. Heaviside and Gibbs shared the motivation to simplify the quaternion expressions and in particular to separate the vector and scalar parts of the mathematics.

Between 1884 and 1890, Heaviside and Hertz derived and simplified Maxwell's equations, arriving independently at the form most familiar today. These formulations were referred to as either the Hertz–Heaviside equations or the Maxwell–Hertz equations. In 1940, Albert Einstein wrote an opinion paper [18] in which he referred to Maxwell's equations, and the term became widely accepted afterward. Table 3.2 illustrates the Heaviside formulations from [19].

Electromagnetic phenomena at the microscopic and the macroscopic levels are governed by Maxwell's equations. The equations can be expressed in both integral and differential forms that are mathematically equivalent, as can be shown by use of Gauss' law and Stokes' theorem. The integral forms are most useful when dealing with macroscopic problems with high degrees of symmetry. In contrast, the differential forms manipulate change and current densities and fields at a point in both space and time.

3.2.2 Modern versions of Maxwell's equations

The modern versions of Maxwell's equations, in both integral and differential (point) forms, are shown in Table 3.3. These are shown in the same order as was used in Table 3.2 for Oliver Heaviside's formulations of James Clerk Maxwell's equations.

Table 3.2 Oliver Heaviside's formulations of James Clerk Maxwell's equations

First circuital law	$curl\ \boldsymbol{H} = 4\pi kE + c\dot{E}$
Second circuital law	$curl\ \boldsymbol{E} = -\dot{\boldsymbol{B}} = -\mu\dot{\boldsymbol{H}}$
Divergence of B	$div\ \boldsymbol{B} = 0$
Divergence of D	$div\ \boldsymbol{D} = \rho$

Table 3.3 Modern versions of Maxwell's equations

Popular title	Integral form	Differential (point) form
Ampere's law	$\oint \boldsymbol{H}\cdot d\boldsymbol{l} = \int \boldsymbol{J}\cdot d\boldsymbol{s} + \frac{d}{dt}\int \boldsymbol{D}\cdot d\boldsymbol{s}$	$\text{curl}\ \boldsymbol{H} = \boldsymbol{J} + \frac{\partial \boldsymbol{D}}{\partial t}$
Faraday's law	$\oint \boldsymbol{E}\cdot d\boldsymbol{l} = -\frac{d}{dt}\int \boldsymbol{B}\cdot d\boldsymbol{s}$	$\text{curl}\ \boldsymbol{E} = -\frac{\partial \boldsymbol{B}}{\partial t}$
Gauss' law for magnetic fields	$\oint \boldsymbol{B}\cdot d\boldsymbol{s} = \boldsymbol{0}$	$\text{div}\ \boldsymbol{B} = 0$
Gauss' law for electric fields	$\oint \boldsymbol{D}\cdot d\boldsymbol{s} = \int \rho dv$	$\text{div}\ \boldsymbol{D} = \rho$

Each of the first two, Ampere's law and Faraday's law, can be converted to and from the integral form to the differential form using Stokes' theorem. In like fashion, each of the last two can be converted to and from the integral form to the differential form using the divergence theorem. The conversion for Ampere's law and for Gauss' law for electric fields will be shown for clarity.

Starting with the differential form of Ampere's law

$$curl\,\boldsymbol{H} = \boldsymbol{J} + \frac{\partial \boldsymbol{D}}{\partial t} \tag{3.65}$$

we integrate both sides over some open surface to obtain

$$\int (curl\,\boldsymbol{H}) \cdot \boldsymbol{ds} = \int \boldsymbol{J} \cdot \boldsymbol{ds} + \int \frac{\partial \boldsymbol{D}}{\partial t} \cdot \boldsymbol{ds} \tag{3.66}$$

We then apply Stokes' theorem to obtain

$$\oint H \cdot \boldsymbol{dl} = \int \boldsymbol{J} \cdot \boldsymbol{ds} + \int \frac{\partial \boldsymbol{D}}{\partial t} \cdot \boldsymbol{ds} \tag{3.67}$$

Assuming we have stationary contours, the $\partial \boldsymbol{D}/\partial t$ can be removed from the integrand and replaced by the ordinary derivative d/dt to yield

$$\oint H \cdot \boldsymbol{dl} = \int \boldsymbol{J} \cdot \boldsymbol{ds} + \frac{d}{dt}\int D \cdot \boldsymbol{ds} \tag{3.68}$$

In similar fashion, starting with the differential form of Gauss' law for electric fields

$$\text{div}\,\boldsymbol{D} = \rho \tag{3.69}$$

we integrate both sides over some volume to obtain

$$\int (div\,\boldsymbol{D}) \cdot \boldsymbol{dv} = \int \rho \cdot \boldsymbol{dv} \tag{3.70}$$

We then apply the divergence theorem to obtain

$$\oint \boldsymbol{D} \cdot \boldsymbol{ds} = \int \rho \cdot \boldsymbol{dv} \tag{3.71}$$

3.2.3 Ampere's law

On April 21, 1820, as part of a series of lectures on electricity and magnetism at the University of Copenhagen, Hans Christiaan Ørsted (sometimes shown as Oersted) planned a lecture on the well-known similarities between electrical and magnetic forces [20]. Ørsted had thought to include as part of this lecture an experiment to demonstrate the effects of a current through a wire conductor on a magnetic compass located in close proximity to the wire conductor. He was unfortunately unable to perform the experiment prior to the lecture, but then during the lecture he decided to go forward with it. Ørsted set up his apparatus consisting of a galvanic battery, an interconnecting wire, and a boxed compass. In this experiment, Ørsted was able to demonstrate a weak response of the compass needle coincident with application of electrical current to the interconnecting wire conductor. Ørsted was occupied with other duties and was unable to return to further investigation of his observations for three months, at which time he had better facilities and was able to assemble several witnesses to the repeat of his earlier experiments.

Ørsted's observations were duly published in *the Journal for Chemistry and Physics* on July 21, 1820. Copies were sent to numerous contemporaries, including Francois Arago and Andre-Marie Ampere. Arago was able to quickly verify Ørsted's results experimentally, reported as much to the Academie Royale des Sciences on September 4, 1820, and then successfully demonstrated them convincingly on September 11, 1820 [21,22]. Ampere was inflamed by Ørsted's observations and Arago's demonstrations and threw himself into an in-depth study of the observed phenomena. Ampere was able to successfully demonstrate two parallel wires carrying electric currents repel or attract each other, dependent on the direction of current flow in the wires. Most importantly, Ampere established the mathematical relationship now known as Ampere's force law describing his observations of the interactions between two current carrying wires [23,24].

Upon graduation from Cambridge in 1854, James Clerk Maxwell embarked on what would become a quest for a complete mathematization of electrical and magnetic phenomena as were then known at the time. Maxwell first read Faraday's three volume "Experimental Researches in Electricity" that contained all of Faraday's published work up to that date, followed by an exhaustive examination of the work of William Thomson (Lord Kelvin) and most, if not all, of the remaining works of scientists and researchers in the field of electricity and magnetism, including Ørsted, Ampere, Gauss, Weber, Helmholtz, and many others.

After this period of study, Maxwell's first published effort to describe electromagnetic phenomena was his paper "On Faraday's Lines of Force", published in 1855 [25]. In this paper, for the first time, Maxwell writes three equations directly relating the total intensity of magnetizing force in a closed curve embracing an electric current

$$a_2 = \frac{d\beta_1}{dz} - \frac{d\gamma_1}{dy}; \quad b_2 = \frac{d\gamma_1}{dx} - \frac{d\alpha_1}{dz}; \quad c_2 = \frac{d\alpha_1}{dy} - \frac{d\beta_1}{dx} \tag{3.72}$$

In these equations, the Greek letter variables α, β, and γ represent the x, y, and z components, respectively, of the magnetizing force of the encircled current. These equations underwent major changes in Maxwell's next paper, a four-part thesis titled "On Physical Lines of Force", with the addition of his concept of displacement current [26–29]. Maxwell returned to these equations in his third paper "A Dynamical Theory of the Electromagnetic Field", ultimately including them in his final form in his seminal work "A Treatise on Electricity and Magnetism". We know the modified form of these equations today as Ampere's law.

The integral form of Ampere's law is given by

$$\oint \boldsymbol{H} \cdot d\boldsymbol{l} = \int \boldsymbol{J} \cdot d\boldsymbol{s} + \frac{d}{dt}\int \boldsymbol{D} \cdot d\boldsymbol{s} \tag{3.73}$$

The integral on the left-hand side describes the circulation of a magnetic field about a contour C. This integral is also referred to as the definition of the magnetomotive force around the contour C. The first integral on the right-hand side represents a conduction current density through the surface bounded by the contour C. The second integral on the right-hand side represents the change in electric flux through the surface bounded by the contour C and is one of the primary contributions of James Clerk Maxwell to the study of electromagnetic phenomena.

The second integral in (3.73) integral may also be expressed as

$$\varepsilon_0 \frac{d}{dt}\int \boldsymbol{E} \cdot d\boldsymbol{s} = \frac{dQ}{dt} \tag{3.74}$$

Equation (3.74) has the units of coulombs per second, or amperes. For historical reasons, the product of the permittivity and the change of electric flux per unit time through a surface is called the "displacement current" even though no charge crosses the surface.

Historically, the interpretation of the second right-hand integral in (3.73) has been that it represents an ordinary conduction current, much like the current density term in the first right-hand integral. At the time that Maxwell introduced his concept of displacement current, the scientific community still held true that an ether existed through which action at a distance could be explained. Maxwell's displacement current term was associated with mechanical changes in the ether. Modern physics has abandoned the ether such that Maxwell's original concept of a displacement current that relied on mechanical changes in the ether is no longer strictly valid.

Maxwell's equations express the relationship between variables that describe rates of change of field quantities and variables that describe charge and current densities at the same point in time. Because of this, Maxwell's equations are not capable of establishing any sort of causal relationship between field quantities. To account for a causal relationship, it is necessary to establish a relationship wherein present events are exclusively determined by past events. The use of variables that are dependent on previous, or retarded, time values is thus required if a causal relationship is desired. Details regarding retarded time values are shown in Chapter 5.

In consideration of the foregoing reasoning, (3.73) may be loosely interpreted in the following manner: the circulation of a magnetic field is determined by the contributions of a changing current density composed of moving charges in the present time sense and moving charges in the past time sense responsible for a time-varying change in a charge distribution. Stated more directly, the circulation of a magnetic field as described by (3.73) is caused by changing current density.

3.2.4 Faraday's law

In the latter part of the year 1831, Michael Faraday conducted a series of experiments that definitively showed the relationship between electrical currents and magnetic fields [30,31]. However, lacking any sort of mathematical basis or theory left his work in the shadows for several years, until Maxwell revisited it in 1854 and gave it a mathematical framework.

As discussed in the section on Ampere's law, Maxwell's first published effort to describe electromagnetic phenomena was his paper "On Faraday's Lines of Force", published in 1855. In this paper, Maxwell fully embraced Faraday's concepts of the workings of electromagnetic induction, referred to by Faraday as the "electro-tonic state". Maxwell proceeds to discuss several mathematical theorems, and after some additional discussion regarding the work done by changes in magnetic effects related to permanent magnets and static currents, he writes these three equations:

$$\alpha_2 = -\frac{1}{4\pi}\frac{d\alpha_0}{dt}; \quad \beta_2 = -\frac{1}{4\pi}\frac{d\beta_0}{dt}; \quad \gamma_2 = -\frac{1}{4\pi}\frac{d\gamma_0}{dt} \tag{3.75}$$

In these equations, the Greek letter variables α_2, β_2, and γ_2 represent electromotive forces. The Greek letter variables α_0, β_0, and γ_0 are the mathematical representation of Maxwell's/Faraday's components of the "Electro-tonic" intensity. Otherwise, these functions remain undefined in Maxwell's paper.

As did the equations that eventually became Ampere's law, these equations will undergo major evolution in Maxwell's second and third papers and will ultimately appear in his Treatise. We know the modified form of these equations today as Faraday's law.

The integral form of Faraday's law is given by

$$\oint \boldsymbol{E} \cdot dl = -\frac{d}{dt}\int \boldsymbol{B} \cdot ds \tag{3.76}$$

The integral on the left-hand side describes the circulation of an electric field about a contour C. This integral is also referred to as the definition of the electromotive force, or emf, around the contour C. The integral on the right-hand side represents the change in magnetic flux through the surface bounded by the contour C.

Equation (3.76) may be expressed as

$$\text{emf} = -\frac{d\psi_m}{dt} \tag{3.77}$$

The emf in (3.77) can be thought of as a voltage source inserted in a loop whose perimeter is contour C, the magnitude of which is equal to the magnitude of the time rate of change of the magnetic flux passing through the loop. Stated another way, the emf induced in a loop or circuit is measured by the decrease of the magnetic flux that passes through that loop or circuit.

The emf in (3.77) should be interpreted as a distributed-parameter quantity and cannot be localized or treated as a lumped element unless the dimensions of the loop are electrically small. It is very important to determine the polarity of this emf. The negative sign in Faraday's law represents what is colloquially referred to as Lenz's law. Stated simply, Lenz' law states that the direction of an induced current (or polarity of an induced emf) is always such as to oppose the change in the circuit or the magnetic field that produces it.

Entirely similar to the reasoning provided at the end of the section on Ampere's law, (3.76) may be interpreted in the following manner: The circulation of an electric field is determined by the contributions of changing current density that has produced a time-variant magnetic field.

3.2.5 Gauss' laws

Two of Maxwell's equations are based on Gauss' mathematical relationships involving potential theory, developed in studies related to terrestrial magnetism. The relationships are named after Gauss in acknowledgement of the origination of the mathematics. These are directly based on the basic vector calculus relationship discussed in (3.57).

The first of these is Gauss' law for magnetic fields. The integral form of this equation is given by

$$\oint \boldsymbol{B} \cdot d\boldsymbol{s} = 0 \tag{3.78}$$

Equation (3.78) essentially states that magnetic field lines form closed paths. The traditional interpretation of this relationship is that no magnetic monopoles exist. Stated another way, a magnetic field is not created by magnetic charges. Although much effort has been expended in the search for magnetic monopoles, to date no one has found one.

The second of Maxwell's equations based on Gauss' potential theory is Gauss' law for electric fields. The integral form of this equation is given by

$$\oint \boldsymbol{D} \cdot d\boldsymbol{s} = \int \rho_v dv \tag{3.79}$$

In contrast to (3.78), (3.79) states that an electric field is created by the presence of a charge density. The quantity ρ_v represents the volume charge density, measured in coulombs per cubic metre (C/m^3). This relationship states the net density of electric flux crossing out of a closed volume bounded by the surface S is equal to the net positive charge within the volume. Recall that electric field lines originate on positive charges and terminate on negative charges. An important aspect of this relationship is to recognize that the source of the electric flux density is the presence of free positive charge in the volume of interest. Gauss' law for electric fields plays a central role in cable shielding.

3.2.6 Equation of continuity

Electrical charge is a physical property of subatomic particles that constitute matter. By convention electrons carry a negative charge, and protons carry a positive charge. Electrical charge may be moved about or redistributed by the presence of an electromagnetic field, but the net charge of an isolated system is a constant. Electrical charge can neither be created nor destroyed. This is a statement of the law of conservation of electrical charge.

Any current leaving a region or volume V enclosed by a surface S will decrease the amount of charge within that volume. The reverse is true that any current entering a region or volume V enclosed by a surface S will increase the amount of charge within that volume.

We can mathematically state for a current leaving a volume

$$I = \oint \boldsymbol{J} \cdot d\boldsymbol{s} = -\frac{dQ}{dt} = -\frac{d}{dt}\int \rho_v dv \tag{3.80}$$

Equation (3.80) states that the current $\boldsymbol{I}$ leaving a volume is equivalent to the net outward flux of the current density vector $\boldsymbol{J}$ through the surface S and is also equal to the rate of decrease of the electrical charge $\boldsymbol{Q}$ contained within the volume V. We can apply the divergence theorem to (3.80) to convert the surface integral of the current density $\boldsymbol{J}$ to the volume integral of the divergence of $\boldsymbol{J}$. We can then restate (3.80) for a stationary volume as

$$\int \nabla \cdot \boldsymbol{J} dv = -\int \frac{\partial \rho_v}{\partial t} dv \tag{3.81}$$

Note that partial differentiation is needed here as the charge density ρ_v may be a function of both time and space. Irrespective of the choice of volume, we can now write the equation of continuity as

$$\nabla \cdot \boldsymbol{J} = -\frac{\partial \rho_v}{\partial t} \tag{3.82}$$

The equation of continuity can be derived from Ampere's law. If we apply the divergence theorem to the point form of Ampere's law, we obtain

$$\nabla \cdot \text{curl}\, \boldsymbol{H} = \nabla \cdot \left(\boldsymbol{J} + \frac{\partial \boldsymbol{D}}{\partial t} \right) \tag{3.83}$$

Since the divergence of the curl of any vector is equal to zero, (3.83) can then be rearranged as

$$\nabla \cdot \boldsymbol{J} = -\left(\nabla \cdot \frac{\partial \boldsymbol{D}}{\partial t} \right) \tag{3.84}$$

By substituting Gauss' law for electric fields into (3.84), we obtain

$$\nabla \cdot \boldsymbol{J} = -\frac{\partial \rho_v}{\partial t} \tag{3.85}$$

which is identical to (3.82).

3.2.7 Constitutive relations

The vector quantities representing the electric field, the electric flux density, the magnetic field, and the magnetic flux density in Maxwell's equations are interrelated by the characteristics of the region in which these vector quantities exist. The electric field and the electric flux density are related by

$$\boldsymbol{D} = \varepsilon \boldsymbol{E} \tag{3.86}$$

and the magnetic field and the magnetic flux density are related by

$$\boldsymbol{B} = \mu \boldsymbol{H} \tag{3.87}$$

An additional relationship exists between the electric field and current density,

$$\boldsymbol{J}_c = \sigma \boldsymbol{E} \tag{3.88}$$

where $\boldsymbol{J}_c$ emphasizes that in this relationship, the current density is a conduction current density caused by the presence of an electric field in a conducting medium. Note the total current density can be expressed as the sum of an impressed source $\boldsymbol{J}_0$ and the conduction current density $\boldsymbol{J}_c$, or $\boldsymbol{J} = \boldsymbol{J}_0 + \boldsymbol{J}_c$.

The variables, sometimes known as the constitutive parameters, ε, μ, and σ, are the permittivity measured in farads per metre (F/m), the permeability measured in henrys per metre (H/m), and the conductivity measured in siemens per metre (S/m), respectively, of the region of interest. The siemens is the reciprocal of the Ω, and is sometimes referred to as the mho.

Regions in which the field vectors are related in a simple fashion as shown in (3.86), (3.87), and (3.88) are referred to as linear, isotropic, and homogeneous. A non-linear region is one in which $\boldsymbol{D}$ and $\boldsymbol{J}_c$, either or both, vary as a function of the magnitude of $\boldsymbol{E}$, and/or $\boldsymbol{B}$ varies as a function of the magnitude of $\boldsymbol{H}$. An anisotropic region is one in which $\boldsymbol{D}$ and $\boldsymbol{J}_c$, either or both, are not parallel to $\boldsymbol{E}$, and/or $\boldsymbol{B}$ is not parallel to $\boldsymbol{H}$. An inhomogeneous region is one in which one or more of the constitutive parameters are functions of position. Dielectrics, such as wire sheaths, are generally anisotropic.

Equations (3.86), (3.87), and (3.88) may be substituted into Maxwell's equations to yield the expressions in Table 3.3.

The emphasis in this text will be on electromagnetic fields in linear, isotropic, and homogeneous media. Where possible, in cases where these conditions are not strictly true, they will be approximated as being such.

In January 1862, Maxwell published the third part of his paper "On Physical Lines of Force". In this paper, Maxwell shows his calculations of the ratio of the energies of the magnetic and electric fields and concludes that "... the velocity of propagation of these fields equals the results of Kohlrausch, Weber, and Fizeau, that is to say, the velocity of light". Maxwell, in his third paper, "A Dynamical Theory of the Electromagnetic Field", gave full credit to Faraday's paper "Thoughts on Ray Vibrations" [32,33] as being the source of Maxwell's theory of light.

The free space values of permittivity and permeability are

$$\varepsilon_0 \cong \frac{1}{36\pi \times 10^9}\ \mathrm{F/m} \tag{3.89}$$

$$\mu_0 = 4\pi \times 10^{-7}\ \mathrm{H/m} \tag{3.90}$$

The velocity of light is given by

$$c \cong \frac{1}{\sqrt{\varepsilon_0 \mu_0}} = 300 \times 10^6\ \mathrm{m/s} \tag{3.91}$$

Based on measurements of the speed of light between 1975 and 1983, performed using laser interferometry, and by international agreement, the length of the metre was established as being equal to length of the path travelled by light in a vacuum in a time interval of $(1/299792458)$s [34]. The corresponding velocity of light is then 299792458 m/s, a difference of 0.069% compared to the value shown in (3.91).

The permittivity and permeability of materials often differ from those of free space, and are normally expressed in relative terms as

$$\varepsilon = \varepsilon_r \varepsilon_0 \tag{3.92}$$

$$\mu = \mu_r \mu_0 \tag{3.93}$$

where ε_r is the relative permittivity (often called the dielectric constant) and μ_r is the relative permeability.

3.2.8 Vector wave equations

Implicit in Maxwell's theory and demonstrated by Hertz, electromagnetic waves propagate through free space. Electromagnetic waves propagate through various materials aside from free space, along wires through the surrounding dielectric, along transmission line structures, and through various materials including metals. Maxwell's equations themselves do not describe this wave structure or motion, but relationships that do describe wave structure and motion can be derived from Maxwell's equations.

The simplest derivation requires a few simplifying assumptions. The field vectors are assumed to be in a linear, isotropic, homogeneous region. The region is assumed to have a zero net charge, and any currents that are in the region are conduction currents.

The point forms of Maxwell's equations in accordance with these assumptions are the same as those shown in Table 3.4. Repeating Ampere's law as a starting point, we have

$$\mathrm{curl}\,\boldsymbol{H} = \sigma \boldsymbol{E} + \varepsilon \frac{\partial \boldsymbol{E}}{\partial t} \tag{3.94}$$

and Gauss' law for electric fields, which under the foregoing assumption of zero net charge is

$$\mathrm{div}\,\boldsymbol{E} = 0 \tag{3.95}$$

Table 3.4 Maxwell's equations in linear, isotropic, and homogeneous regions

Popular title	Integral form	Differential (point) form
Ampere's law	$\oint \boldsymbol{H}\cdot d\boldsymbol{l} = \sigma \int \boldsymbol{E}\cdot d\boldsymbol{s} + \varepsilon \frac{d}{dt}\int \boldsymbol{E}\cdot d\boldsymbol{s}$	curl $\boldsymbol{H} = \sigma\boldsymbol{E} + \varepsilon\frac{\partial \boldsymbol{E}}{\partial t}$
Faraday's law	$\oint \boldsymbol{E}\cdot d\boldsymbol{l} = -\mu\frac{d}{dt}\int \boldsymbol{H}\cdot d\boldsymbol{s}$	curl $\boldsymbol{E} = -\mu\frac{\partial \boldsymbol{H}}{\partial t}$
Gauss' law for magnetic fields	$\oint \boldsymbol{H}\cdot d\boldsymbol{s}=0$	div $\boldsymbol{H} = 0$
Gauss' law for electric fields	$\oint \boldsymbol{E}\cdot d\boldsymbol{s}=\frac{1}{\varepsilon}\int \rho dv$	div $\boldsymbol{E} = \frac{\rho}{\varepsilon}$

If we now take the curl of Faraday's law and substitute Ampere's law, we obtain

$$\text{curl curl } \boldsymbol{E} = -\mu \text{ curl}\left(-\frac{\partial \boldsymbol{H}}{\partial t}\right) = -\mu\frac{\partial}{\partial t}(\text{curl } \boldsymbol{H}) \tag{3.96}$$

$$= -\mu\sigma\frac{\partial \boldsymbol{E}}{\partial t} - \mu\varepsilon\frac{\partial^2 \boldsymbol{E}}{\partial t^2} \tag{3.97}$$

In like fashion, if we take the curl of Ampere's law and substitute Faraday's law, we obtain

$$\text{curl curl } \boldsymbol{H} = \text{curl}\left(\sigma\boldsymbol{E} + \varepsilon\frac{\partial \boldsymbol{E}}{\partial t}\right) = \sigma(\text{curl } \boldsymbol{E}) + \varepsilon\frac{\partial}{\partial t}(\text{curl } \boldsymbol{E}) \tag{3.98}$$

$$= -\mu\sigma\frac{\partial \boldsymbol{H}}{\partial t} - \mu\varepsilon\frac{\partial^2 \boldsymbol{H}}{\partial t^2} \tag{3.99}$$

Using the del symbol and the vector identity for the Laplacian

$$\nabla \times \nabla \times \boldsymbol{A} = \nabla(\nabla\cdot\boldsymbol{A}) - \nabla\cdot(\nabla\boldsymbol{A}) = \nabla(\nabla\cdot\boldsymbol{A}) - \nabla^2\boldsymbol{A} \tag{3.100}$$

and noting that div $\boldsymbol{E} = 0$, (3.97) can be restated as

$$\nabla^2\boldsymbol{E} = \mu\sigma\frac{\partial \boldsymbol{E}}{\partial t} + \frac{1}{c^2}\frac{\partial^2 \boldsymbol{E}}{\partial t^2} \tag{3.101}$$

and similarly noting that div $\boldsymbol{H} = 0$, (3.99) can be restated as

$$\nabla^2\boldsymbol{H} = \mu\sigma\frac{\partial \boldsymbol{H}}{\partial t} + \frac{1}{c^2}\frac{\partial^2 \boldsymbol{H}}{\partial t^2} \tag{3.102}$$

Equations (3.101) and (3.102) are three-dimensional homogeneous vector wave equations. These equations describe wave propagation through a source-free region. In free space, the conductivity $\sigma = 0$, and (3.101) and (3.102) reduce to

$$\nabla^2\boldsymbol{E} = \frac{1}{c^2}\frac{\partial^2 \boldsymbol{E}}{\partial t^2} \tag{3.103}$$

$$\nabla^2 \boldsymbol{H} = \frac{1}{c^2}\frac{\partial^2 \boldsymbol{H}}{\partial t^2} \tag{3.104}$$

A more general form of (3.103), assuming the electric field is a function of z and t only, is given by

$$\frac{1}{c^2}\frac{\partial^2 \boldsymbol{E}}{\partial t^2} - \frac{\partial^2 \boldsymbol{E}}{\partial z^2} = \frac{\partial^2 \boldsymbol{E}}{\partial t^2} - \frac{1}{\mu_0 \varepsilon_0}\frac{\partial^2 \boldsymbol{E}}{\partial z^2} = 0 \tag{3.105}$$

The wave equation of the form in (3.105) has solutions of the form

$$\boldsymbol{E} = \boldsymbol{E}_0(z \pm vt) \tag{3.106}$$

Any arbitrary function f whose argument is written as shown in (3.106) and whose derivatives are mathematically well behaved will satisfy (3.105). The solution $(z - vt)$ represents a wave travelling in the positive z-direction. Travelling electromagnetic plane waves are most often represented as sinusoidal waves. In the Cartesian coordinate system, such waves may be represented as

$$\boldsymbol{E} = \boldsymbol{E}_0 \sin\left(k_x x + k_y y + k_z z - \omega t\right) \tag{3.107}$$

$$\boldsymbol{E} = \boldsymbol{E}_0 \sin\left(\boldsymbol{k} \cdot \boldsymbol{r} - \omega t\right) \tag{3.108}$$

where the vector $\boldsymbol{k}$ is known as the wavevector or wavenumber, whose direction is that of the wave's propagation, and $\boldsymbol{r}$ represents a vector from the origin to the plane surface of the wavefront. Since we are restricting ourselves to wave motion in the z-direction only, (3.108) can be rewritten as

$$\boldsymbol{E} = \boldsymbol{E}_0 \sin\left(kz - \omega t\right) \tag{3.109}$$

and the corresponding magnetic field

$$\boldsymbol{H} = \boldsymbol{H}_0 \sin\left(kz - \omega t\right) \tag{3.110}$$

Equation (3.109) is a solution to (3.105), and (3.110) is a solution to the corresponding equation for the magnetic field. It is now necessary to show that these same equations also satisfy Maxwell's equations for divergence and curl.

We can substitute (3.109) into (3.95) to yield

$$\nabla \cdot \boldsymbol{E} = \nabla \cdot \left[\boldsymbol{E}_0 \sin\left(k_x x + k_y y + k_z z - \omega t\right)\right] \tag{3.111}$$

$$= \left(\mathbf{k} \cdot \boldsymbol{E}_0\right)\left[\cos\left(k_x x + k_y y + k_z z - \omega t\right)\right] \tag{3.112}$$

Substituting (3.110) into Gauss' law for magnetic fields yields a similar result

$$\nabla \cdot \boldsymbol{H} = \nabla \cdot \left[\boldsymbol{H}_0 \sin\left(k_x x + k_y y + k_z z - \omega t\right)\right] \tag{3.113}$$

$$= \left(\mathbf{k} \cdot \boldsymbol{H}_0\right)\left[\cos\left(k_x x + k_y y + k_z z - \omega t\right)\right] \tag{3.114}$$

By inspection, for (3.112) and (3.114) to hold for all time and space,

$$(\mathbf{k} \cdot \boldsymbol{E}_0) = (\mathbf{k} \cdot \boldsymbol{H}_0) = 0$$

which shows that the divergence relationships for the electric and magnetic fields require that each is in a direction at a right-angle to the direction of propagation.

Using (3.94) and assuming the region has no conductivity such that no conduction currents exist, and substituting (3.109) and (3.110) into (3.94), we have

$$\text{curl}\,\boldsymbol{H} = \varepsilon\frac{\partial \boldsymbol{E}}{\partial t}$$

$$\text{curl}\left[\boldsymbol{H}_0\sin\left(k_x x + k_y y + k_z z - \omega t\right)\right] = \varepsilon\frac{\partial}{\partial t}\left[\boldsymbol{E}_0\sin\left(k_x x + k_y y + k_z z - \omega t\right)\right]$$

$$(\boldsymbol{k} \times \boldsymbol{H}_0)\left[\cos\left(k_x x + k_y y + k_z z - \omega t\right)\right] = \frac{-k}{\eta}\left[\boldsymbol{E}_0\cos\left(k_x x + k_y y + k_z z - \omega t\right)\right]$$

$$\left(\widehat{\boldsymbol{k}} \times \boldsymbol{H}_0\right)\left[\cos\left(k_x x + k_y y + k_z z - \omega t\right)\right] = \frac{-\boldsymbol{E}_0}{\eta}\left[\cos\left(k_x x + k_y y + k_z z - \omega t\right)\right]$$

where $\widehat{\boldsymbol{k}} = \frac{\boldsymbol{k}}{|\boldsymbol{k}|}$ is a unit vector in the direction of $\boldsymbol{k}$, and $\eta \approx 120\pi\ \Omega$, the impedance of free space.

$$\left|-\widehat{\boldsymbol{k}} \times \boldsymbol{H}_0\right| = \left|\frac{\boldsymbol{E}_0}{\eta}\right| \tag{3.115}$$

or

$$|\boldsymbol{H}_0| = \left|\frac{\boldsymbol{E}_0}{\eta}\right| \tag{3.116}$$

since $\left|-\widehat{\boldsymbol{k}}\right| = 1$.

Equation (3.115) shows the electric field, the magnetic field, and the direction of propagation form an orthogonal set of vector relationships.

Note this orthogonality applies strictly to a plane wave in free space. Waves that propagate in this fashion are known as Transverse ElectroMagnetic (TEM) waves. Electric and magnetic fields may not be orthogonal to each other and the direction of propagation in regions close to localized sources, in cases of oblique incidence where plane waves are refracted into conductive materials splitting the electric and magnetic fields across planes of constant amplitude at an angle with respect to planes of constant phase, or in regions of anisotropy or inhomogeneity. In such instances, one or the other field propagates as a longitudinal component. Primary examples of this occur in waveguides, wherein Transverse Electric (TE) or Transverse Magnetic (TM) fields exist and propagate.

Before proceeding further, it is useful to express Maxwell's equations in terms of phasor quantities, as shown in Table 3.5. It is assumed here that we will be primarily interested in sinusoidal steady-state behaviour of electromagnetic fields, also referred to as time harmonic behaviour. Time harmonic forms can be

Table 3.5 Phasor form of Maxwell's equations in linear, isotropic, and inhomogeneous regions

Popular title	Integral form	Differential (point) form
Ampere's law	$\oint \widehat{\boldsymbol{H}} \cdot d\boldsymbol{l} = (\sigma + j\omega\varepsilon) \int \boldsymbol{E} \cdot d\boldsymbol{s}$	curl $\widehat{\boldsymbol{H}} = (\sigma + j\omega\varepsilon)\widehat{\boldsymbol{E}}$
Faraday's law	$\oint \widehat{\boldsymbol{E}} \cdot d\boldsymbol{l} = -j\omega\mu \int \widehat{\boldsymbol{H}} \cdot d\boldsymbol{s}$	curl $\widehat{\boldsymbol{E}} = -j\omega\mu\widehat{\boldsymbol{H}}$
Gauss' law for magnetic fields	$\oint \widehat{\boldsymbol{H}} \cdot d\boldsymbol{s} = \boldsymbol{0}$	div $\widehat{\boldsymbol{H}} = 0$
Gauss' law for electric fields	$\oint \widehat{\boldsymbol{E}} \cdot d\boldsymbol{s} = \frac{1}{\varepsilon} \int \widehat{\rho} dv$	div $\boldsymbol{E} = \frac{\widehat{\rho}}{\varepsilon}$

expressed as complex quantities. Phasor quantities provide a shorthand means to represent complex quantities. Use of phasor quantities will simplify computations and presentation of concepts such as power flow and media impedance. The basis for phasor notation is well defined in multiple sources, so that we will simply introduce Maxwell's equations in their phasor form without additional material.

The hat ($\widehat{X}$) notation shown in this table will be used to denote complex quantities in following material.

3.2.9 Energy in the electromagnetic field

In a previous section, it was shown that propagating electric and magnetic fields in free space are orthogonal to the direction of propagation. These fields contain energy, and it is entirely reasonable to think that energy must flow in some direction, rather than just appear and disappear at one point or another, or from one point to the next. In 1884, John Henry Poynting published a paper discussing this energy flow [35]. Heaviside published almost identical results in February 1885 [36,37], not realizing that Poynting had published previously in the Philosophical Transactions. Poynting provided a more generally applicable description and the mathematics to accompany it. Heaviside later gave full credit to Poynting, and the relationship that Poynting developed now bears his name.

Poynting defined the vector $\boldsymbol{S}$ as the cross product of propagating $\boldsymbol{E}$ and $\boldsymbol{H}$ fields, with the direction of positive flow determined by "crossing" the $\boldsymbol{E}$ vector into the $\boldsymbol{H}$ vector.

$$\boldsymbol{S} = \boldsymbol{E} \times \boldsymbol{H} \tag{3.117}$$

The units of $\boldsymbol{S}$ are watts per square metre, and hence $\boldsymbol{S}$ describes the instantaneous power flow per unit area.

Poynting also explored the divergence of $\boldsymbol{S}$. By use of the vector identity

$$\nabla \cdot (\boldsymbol{A} \times \boldsymbol{B}) = \boldsymbol{B} \cdot (\nabla \times \boldsymbol{A}) - \boldsymbol{A} \cdot (\nabla \times \boldsymbol{B}) \tag{3.118}$$

the divergence of $\boldsymbol{S}$ may be written as

$$\nabla \cdot \boldsymbol{S} = \nabla \cdot (\boldsymbol{E} \times \boldsymbol{H}) = \boldsymbol{H} \cdot (\text{curl } \boldsymbol{E}) - \boldsymbol{E} \cdot (\text{curl } \boldsymbol{H}) \tag{3.119}$$

Substituting Ampere's and Faraday's laws into (3.119) yields

$$\nabla \cdot \boldsymbol{S} = \mathbf{H} \cdot \left(-\mu \frac{\partial \boldsymbol{H}}{\partial t} \right) - \boldsymbol{E} \cdot \left(\sigma \boldsymbol{E} + \varepsilon \frac{\partial \boldsymbol{E}}{\partial t} \right) \tag{3.120}$$

$$= -\mathbf{H} \cdot \frac{\partial \boldsymbol{B}}{\partial t} - \boldsymbol{E} \cdot \boldsymbol{J} - \boldsymbol{E} \cdot \frac{\partial \boldsymbol{D}}{\partial t} \tag{3.121}$$

The units of (3.121) are watts per cubic metre. (3.121) is known as the point form of Poynting's theorem. Integrating (3.121) over a volume and applying the divergence theorem yields

$$-\oint_s \boldsymbol{S} \cdot ds = \int_v (\boldsymbol{E} \cdot \boldsymbol{J}) dv + \int_v \left(\mathbf{H} \cdot \frac{\partial \boldsymbol{B}}{\partial t} + \boldsymbol{E} \cdot \frac{\partial \boldsymbol{D}}{\partial t} \right) dv \tag{3.122}$$

The units of (3.122) are watts. Equation (3.122) is known as the integral form of Poynting's theorem and this states that the inward flux of the Poynting vector through a closed surface is the sum of power dissipated in the volume enclosed by the surface and the time rate of change of the energy stored in that same volume.

Equations (3.117), (3.121) and (3.122) are all instantaneous representations of electromagnetic energy flow. When considering time harmonic conditions, we will be more interested in the time average of the power, rather than the instantaneous power. The phasor version of the Poynting vector is given by

$$\widehat{\boldsymbol{S}} = \widehat{\boldsymbol{E}} \times \widehat{\boldsymbol{H}}^* \tag{3.123}$$

where $\widehat{\boldsymbol{H}}^*$ denotes the complex conjugate of the phasor $\widehat{\boldsymbol{H}}$. The density of the time average power is given by

$$\boldsymbol{S}_{av} = \frac{1}{2} \mathcal{R}e\, \widehat{\boldsymbol{S}} = \frac{1}{2} \mathcal{R}e \left(\widehat{\boldsymbol{E}} \times \widehat{\boldsymbol{H}}^* \right) = \frac{1}{2} \mathcal{R}e \left(\widehat{\boldsymbol{E}}^* \times \widehat{\boldsymbol{H}} \right) \tag{3.124}$$

The units of (3.124) are watts per square metre.

Assuming the constitutive parameters μ, ε, and σ are constant and are not functions of frequency, (3.124) can be used to determine the time-averaged power dissipated by ohmic losses, the time-averaged magnetic stored energy, and the time-averaged electric stored energy.

We start by taking the dot product of Faraday's law, $\left(\text{curl}\, \widehat{\boldsymbol{E}} = -j\omega\mu\widehat{\boldsymbol{H}} \right)$, by $\widehat{\boldsymbol{H}}^*$, then take the conjugate of Ampere's law, including the total current density $\widehat{\boldsymbol{J}} = \widehat{\boldsymbol{J}}_0 + \widehat{\boldsymbol{J}}_c$, $\left(\text{curl}\, \widehat{\boldsymbol{H}} = \widehat{\boldsymbol{J}}_0 + (\sigma + j\omega\varepsilon)\widehat{\boldsymbol{E}} \right)$ and form the dot product of that result by $\widehat{\boldsymbol{E}}$, and then finally subtract the two results to yield

$$\widehat{\boldsymbol{H}}^* \cdot \left(\nabla \times \widehat{\boldsymbol{E}} = -j\omega\mu\widehat{\boldsymbol{H}} \right) = \widehat{\boldsymbol{H}}^* \cdot \nabla \times \widehat{\boldsymbol{E}} = \widehat{\boldsymbol{H}}^* \cdot -j\omega\mu\widehat{\boldsymbol{H}} \tag{3.125}$$

$$\widehat{\boldsymbol{E}} \cdot \left(\nabla \times \widehat{\boldsymbol{H}}^{*} = \widehat{\boldsymbol{J}}_{0}^{*} + (\sigma + j\omega\varepsilon)\widehat{\boldsymbol{E}}^{*} \right) = \widehat{\boldsymbol{E}} \cdot \nabla \times \widehat{\boldsymbol{H}}^{*}$$
$$= \widehat{\boldsymbol{E}} \cdot \widehat{\boldsymbol{J}}_{0}^{*} + \widehat{\boldsymbol{E}} \cdot \sigma \widehat{\boldsymbol{E}}^{*} - \widehat{\boldsymbol{E}} \cdot j\omega\varepsilon\widehat{\boldsymbol{E}}^{*} \tag{3.126}$$

Combining

$$\widehat{\boldsymbol{H}}^{*} \cdot \nabla \times \widehat{\boldsymbol{E}} - \widehat{\boldsymbol{E}} \cdot \nabla \times \widehat{\boldsymbol{H}}^{*} = \widehat{\boldsymbol{H}}^{*} \cdot -j\omega\mu\widehat{\boldsymbol{H}} - \widehat{\boldsymbol{E}} \cdot \sigma\widehat{\boldsymbol{E}}^{*} + \widehat{\boldsymbol{E}} \cdot j\omega\varepsilon\widehat{\boldsymbol{E}}^{*} - \widehat{\boldsymbol{E}} \cdot \widehat{\boldsymbol{J}}_{0}^{*} \tag{3.127}$$

Using (3.118), we can write (3.127) as

$$-\nabla \cdot \left(\widehat{\boldsymbol{E}} \times \widehat{\boldsymbol{H}}^{*} \right) = \sigma \widehat{\boldsymbol{E}}^{*} \cdot \widehat{\boldsymbol{E}} + j\omega \left(\mu \widehat{\boldsymbol{H}} \cdot \widehat{\boldsymbol{H}}^{*} - \varepsilon \widehat{\boldsymbol{E}} \cdot \widehat{\boldsymbol{E}}^{*} \right) + \widehat{\boldsymbol{E}} \cdot \widehat{\boldsymbol{J}}_{0}^{*} = -\nabla \cdot \boldsymbol{S} \tag{3.128}$$

Recasting (3.128) into terms of (3.124) yields

$$-\frac{1}{2}\widehat{\boldsymbol{E}} \cdot \widehat{\boldsymbol{J}}_{0}^{*} = \frac{1}{2}\sigma\widehat{\boldsymbol{E}}^{*} \cdot \widehat{\boldsymbol{E}} + j2\omega \left(\frac{\dfrac{\mu\widehat{\boldsymbol{H}} \cdot \widehat{\boldsymbol{H}}^{*}}{4} - \varepsilon\widehat{\boldsymbol{E}} \cdot \widehat{\boldsymbol{E}}^{*}}{4} \right) + \nabla \cdot \boldsymbol{S} \tag{3.129}$$

From which we can see that

1. The time-averaged power dissipated is

$$\frac{1}{2}\sigma\widehat{\boldsymbol{E}}^{*} \cdot \widehat{\boldsymbol{E}} = \sigma\langle|\boldsymbol{E}|^{2}\rangle \tag{3.130a}$$

2. The time-averaged magnetic stored energy is

$$\frac{1}{4}\mu\widehat{\boldsymbol{H}} \cdot \widehat{\boldsymbol{H}}^{*} = \frac{1}{2}\mu\langle|\boldsymbol{H}|^{2}\rangle \tag{3.130b}$$

3. The time-averaged electric stored energy is

$$\frac{1}{4}\varepsilon\widehat{\boldsymbol{E}} \cdot \widehat{\boldsymbol{E}}^{*} = \frac{1}{2}\varepsilon\langle|\boldsymbol{E}|^{2}\rangle \tag{3.130c}$$

4. The complex power delivered by the source to the field is

$$-\frac{1}{2}\widehat{\boldsymbol{E}} \cdot \widehat{\boldsymbol{J}}_{0}^{*} \tag{3.130d}$$

3.2.10 Boundary conditions

In infinite and unbounded homogeneous media, solutions to Maxwell's equations are relatively easy to obtain. In most practical situations, solutions are required where various materials exhibiting different constituent parameters are involved. Under such circumstances, knowledge of the materials involved and how they change at the associated boundaries of the regions within which the fields exist is necessary for the solution process. These boundary conditions must be satisfied by the differential relationships embodied in Maxwell's equations.

The integral forms of Maxwell's equations are assumed to hold across regions within which discontinuous materials interface. The integral forms of the curl relationships of Maxwell's equations are used to establish the tangential components of the boundary conditions, and the integral forms of the divergence relationships are used to establish the normal components of the boundary conditions.

As an example, Figure 3.3 illustrates a construction to support determination of the tangential components of the electric field. In the figure is shown a rectangular contour, C_0, with sides Δh and Δw. The area enclosed by C_0 is not shown but is identified as S_0. The positive unit vector normal to the surface S is $\boldsymbol{n}$, directed from medium #1 into medium #2, and a unit vector tangential to the surface S is $\boldsymbol{u}$. A unit vector $\boldsymbol{n_0}$ normal to the area enclosed by C_0 is then given by

$$\boldsymbol{n}_0 = \boldsymbol{n} \times \boldsymbol{u} \tag{3.131}$$

Let $\boldsymbol{u}_1$ and $\boldsymbol{u}_2$ be unit vectors in the anticlockwise direction on the upper and lower sides of the rectangle formed by contour C_0.

Applying Faraday's law, we have

$$\oint \boldsymbol{E} \cdot d\boldsymbol{l} + \mu \frac{d}{dt} \int (\boldsymbol{H} \cdot \boldsymbol{n}_0) ds = 0$$

which can be approximated, neglecting higher order differentials, by

$$(\boldsymbol{E}_1 \cdot \boldsymbol{u}_1 + \boldsymbol{E}_2 \cdot \boldsymbol{u}_2)\Delta w \cong \left(-\mu \frac{\partial \boldsymbol{H}}{\partial t} \cdot \boldsymbol{n}_0 \right) \Delta h \Delta w$$

Since $\boldsymbol{u}_2 = -\boldsymbol{u}_1$, and allowing $\Delta w \rightarrow 0$, we have

$$(\boldsymbol{E}_1 \cdot \boldsymbol{u}_1 + \boldsymbol{E}_2 \cdot -\boldsymbol{u}_1)\Delta w = 0 \tag{3.132a}$$

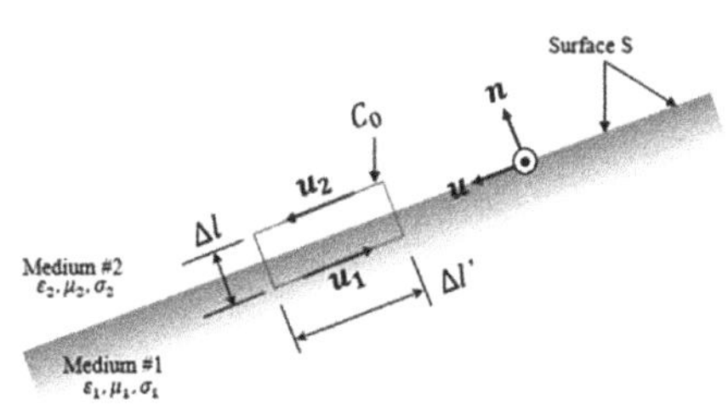

Figure 3.3 Construction for tangential components

or

$$E_{1_t} = E_{2_t} \tag{3.132b}$$

A second example is provided by Figure 3.4, which illustrates a construction to support determination of the normal components of the magnetic flux density. In the figure is shown a right circular cylindrically shaped pillbox with height Δl and end-cap surface area Δa.

Using the divergence relationship

$$\oint (\boldsymbol{n} \cdot \boldsymbol{B}) da = 0 \tag{3.133}$$

Neglecting differentials of higher order, we have an approximation

$$(\boldsymbol{B} \cdot \boldsymbol{n}_1 + \boldsymbol{B} \cdot \boldsymbol{n}_2)\Delta a \cong 0$$

The contribution of the walls is directly proportional to the height of the cylinder. If we let Δl and Δa shrink to an infinitesimal size and recognizing that $\boldsymbol{n}_1 = -\boldsymbol{n}$, $\boldsymbol{B}$ can be assumed to be the same on each side of the interface.

$$\boldsymbol{n} \cdot (\boldsymbol{B}_2 - \boldsymbol{B}_1) = 0 \tag{3.134a}$$

or

$$B_{2n} = B_{1n} \tag{3.134b}$$

Similar calculations can be used to establish boundary conditions for the magnetic field and the electric flux density. Without working through further derivations, four statements can be made regarding boundary conditions:

1. The tangential component of an $\boldsymbol{E}$ field is continuous across an interface.
2. The tangential component of an $\boldsymbol{H}$ field is discontinuous across an interface where a surface current exists.
3. The normal component of a $\boldsymbol{D}$ field is discontinuous across an interface where a surface charge exists.
4. The normal component of a $\boldsymbol{B}$ field is continuous across an interface.

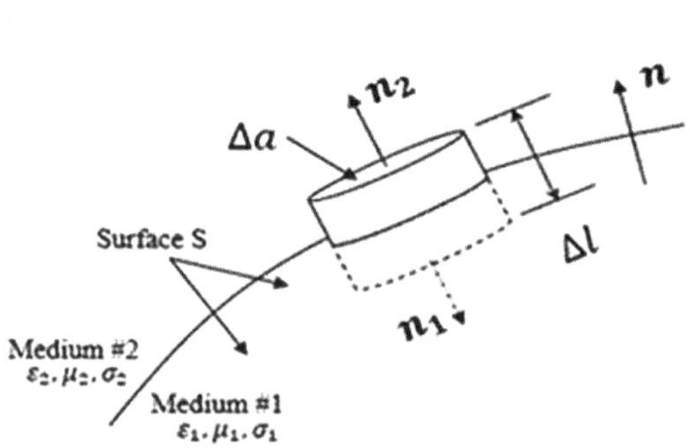

Figure 3.4 Construction for normal components

Table 3.6 Boundary conditions for material interfaces

Component	Boundary condition
	General case
E Field, tangential	$\boldsymbol{n} \times (\boldsymbol{E}_2 - \boldsymbol{E}_1) = 0$
D Field, tangential	$\boldsymbol{n} \times \left(\boldsymbol{D}_2 - \frac{\varepsilon_2}{\varepsilon_1}\boldsymbol{D}_1\right) = 0$
H Field, tangential	$\boldsymbol{n} \times (\boldsymbol{H}_2 - \boldsymbol{H}_1) = \boldsymbol{J}_s$
B Field, tangential	$\boldsymbol{n} \times \left(\boldsymbol{B}_2 - \frac{\mu_2}{\mu_1}\boldsymbol{B}_1\right) = \mu_2\boldsymbol{J}_s$
E Field, normal	$\boldsymbol{n} \cdot \left(\boldsymbol{E}_2 - \frac{\varepsilon_1}{\varepsilon_2}\boldsymbol{E}_1\right) = \frac{\rho_s}{\varepsilon_2}$
D Field, normal	$\boldsymbol{n} \cdot (\boldsymbol{D}_2 - \boldsymbol{D}_1) = \rho_s$
H Field, normal	$\boldsymbol{n} \cdot \left(\boldsymbol{H}_2 - \frac{\mu_1}{\mu_2}\boldsymbol{H}_1\right) = 0$
B Field, normal	$\boldsymbol{n} \cdot (\boldsymbol{B}_2 - \boldsymbol{B}_1) = 0$
	Selected other cases
	$\sigma = \infty$ in Medium 2
E Field, tangential	$E_{1_t} = 0$
H Field, tangential	$H_{1_t} = J_s$; $H_{2_t} = 0$
D Field, normal	$D_{1_n} = \rho_s$
	$\mu = \infty$ in Medium 2
H Field, tangential	$H_{1_t} = 0$

Table 3.6 provides a summary of boundary conditions for electric and magnetic fields.

3.2.11 Lossless and lossy media

The wave equations in (3.101) and (3.102) can be restated in terms of phasor quantities as

$$\nabla^2\widehat{\boldsymbol{E}} = \mu\sigma\left(j\omega\widehat{\boldsymbol{E}}\right) + \mu\varepsilon(j\omega)^2\widehat{\boldsymbol{E}} = j\omega\mu(\sigma + j\omega\varepsilon)\widehat{\boldsymbol{E}} \tag{3.135}$$

$$\nabla^2\widehat{\boldsymbol{H}} = j\omega\mu(\sigma + j\omega\varepsilon)\widehat{\boldsymbol{H}} \tag{3.136}$$

Here, the quantity $j\omega\mu(\sigma + j\omega\varepsilon)$ will be referred to as γ^2. The positive square root of γ^2 is known as the propagation constant, and is a complex number normally expressed as

$$\gamma = \sqrt{j\omega\mu(\sigma + j\omega\varepsilon)} = \alpha + j\beta \tag{3.137}$$

where α is known as the attenuation constant and β is known as the phase constant or the wavenumber. The units of α are Nepers per metre (Np/m). 1 Np/m is equivalent to 8.69 dB/m loss. The units of β are radians per metre (rad/m). The phase constant represents the amount of phase shift that occurs as the wave travels over the distance of 1 metre.

The quantity $(\sigma + j\omega\varepsilon)$ may be replaced by $j\omega\varepsilon[1 - j(\sigma/\omega\varepsilon)]$. The term $\sigma/\omega\varepsilon$ in this expression is known as the loss tangent of the medium. The loss tangent is a well-documented parameter for many different materials. The name loss tangent derives from the ratio of the conduction current density to the displacement current density in the material, where

$$\tan\phi = \frac{\left|\widehat{\boldsymbol{J}}_c\right|}{\left|\widehat{\boldsymbol{J}}_d\right|} = \frac{\sigma}{\omega\varepsilon} \tag{3.138}$$

The loss tangent provides a convenient method for distinguishing between dielectrics and conductors, as well as determining the attenuation and phase constants and the velocity of propagation in the medium.

The ratio of the electric field to the magnetic field in a plane wave is known as the intrinsic impedance of the medium through which the fields are propagating. This impedance is given by

$$\eta = \frac{j\omega\mu}{\gamma} = \sqrt{\frac{j\omega\mu}{\sigma + j\omega\varepsilon}} = \sqrt{\frac{j\omega\mu}{j\omega\varepsilon[1 - j(\sigma/\omega\varepsilon)]}} = \sqrt{\frac{\mu}{\varepsilon}}\left(\frac{1}{\sqrt{[1 - j(\sigma/\omega\varepsilon)]}}\right) \tag{3.139}$$

The units of (3.139) are ohms; (3.139) is also known as the wave impedance for plane waves and TEM waves.

Lossless media tend to be materials considered as poor conductors. In poor conductors, σ is typically very small $(\sigma \approx 0)$ such that $\sigma/\omega\varepsilon \ll 1$, and $\gamma \cong \sqrt{j\omega\mu(j\omega\varepsilon)} = j\omega\sqrt{\mu\varepsilon}$.

Equations (3.135) and (3.136) reduce to

$$\nabla^2\widehat{\boldsymbol{E}} = \gamma^2\widehat{\boldsymbol{E}} = -\omega^2\mu\varepsilon\widehat{\boldsymbol{E}} \tag{3.140}$$

$$\nabla^2\widehat{\boldsymbol{H}} = \gamma^2\widehat{\boldsymbol{H}} = -\omega^2\mu\varepsilon\widehat{\boldsymbol{H}} \tag{3.141}$$

Equation (3.139) reduces to

$$\eta = \frac{j\omega\mu}{\gamma} = \frac{j\omega\mu}{j\omega\sqrt{\mu\varepsilon}} = \sqrt{\frac{\mu}{\varepsilon}} \cong 120\pi \text{ (in free space)} \tag{3.142}$$

and α and β become

$$\alpha \cong \frac{\boldsymbol{\sigma}}{\mathbf{2}}\sqrt{\frac{\mu}{\varepsilon}} \cong 0 \tag{3.143a}$$

$$\beta \cong \omega\sqrt{\mu\varepsilon} \tag{3.143b}$$

β is directly related to the wavelength, λ, of the wave in the medium and can be restated as

$$\beta = \frac{2\pi}{\lambda} \tag{3.144}$$

β is also directly related to the velocity, u, of the wave in the medium. The velocity of the wave in the medium is given by $1/\sqrt{\mu\varepsilon}$. Using this, β can be restated as

$$\beta \cong \frac{\omega}{u} \tag{3.145}$$

Equations (3.144) and (3.145) can also be restated as

$$\lambda = \frac{2\pi}{\beta} = \frac{u}{f} \tag{3.146}$$

Equation (3.146) indicates that as the frequency increases, the wavelength becomes shorter. If μ, or ε, or both, are greater than their corresponding values in free space, the velocity u will be less than c.

In a lossy medium, $\sigma > 0$, and (3.135) and (3.136) do not reduce

$$\nabla^2\widehat{\boldsymbol{E}} = \gamma^2\widehat{\boldsymbol{E}} = \mu\sigma\left(j\omega\widehat{\boldsymbol{E}}\right) + \mu\varepsilon(j\omega)^2\widehat{\boldsymbol{E}} = j\omega\mu(\sigma + j\omega\varepsilon)\widehat{\boldsymbol{E}} \tag{3.147}$$

$$\nabla^2\widehat{\boldsymbol{H}} = \gamma^2\widehat{\boldsymbol{H}} = j\omega\mu(\sigma + j\omega\varepsilon)\widehat{\boldsymbol{H}} \tag{3.148}$$

γ is the same as (3.137).

$$\gamma = \sqrt{j\omega\mu(\sigma + j\omega\varepsilon)} = \alpha + j\beta \tag{3.149}$$

and (3.139) does not reduce

$$\eta = \frac{j\omega\mu}{\gamma} = \sqrt{\frac{j\omega\mu}{(\sigma + j\omega\mu\varepsilon)}} = \sqrt{\frac{\mu}{\varepsilon}}\left(\frac{1}{\sqrt{[1 - j(\sigma/\omega\varepsilon)]}}\right) \tag{3.150}$$

For the lossy medium, α and β can be found as follows. Squaring both sides of (3.149) and equating the real and imaginary parts of the result yields

$$\beta^2 - \alpha^2 = \omega^2\mu\varepsilon \tag{3.151a}$$

$$\alpha\beta = \frac{\omega\mu\sigma}{2} \tag{3.151b}$$

Solving the system of (3.151) for α and β, and constraining the solution such that $\alpha \geq 0$ and $\beta \geq 0$ when $\omega \geq 0$ yields

$$\beta = \omega\sqrt{\mu\varepsilon}\left\{\frac{1}{2}\left[\sqrt{1 + \left(\frac{\sigma}{\omega\varepsilon}\right)^2} + 1\right]\right\}^{1/2} \geq \omega\sqrt{\mu\varepsilon} \tag{3.152}$$

$$\alpha = \omega\sqrt{\mu\varepsilon}\left\{\frac{1}{2}\left[\sqrt{1 + \left(\frac{\sigma}{\omega\varepsilon}\right)^2} - 1\right]\right\}^{1/2} = \frac{(\sigma/2)\sqrt{\mu\varepsilon}}{\left\{\frac{1}{2}\left[\sqrt{1 + \left(\frac{\sigma}{\omega\varepsilon}\right)^2} + 1\right]\right\}^{1/2}} \leq \beta \tag{3.153}$$

A very important consideration in the design of electromagnetic shielding is the passage of waves across the boundary between two media. Note that throughout the following discussions on the incidence of electromagnetic waves on boundaries, the Cartesian system is used. This makes sense, as regardless of the geometry, at the locations of incidence, reflection, and refraction of the propagation vector, the surfaces of boundaries may be considered as planar. For wire and cable shielding, this also makes complete sense. If the radius of the wire or cable shield is large with respect to the thickness of the shield, the wave propagation into and through the shield may be regarded as substantially plane. This allows us to focus on normal incidence scenarios.

Figure 3.5 illustrates normal incidence of a linearly polarized TEM wave moving through one lossless dielectric material and into another lossless dielectric material, where both media are dissipationless ($\sigma_1 = \sigma_2 = 0$), isotropic, and homogeneous. An example of this scenario for wire and cable is an unshielded wire with a dielectric jacket covering exposed to the environment. This example

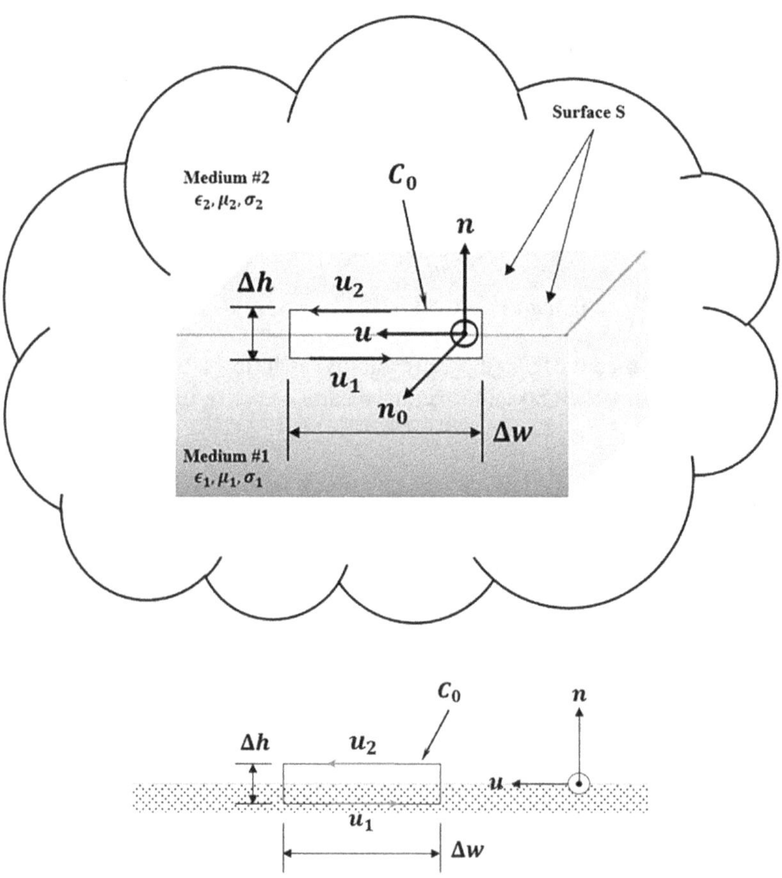

Figure 3.5 Normal incidence of plane wave at dielectric/dielectric boundary

highlights the potential power loss to the surrounding environment through the dielectric jacket. Generally, the two media will exhibit different constitutive parameters (ε_1, μ_1; ε_2, μ_2) and therefore will exhibit different impedance characteristics, thus creating a discontinuity at the interface. The incident wave will be partially transmitted and partially reflected at the interface by the presence of this discontinuity. Analysis proceeds by assuming the wave is travelling in the $+z$-direction in accordance with Poynting's theorem. z is negative valued in medium #1.

The incident wave may be represented as

$$\boldsymbol{E}_i = \boldsymbol{u}_x E_{i0} e^{-j\beta_1 z} \tag{3.154}$$

$$\boldsymbol{H}_i = \frac{1}{\eta_1}\boldsymbol{u}_z \times \boldsymbol{E}_i = \boldsymbol{u}_y \frac{E_{i0}}{\eta_1} e^{-j\beta_1 z} \tag{3.155}$$

where E_{i0} = the magnitude of the incident electric field at $z = 0$

- β_1 = the phase constant of medium #1
- η_1 = the intrinsic impedance of medium #1

The reflected wave is given by

$$\boldsymbol{E}_r = \boldsymbol{u}_x E_{r0} e^{j\beta_1 z} \tag{3.156}$$

$$\boldsymbol{H}_r = -\frac{1}{\eta_1}\boldsymbol{u}_z \times \boldsymbol{E}_r = -\boldsymbol{u}_y \frac{E_{r0}}{\eta_1} e^{j\beta_1 z} \tag{3.157}$$

The transmitted wave is given by

$$\boldsymbol{E}_t = \boldsymbol{u}_x E_{t0} e^{-j\beta_2 z} \tag{3.158}$$

$$\boldsymbol{H}_t = \frac{1}{\eta_2}\boldsymbol{u}_z \times \boldsymbol{E}_t = \boldsymbol{u}_y \frac{E_{t0}}{\eta_2} e^{-j\beta_2 z} \tag{3.159}$$

The boundary conditions from Table 3.6 require that the tangential components of the electric and magnetic fields are continuous across the interface. Thus,

$$\boldsymbol{E}_i(0) + \boldsymbol{E}_r(0) = \boldsymbol{E}_t(0), \text{ or } E_{i0} + E_{r0} = E_{t0} \tag{3.160a}$$

$$\boldsymbol{H}_i(0) + \boldsymbol{H}_r(0) = \boldsymbol{H}_t(0), \text{ and } \frac{1}{\eta_1}(E_{i0} - E_{r0}) = \frac{E_{t0}}{\eta_2} \tag{3.160b}$$

Solving equations (3.160) for E_{r0} and E_{t0} yields

$$E_{r0} = E_{i0}\left(\frac{\eta_2 - \eta_1}{\eta_2 + \eta_1}\right) \tag{3.161a}$$

$$E_{t0} = E_{i0}\left(\frac{2\eta_2}{\eta_2 + \eta_1}\right) \tag{3.161b}$$

The ratios between these two expressions are given by

$$\frac{E_{r0}}{E_{i0}} = \left(\frac{\eta_2 - \eta_1}{\eta_2 + \eta_1}\right) = \Gamma \tag{3.162}$$

$$\frac{E_{t0}}{E_{i0}} = \left(\frac{2\eta_2}{\eta_2 + \eta_1}\right) = \mathrm{T} \tag{3.163}$$

where

Γ = the reflection coefficient, the ratio at the interface of the complex reflected electric field to the complex incident electric field.

T = the transmission coefficient, the ratio at the interface of the complex transmitted electric field to the complex incident electric field.

The coefficients Γ and T are related by

$$1 + \Gamma = \mathrm{T} \tag{3.164}$$

Using (3.130a), the time-averaged power carried in the $+z$ direction in medium #1 is

$$\langle S_{z1} \rangle = \frac{|E_{i0}|^2 - |E_{r0}|^2}{2\eta_1} = \frac{|E_{i0}|^2}{2\eta_1}\left(1 - |\Gamma|^2\right) \tag{3.165}$$

The corresponding time-averaged power carried in the $+z$ direction in medium #2 is

$$\langle S_{z2} \rangle = \frac{|E_{t0}|^2}{2\eta_2} \tag{3.166}$$

Since the interface is lossless, $\langle S_{z1} \rangle = \langle S_{z2} \rangle$, such that

$$\frac{|E_{i0}|^2}{2\eta_1}\left(1 - |\Gamma|^2\right) = \frac{|E_{t0}|^2}{2\eta_2} \tag{3.167}$$

$$\left(1 - |\Gamma|^2\right) = \frac{2\eta_1}{2\eta_2}\frac{|E_{t0}|^2}{|E_{i0}|^2} = \frac{\eta_1}{\eta_2}|\mathrm{T}|^2 \tag{3.168}$$

Equation (3.168) indicates the transmitted power, modified by the ratio of the impedances of the two media, is equal to the incident power, less the reflected power.

Examination of the foregoing shows a close analogy to the analysis used for transmission lines, an observation used to great advantage by Schelkunoff in formulating his shielding equations.

We turn our attention now to an interface between a lossless dielectric and a lossy medium characterized by $[(\sigma/\omega\varepsilon) \gg 1]$. The presence of a finite conductivity in one of the two interfacing media introduces a large degree of complexity in the mathematical relationships between incident, reflected, and transmitted (refracted) waves. The complexity can be bounded by considering two separate cases.

Case 1 is defined as $\frac{\sigma}{\omega\varepsilon} \ll 1$, where

$$\beta = \omega\sqrt{\mu\varepsilon}\left\{\frac{1}{2}\left[\sqrt{1+\left(\frac{\sigma}{\omega\varepsilon}\right)^2}+1\right]\right\}^{1/2} \cong \omega\sqrt{\mu\varepsilon} \tag{3.169}$$

$$\alpha = \omega\sqrt{\mu\varepsilon}\left\{\frac{1}{2}\left[\sqrt{1+\left(\frac{\sigma}{\omega\varepsilon}\right)^2}-1\right]\right\}^{1/2} \cong \frac{\sigma}{2}\sqrt{\frac{\mu}{\varepsilon}} \tag{3.170}$$

Case 1 applies in instances where the conduction current is much less than the displacement current in the lossy medium, possibly an imperfect dielectric or a magnetic medium exhibiting a low electrical conductivity.

Case 2 is defined as $\frac{\sigma}{\omega\varepsilon} \gg 1$, where

$$\beta = \omega\sqrt{\mu\varepsilon}\left\{\frac{1}{2}\left[\sqrt{1+\left(\frac{\sigma}{\omega\varepsilon}\right)^2}+1\right]\right\}^{1/2} \cong \sqrt{\frac{\omega\mu\sigma}{2}} \tag{3.171}$$

$$\alpha = \omega\sqrt{\mu\varepsilon}\left\{\frac{1}{2}\left[\sqrt{1+\left(\frac{\sigma}{\omega\varepsilon}\right)^2}-1\right]\right\}^{1/2} \cong \sqrt{\frac{\omega\mu\sigma}{2}} \tag{3.172}$$

Case 2 applies in instances where the displacement current is much less than the conduction current in the lossy medium, possibly an imperfect conductor such as copper or aluminium.

In either case, the angle of the propagation constant can be found as

$$\tan^{-1}\frac{\alpha}{\beta} = \frac{\left\{\frac{1}{2}\left[\sqrt{1+\left(\frac{\sigma}{\omega\varepsilon}\right)^2}-1\right]\right\}^{1/2}}{\left\{\frac{1}{2}\left[\sqrt{1+\left(\frac{\sigma}{\omega\varepsilon}\right)^2}+1\right]\right\}^{1/2}} \tag{3.173}$$

The angle of the propagation constant is equal to the phase angle between the magnetic and electric fields.

For Case 1, we have

$$\tan^{-1}\frac{\alpha}{\beta} = \frac{\frac{\sigma}{2}\sqrt{\frac{\mu}{\varepsilon}}}{\omega\sqrt{\mu\varepsilon}} = \frac{\sigma}{2\omega\varepsilon} \tag{3.174}$$

which clearly indicates the angle of the propagation vector is a function of the applied frequency, and the dielectric constant and conductivity of the medium. In a non-conducting medium ($\sigma = 0$), the two fields are in phase.

For Case 2, we have

$$\tan^{-1}\frac{\alpha}{\beta} = \frac{\sqrt{\frac{\omega\mu\sigma}{2}}}{\sqrt{\frac{\omega\mu\sigma}{2}}} = 1 \tag{3.175}$$

$$\gamma = \sqrt{j\omega\mu(\sigma + j\omega\varepsilon)} \cong \sqrt{j\omega\mu\sigma} = (1+j)\sqrt{\frac{\omega\mu\sigma}{2}} \tag{3.176}$$

and (3.150) reduces to

$$\eta = \frac{j\omega\mu}{\gamma} = \sqrt{\frac{j\omega\mu}{(\sigma + j\omega\varepsilon)}} = \sqrt{\frac{\mu}{\varepsilon}}\left(\frac{1}{\sqrt{[1 - j(\sigma/\omega\varepsilon)]}}\right) = (1 + j)\sqrt{\frac{\omega\mu}{2\sigma}} \quad (3.177)$$

Equations (3.175) and (3.176) clearly indicate in Case 2 the angle of the propagation vector is a constant in the medium, regardless of frequency. For a "good" conductor, represented by Case 2, the phase angle between the electric and magnetic fields is equal to 45°, or $\pi/4$ radians. That is, the electric field leads the magnetic field by 45°, or $\pi/4$ radians. Note that most practical electrical conductors, so-called "good" conductors, will exhibit this behaviour. The following discussion assumes we have a "good" conductor as the lossy material.

Figure 3.6 illustrates normal incidence of a linearly polarized TEM wave moving through a lossless dielectric material and into a "good" conductor. An example of this scenario for wire and cable is a shielded wire with a dielectric jacket covering and a shield covering the jacket, the shielded wire or cable exposed to the environment. As before, this example highlights the potential power loss to the surrounding environment, this time into the shield. The lossless material is dissipationless ($\sigma_1 = 0$). Both materials are isotropic and homogeneous, and both materials may exhibit different constitutive parameters (ε_1, μ_1; ε_2, μ_2). Therefore, each of the two materials may exhibit different impedance characteristics, thus creating a discontinuity at the interface. The incident wave will be partially transmitted and partially reflected at the interface by the presence of this discontinuity. Analysis to follow is constrained by Case 2, previously discussed, and proceeds by assuming the wave is travelling in the $+z$-direction in accordance with Poynting's theorem. z is negatively valued in medium #1.

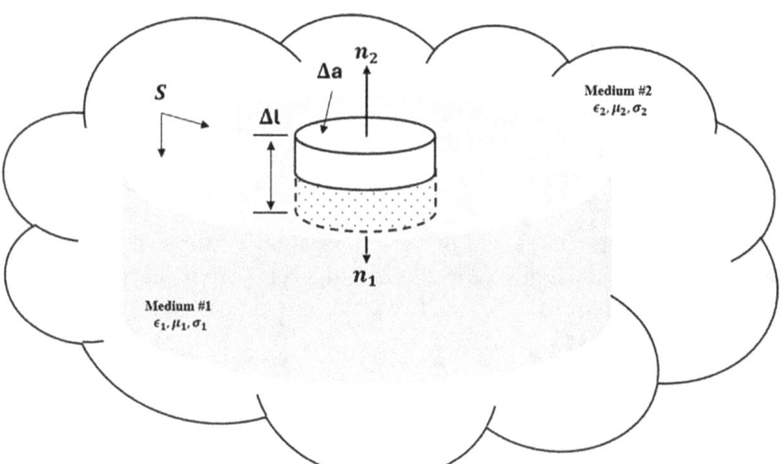

Figure 3.6 Normal incidence of plane wave at dielectric/"good" conductor boundary

The incident wave may be represented as

$$\boldsymbol{E}_i = \boldsymbol{u}_x E_{i0} e^{-j\beta_1 z} \tag{3.178}$$

$$\boldsymbol{H}_i = \frac{1}{\eta_1}\boldsymbol{u}_z \times \boldsymbol{E}_i = \boldsymbol{u}_y \frac{E_{i0}}{\eta_1} e^{-j\beta_1 z} \tag{3.179}$$

The reflected wave is given by

$$\boldsymbol{E}_r = \boldsymbol{u}_x E_{r0} e^{j\beta_1 z} \tag{3.180}$$

$$\boldsymbol{H}_r = -\frac{1}{\eta_1}\boldsymbol{u}_z \times \boldsymbol{E}_r = -\boldsymbol{u}_y \frac{E_{r0}}{\eta_1} e^{j\beta_1 z} \tag{3.181}$$

The transmitted wave decays as it moves deeper into the conductor, and is given by

$$\boldsymbol{E}_t = \boldsymbol{u}_x E_{t0} e^{-\alpha_2 z} e^{-j\beta_2 z} \tag{3.182}$$

$$\boldsymbol{H}_t = \frac{1}{\tilde{\eta}_2}\boldsymbol{u}_z \times \boldsymbol{E}_t = \boldsymbol{u}_y \frac{E_{t0}}{\tilde{\eta}_2} e^{-\alpha_2 z} e^{-j\beta_2 z} \tag{3.183}$$

where $\tilde{\eta}_2 = (1+j)\sqrt{\frac{\omega\mu_2}{2\sigma_2}}$ represents the complex impedance of medium #2.

The boundary conditions from Table 3.6 require that the tangential components of the electric and magnetic fields are continuous across the interface. Thus,

$$\boldsymbol{E}_i(0) + \boldsymbol{E}_r(0) = \boldsymbol{E}_t(0), \text{ or } E_{i0} + E_{r0} = E_{t0} \tag{3.184a}$$

$$\boldsymbol{H}_i(0) + \boldsymbol{H}_r(0) = \boldsymbol{H}_t(0), \text{ or } \frac{1}{\eta_1}(E_{i0} - E_{r0}) = \frac{E_{t0}}{\tilde{\eta}_2} \tag{3.184b}$$

Solving equations (3.184) for E_{r0} and E_{t0} yields

$$E_{r0} = E_{i0}\left(\frac{\tilde{\eta}_2 - \eta_1}{\tilde{\eta}_2 + \eta_1}\right) \tag{3.185a}$$

$$E_{t0} = E_{i0}\left(\frac{2\tilde{\eta}_2}{\tilde{\eta}_2 + \eta_1}\right) \tag{3.185b}$$

The ratios between these two expressions are given by

$$\frac{E_{r0}}{E_{i0}} = \left(\frac{\tilde{\eta}_2 - \eta_1}{\tilde{\eta}_2 + \eta_1}\right) = \frac{(\tilde{\eta}_2/\eta_1) - 1}{(\tilde{\eta}_2/\eta_1) + 1} = \tilde{\Gamma} \tag{3.186}$$

$$\frac{E_{t0}}{E_{i0}} = \left(\frac{2\tilde{\eta}_2}{\tilde{\eta}_2 + \eta_1}\right) = \tilde{\mathrm{T}} \tag{3.187}$$

Consider we know

$$E_{io}/H_{i0} = \eta_1; \text{ and } E_{ro}/H_{r0} = -\eta_1; \text{ and } E_{to}/H_{t0} = \tilde{\eta}_2 \tag{3.188}$$

Given our selection of Case 2 and assuming $|\tilde{\eta}_2| \ll |\eta_1|$, (3.188) enables us to write (3.187) as

$$\frac{E_{t0}}{E_{i0}} \cong \left(\frac{2\tilde{\eta}_2}{\eta_1}\right) \Rightarrow E_{t0} \cong 2E_{i0}\left(\frac{\tilde{\eta}_2}{\eta_1}\right) \tag{3.189a}$$

$$\frac{E_{t0}}{E_{i0}} \cong \left(\frac{2\tilde{\eta}_2}{\eta_1}\right) \Rightarrow H_{t0}\tilde{\eta}_2 \cong \left(\frac{2\tilde{\eta}_2}{\eta_1}\right)H_{i0}\eta_1 \Rightarrow H_{t0} \cong 2H_{i0} \tag{3.189b}$$

Equation (3.189a) indicates the transmitted electric field intensity at the interface is twice that of the incident field and is very small in the conductor, reduced by the ratio of the conductor impedance to that of the dielectric. Equation (3.189b) indicates the transmitted magnetic field intensity at the interface is twice that of the incident field at the interface and indicates further the presence of a surface current density at the interface.

It is interesting to note the relationship between the two fields by comparing them in a ratio.

$$\frac{\frac{E_{t0}}{\tilde{\eta}_2}e^{-\alpha_2 z}e^{-j\beta_2 z}}{E_{t0}e^{-\alpha_2 z}e^{-j\beta_2 z}} = \frac{1}{\tilde{\eta}_2} \tag{3.190}$$

$$\frac{H_{t0}}{E_{t0}} = \frac{1}{|\tilde{\eta}_2|} = \sqrt{\frac{\sigma_2}{\omega\mu_2}} \tag{3.191}$$

The transmitted magnetic field intensity is proportionately larger than the transmitted electric field intensity as a function of the ratio of the conductivity of the conductor and the frequency of the wave multiplied by the permeability of the conductor.

Consider further that

$$|\tilde{\eta}_2|/\eta_1 = \frac{\sqrt{\omega\mu_2/\sigma_2}}{\sqrt{\mu_0/\varepsilon_0}} = \sqrt{\frac{\omega\mu_2\varepsilon_0}{\mu_0\sigma_2}} \tag{3.192}$$

Assuming both media are non-magnetic, and based on our selection of Case 2, we have

$$|\tilde{\eta}_2|/\eta_1 = \sqrt{\frac{\omega\varepsilon_0}{\sigma_2}} \ll 1 \tag{3.193}$$

Avoiding for the moment the simpler conclusion that for Case 2, (3.186) is equal to -1 (this would be true if medium #2 were a perfect electrical conductor), using the binomial theorem we can expand (3.186) to yield an estimate, as follows:

$$\begin{aligned}\tilde{\Gamma} &= (\tilde{\eta}_2/\eta_1 - 1)(\tilde{\eta}_2/\eta_1 + 1)^{-1} \\ &= (\tilde{\eta}_2/\eta_1 - 1)\left[1 - (\tilde{\eta}_2/\eta_1) + (\tilde{\eta}_2/\eta_1)^2 - (\tilde{\eta}_2/\eta_1)^3 + \cdots\right] \\ &= -1 + 2(\tilde{\eta}_2/\eta_1) - 2(\tilde{\eta}_2/\eta_1)^2 + 2(\tilde{\eta}_2/\eta_1)^3 - \cdots \end{aligned} \tag{3.194}$$

so that

$$\tilde{\Gamma} = -1 + 2(\tilde{\eta}_2/\eta_1) \text{ for } |\tilde{\eta}_2|/\eta_1 = \sqrt{\frac{\omega\varepsilon_0}{\sigma_2}} \ll 1 \tag{3.195}$$

As expected, assuming that $|\tilde{\eta}_2| \ll |\eta_1|$, (3.195) indicates the reflection coefficient for Case 2 asymptotically approaches -1 as the conductivity of medium #2 increases.

Although the electric field is quite small inside the conductor, it still contributes to a surface current density. Inside the conductor, we have a conduction current density that, like the electric and magnetic fields, decays as it moves deeper into the conductor

$$\boldsymbol{J}_c = \boldsymbol{u}_x\sigma_2 2E_{i0}\left(\frac{\tilde{\eta}_2}{\eta_1}\right)e^{-\alpha_2 z}e^{-j\beta_2 z} = \boldsymbol{u}_x\sigma_2 2E_{i0}\left(\frac{\tilde{\eta}_2}{\eta_1}\right)e^{-\gamma_2 z} \tag{3.196}$$

We can determine the surface current density by finding the current through an area oriented perpendicularly to the incident electric field and the boundary surface at $z = 0$. We can define this area as having a length l along the boundary surface with a depth d into the conductor material. The total conduction current through this area is then given by

$$I = \iint \sigma_2 \boldsymbol{E}_{t0} \cdot d\boldsymbol{S} = 2l\sigma_2 H_{i0}\tilde{\eta}_2 \int_0^d e^{-\gamma_2 z}dz \tag{3.197}$$

$$\cong \frac{2l\sigma_2 H_{i0}\tilde{\eta}_2}{\gamma_2} = \frac{2l\sigma_2 H_{i0}\left(\frac{j\omega\mu_2}{\sigma_2}\right)}{\sqrt{j\omega\mu_2\sigma_2}} = 2lH_{i0} \tag{3.198}$$

Dividing by l converts (3.198) to the desired surface current density per square metre expression

$$|\boldsymbol{J}_s| \cong 2H_{i0} \tag{3.199}$$

Equation (3.199) confirms the (expected) presence of a surface current density at the interface. As we have found, the magnetic field intensity at the conducting interface is twice the value of the incident magnetic field intensity and is equal in magnitude to the surface current density. Note this current is induced by the magnetic field at the interface and will circle the magnetic field vector at the interface and inside the conductor such that they flow tangentially in the same direction as the incident electric field at the interface but in the opposite direction inside the conductor, forming what are known as eddy currents.

Using (3.198), the fraction of power reflected is approximately

$$|\tilde{\Gamma}|^2 \cong 1 - 4\sqrt{\frac{\omega\varepsilon_1}{2\sigma_2}} \tag{3.200}$$

The fraction transmitted into the conductor is

$$1 - |\tilde{\Gamma}|^2 \cong 4\sqrt{\frac{\omega\varepsilon_1}{2\sigma_2}} \tag{3.201}$$

The time-averaged power flowing in the $+z$ direction into the conductor per square metre, which is equal to the incident power times the fraction transmitted, is then given by

$$\langle S_{z2}\rangle = \frac{1}{2}|E_{i0}|^2\left(1 - |\tilde{\Gamma}|^2\right) = \frac{1}{2}|H_{i0}|^2\eta_1\left(1 - |\tilde{\Gamma}|^2\right) \tag{3.202}$$

$$= \frac{1}{2}|H_{i0}|^2\eta_1\left(4\sqrt{\frac{\omega\varepsilon_1}{2\sigma_2}}\right) = \frac{1}{2}|H_{i0}|^2\left(\sqrt{\frac{\mu_1}{\varepsilon_1}}\right)\left(4\sqrt{\frac{\omega\varepsilon_1}{2\sigma_2}}\right) \tag{3.203}$$

$$= \frac{1}{2}|H_{i0}|^2\left(\sqrt{\frac{\mu_1}{\varepsilon_1}}\right)\left(4\sqrt{\frac{\omega\varepsilon_1}{2\sigma_2}}\right) = \frac{1}{2}|2H_{i0}|^2\sqrt{\frac{\omega\mu_1}{2\sigma_2}} \tag{3.204}$$

Assuming the conductor is non-magnetic, $\mu_1 = \mu_2$, so that

$$= \frac{1}{2}|2H_{i0}|^2\sqrt{\frac{\omega\mu_2}{2\sigma_2}} = \frac{1}{2}|2H_{i0}|^2\mathcal{R}e(\tilde{\eta}_2) \tag{3.205}$$

The corresponding time-averaged power flowing in the $+z$ direction in the conductor per square metre is given by

$$\langle S_{z2}\rangle = \frac{1}{2}\mathcal{R}e\left(E_{t0} \times H_{t0}^*\right) = \frac{1}{2}\left(H_{t0}\mathcal{R}e(\tilde{\eta}_2) \times H_{t0}^*\right) \tag{3.206}$$

$$= \frac{1}{2}|H_{t0}|^2\mathcal{R}e(\tilde{\eta}_2) = \frac{1}{2}|2H_{i0}|^2\mathcal{R}e(\tilde{\eta}_2) \tag{3.207}$$

$$= \frac{1}{2}|2H_{i0}|^2\mathcal{R}e(\tilde{\eta}_2) = \frac{1}{2}|H_{t0}|^2\sqrt{\frac{\omega\mu_2}{2\sigma_2}} \tag{3.208}$$

The term $\sqrt{\frac{\omega\mu_2}{2\sigma_2}}$ can be written as

$$\frac{1}{\sqrt{\frac{2\sigma_2}{\omega\mu_2}}} = \frac{1}{\sigma_2\sqrt{\frac{2}{\omega\mu_2\sigma_2}}} = \frac{1}{\sigma_2/\alpha_2} \tag{3.209}$$

The electric and magnetic fields and the current density all decay as a function of depth as they penetrate a conductor. The attenuation of the waves' magnitude as they penetrate the conductor is directly proportional to the factor $e^{-\alpha z}$, where in this expression, the variable z simply represents the distance of penetration. Over a distance defined by

$$\delta \equiv \frac{1}{\alpha} \tag{3.210}$$

the wave's magnitude will be attenuated by $1/e$, or $\approx 37\%$. Substituting (3.210) into (3.209) yields

$$\delta = \frac{1}{\sqrt{\frac{\omega\mu\sigma}{2}}} = \frac{1}{\sqrt{\pi f \mu\sigma}} \tag{3.211}$$

Equation (3.211) is known as the harmonic skin depth.

Substituting (3.209) and (3.211) into (3.208) yields

$$\langle S_{z2} \rangle = \frac{1}{2}|2H_{i0}|^2\left(\frac{1}{\delta\sigma_2}\right) \tag{3.212}$$

Equation (3.212) indicates that power is dissipated in the conductor by the surface current in inverse proportion to the skin depth multiplied by the conductivity of the conductor.

3.2.12 Skin depth

Skin depth, sometimes referred to as skin effect, is a descriptive term for the attenuation of electromagnetic fields inside a material, typically used in conjunction with discussions of conductors, and directly related to the ability of a conductive material to perform as an electromagnetic shield. The first mathematical discussion of the non-uniformity of alternating-current density in a wire was provided by James Clerk Maxwell in his Treatise in 1873 [38]. The effect of current tending to flow on the surface of a conductor was first described by Horace Lamb in a paper focused on the induction of currents in cylindrical and spherical conductors in November of 1883 [39]. Oliver Heaviside contributed an extensive mathematical literature to the whole subject between the years of 1884 and 1887 in instalments in The Electrician, a popular and widely read source for the burgeoning discipline of all things having to do with electricity in the late nineteenth century. Using his transmission line wave theory, Heaviside found modified expressions for both the resistance and inductance of the lines and discussed the effect of increasing frequency on these parameters due to the change in magnetic and current density field distributions. Heaviside correctly recognized the resistance is found to be increased and the internal inductance is found to be decreased as frequency increases and formed descriptive differential equations with the modified parameters to account for skin and proximity effects in his transmission line model. In addition to his 1884 paper on energy flow, J.H. Poynting contributed further to the mathematics of the subject in his 1885 paper [40]. David Edward Hughes published a Presidential Address to the Society of Telegraph Engineers and Electricians in January of 1886 to report the results of experimental efforts to explore the phenomenon [41]. Although the paper and the associated experimentation suffered from several technical shortcomings that led to eventual discreditation, Hughes' report kindled lengthy and sometimes emotional debate during which skin effect played a central role, forcing contemporary practising electrical engineers to revise their understanding of the phenomenon and to re-evaluate the importance and applicability of sound, careful mathematical reasoning to practical design aspects [42–44]. In August of 1886, Heaviside wrote the first of a series of papers under the heading of "On the Self-Induction of Wires", in which he provided a critique of Hughes' efforts and provided a lengthy mathematical discussion of the topic [45]. In a later article published in 1891 [46], Heaviside wrote:

> It was discovered by mathematical reasoning that when an electric current is started in a wire, it begins entirely upon its skin, intact upon the outside of its skin; and that, in consequence, sufficiently rapidly impressed fluctuations of the current keep to the skin of the wire, and do not sensibly penetrate its interior.
>
> Now very few (if any) unmathematical electricians can understand this fact; many of them neither understand it nor believe it. Even many who do believe it do so, I believe, simply because they are told so, and not because they can in the least feel positive about its truth of their own knowledge. As an eminent practician (sic) remarked, after prolonged skepticism, "When Sir W. Thompson says so, who can doubt it?"

Lord Rayleigh in 1886 first gave the formula for skin effect in an infinitely wide strip [47]. Heinrich Friedrich Weber in 1886, Josef Stefan in 1887, and Oliver Lodge in 1888 contributed further material [48–50]. William Thomson (Lord Kelvin) gave an expression describing the effect using ber-bei functions (also known as Kelvin functions) in 1889 [51]. Heinrich Hertz in 1889 and Sir Joseph John Thomson in 1893 discussed the subject from both the experimental and mathematical standpoints [52,53]. The term "skin effect" was first used by Mr. J. Swinburne during discussions of the Institute of Electrical Engineers in 1891 [54]. All these researchers, and many others who followed in their footsteps, agreed that alternating currents do not flow inside conductors but rather along the outside surface of conductors, with the effect enhanced in proportion to the applied frequency.

Simply put, electromagnetic energy at moderate to high frequencies is not transmitted in the interior of a conductor; it travels in the region surrounding the conductor. The conductor itself acts as a guide for the wave. Currents established at the surface of a conductor propagate in a direction parallel to the direction of the induced current density and are attenuated by ohmic losses. Put succinctly, as an electromagnetic wave propagates along a wire, it will penetrate the surface of the wire as a function of frequency. At low frequencies, the penetration will be greater than at higher frequencies.

In general, in a "good" conductor, the conduction current is far greater than the displacement current, and (3.135) and (3.136) may be reduced to

$$\nabla^2\widehat{\boldsymbol{E}} = j\omega\mu\sigma\widehat{\boldsymbol{E}} \tag{3.213a}$$

$$\nabla^2\widehat{\boldsymbol{H}} = j\omega\mu\sigma\widehat{\boldsymbol{H}} \tag{3.213b}$$

A similar expression can be written for current density

$$\nabla^2\widehat{\boldsymbol{J}} = j\omega\mu\sigma\widehat{\boldsymbol{J}} \tag{3.213c}$$

With δ defined, equations (3.213) become

$$\nabla^2\widehat{\boldsymbol{E}} = j\omega\mu\sigma\widehat{\boldsymbol{E}} = j\frac{2}{\delta^2}\widehat{\boldsymbol{E}} \tag{3.214a}$$

$$\nabla^2\boldsymbol{H} = j\omega\mu\sigma\widehat{\boldsymbol{H}} = j\frac{2}{\delta^2}\widehat{\boldsymbol{H}} \tag{3.214b}$$

$$\nabla^2\widehat{\boldsymbol{J}} = j\omega\mu\sigma\widehat{\boldsymbol{J}} = j\frac{2}{\delta^2}\widehat{\boldsymbol{J}} \tag{3.214c}$$

Skin depth embodies a magnetic diffusion time, defined as

$$\tau_m = \mu\sigma l^2 \tag{3.215}$$

The skin depth is the length (depth into the material) that makes the diffusion time equal to twice the reciprocal angular frequency of the driving field, so that

$$\tau_m = \frac{2}{\omega} = \mu\sigma\delta^2 \tag{3.216}$$

Thus, the diffusion time is equal to the square of the skin depth multiplied by the permeability and conductivity of the material. Stated another way,

$$\tau_m = \frac{2}{\omega} = \frac{T}{\pi} \tag{3.217}$$

As an example, in copper having a conductivity of 5.8×10^7 Siemens/m, the skin depth is on the order of 6.6×10^{-5} metres at 1 MHz. The corresponding magnetic diffusion time is on the order of 0.32×10^{-6} seconds. A full cycle of the frequency is equal to 1.0×10^{-6} seconds. So, the wave will penetrate the material to the skin depth in roughly one third of the period. In this same time, the wave will have decayed to approximately 0.368 ($\cong$ 37%) of its original magnitude.

Clearly as the frequency increases, the skin depth and the magnetic diffusion time decrease, entirely consistent with Heaviside's findings. This behaviour is central to electromagnetic shielding. The ramifications of this in terms of wire resistance and inductance will be explored at length in the next chapter, and these results will play a significant role in wire shielding.

3.2.13 Polarization of plane waves

Plane waves may be characterized by their polarization. By convention, the polarization of an electromagnetic wave describes the time-varying behaviour of the electric field intensity vector at some point in space. Waves may be polarized elliptically, linearly, or circularly. In general, waves are elliptically polarized, with linear and circular polarizations being special cases of elliptical polarization. The shape of the polarization is traced out by the tip of the electric intensity vector as it moves in the direction of propagation. An elliptical polarization results from a wave that appears to rotate about the axis of the propagation vector and whose magnitude varies as a function of the rotation angle. A linear polarization results from a wave that remains fixed in orientation and does not appear to rotate about the axis of the propagation vector. A circular polarization results from a wave that appears to rotate about the axis of the propagation vector and whose magnitude is constant throughout the rotation. Waves may appear to rotate in an anticlockwise direction, also called a left-hand polarization, or a clockwise direction, also called a right-hand polarization. The apparent rotation is caused by the phase difference of the electric field intensity vector components as they oscillate about the direction vector.

As an example, consider a plane wave, travelling in the positive z-direction, that is the sum of two vector fields. The instantaneous electric field is given by

$$\boldsymbol{E}_T = |\boldsymbol{E}_1|\cos(\omega t - \beta z)\boldsymbol{u}_x + |\boldsymbol{E}_2|\cos(\omega t - \beta z + \theta)\boldsymbol{u}_y \tag{3.218}$$

If we were to view this wave looking in the positive z-direction at the position $z = 0$, (3.218) would be

$$\boldsymbol{E}_T = |\boldsymbol{E}_1|\cos(\omega t)\boldsymbol{u}_x + |\boldsymbol{E}_2|\cos(\omega t + \theta)\boldsymbol{u}_y \tag{3.219}$$

Four possibilities exist for this electric field intensity vector:

1. $|\boldsymbol{E}_1| = |\boldsymbol{E}_2|$, and $\theta = 0$
2. $|\boldsymbol{E}_1| \neq |\boldsymbol{E}_2|$, and $\theta = 0$
3. $|\boldsymbol{E}_1| = |\boldsymbol{E}_2|$, and $\theta = -90$
4. $|\boldsymbol{E}_1| \neq |\boldsymbol{E}_2|$, and $\theta = -90$.

For the first possibility, (3.219) becomes

$$\boldsymbol{E}_T = [|\boldsymbol{E}_1|\cos(\omega t)](\boldsymbol{u}_x + \boldsymbol{u}_y) \tag{3.220}$$

Equation (3.220) indicates the electric field intensity vector will oscillate back and forth along a straight line that is oriented at a 45° angle measured from the x-axis.

For the second possibility, (3.219) becomes

$$\boldsymbol{E}_T = [|\boldsymbol{E}_1|\cos(\omega t)]\left(\boldsymbol{u}_x + \frac{|\boldsymbol{E}_2|}{|\boldsymbol{E}_1|}\boldsymbol{u}_y\right) \tag{3.221}$$

Equation (3.221) indicates the electric field intensity vector will oscillate back and forth along a straight line that is oriented at an angle measured from the x-axis whose tangent is equal to $|\boldsymbol{E}_2|/|\boldsymbol{E}_1|$.

For the third possibility, (3.219) becomes

$$\boldsymbol{E}_T = |\boldsymbol{E}_1|(\cos(\omega t)\boldsymbol{u}_x + \sin(\omega t)\boldsymbol{u}_y) \tag{3.222}$$

Equation (3.222) indicates the electric field intensity vector will appear to rotate about the z-axis with a constant magnitude and that will always be oriented at an angle measured from the x-axis whose tangent is equal to $\sin(\omega t)/\cos(\omega t)$. Because $\theta = -90°$ the electric field intensity vector will appear to rotate in the clockwise direction and is said to have right-handed circular polarization.

For the fourth possibility, (3.219) becomes

$$\boldsymbol{E}_T = |\boldsymbol{E}_1|\left(\cos(\omega t)\boldsymbol{u}_x + \frac{|\boldsymbol{E}_2|}{|\boldsymbol{E}_1|}\sin(\omega t)\boldsymbol{u}_y\right) \tag{3.223}$$

Equation (3.223) indicates the electric field intensity vector will appear to rotate about the z-axis in the clockwise direction and will trace out an elliptical path. Because $\theta = -90°$, (3.223) is said to have right-handed elliptical polarization.

References

[1] J.D. Jackson, *Classical Electrodynamics*, 3rd ed., John Wiley & Sons, New York, NY, 1999.

[2] O. Jefimenko, *Electricity and Magnetism: An Introduction to the Theory of Electric and Magnetic Fields*, 2nd ed., Electret Scientific Company, Star City, WV, 1989.

[3] J. Schwinger, L.L. DeRaad, K.A. Milton, and W. Tsai, *Classical Electrodynamics*, Westview Press, Boulder, CO, 1998.

[4] A. Zangwill, *Modern Electrodynamics*, Cambridge University Press, New York, NY, 2012.

[5] E.M. Purcell, and D.J. Morin, *Electricity and Magnetism*, 3rd ed., Cambridge University Press, New York, NY, 2013.

[6] M.A. Heald, and J.B. Marion, *Classical Electromagnetic Radiation*, 3rd ed., Saunders College Publishing Fort Worth, TX, 1994.

[7] J.D. Griffiths, *Introduction to Electrodynamics*, 2nd ed., Prentice Hall, Englewood Cliffs, NJ, 1989.

[8] C.A. Balanis, *Advanced Engineering Electromagnetics*, John Wiley & Sons, New York, NY, 1989.

[9] W.K.H. Panofsky, and M. Phillips, *Classical Electricity and Magnetism*, 2nd ed., Addison-Wesley, Reading, MA, 1990.

[10] S.S. Attwood, *Electric and Magnetic Fields*, 3rd ed., John Wiley & Sons, New York, NY, 1956.

[11] S.A. Schelkunoff, *Electromagnetic Waves*, D. Van Nostrand, New York, NY, 1943.

[12] D.K. Cheng, *Field and Wave Electromagnetics*, 2nd ed., Addison-Wesley, Reading, MA, 1989.

[13] C.R. Paul, K.W. Whites, and S.A. Nasar, *Introduction to Electromagnetic Fields*, 3rd ed., The McGraw-Hill Companies, Boston, MA, 1998.

[14] W.R. Smythe, *Static and Dynamic Electricity*, 3rd ed., The McGraw-Hill Companies, Boston, MA, 1968.

[15] J.C. Maxwell, *A Treatise on Electricity and Magnetism*, Vols. I & II, Oxford at the Clarendon Press, 1873.

[16] J.C. Maxwell, "A Dynamical Theory of the Electromagnetic Field," *Philosophical Transactions of the Royal Society*, vol. 155, pp. 459–512, 1865 [Online]. https://royalsocietypublishing.org/doi/10.1098/rstl.1865.0008 (accessed May 16, 2023).

[17] W.R. Hamilton, "On Quaternions, or On a New System of Imaginaries in Algebra," editor David R. Wilkins, *Philosophical Magazine*, (1844–1850), 2000 [Online]. (accessed May 16, 2023) https://www.maths.tcd.ie/pub/HistMath/People/Hamilton/OnQuat/OnQuat.pdf

[18] A. Einstein, "Considerations Concerning the Fundaments of Theoretical Physics," *Science*, vol. 91, no. 2369, pp. 487–492, 1940. [Online]. https://www.science.org/doi/10.1126/science.91.2369.487 (accessed May 16, 2023).

[19] O. Heaviside, *Electrical Papers, Vol. I*, MacMillan and Co., London, 1892, pp. 447–476.

[20] K. Jelved, A.D. Jackson, and O. Knudson (Eds & Trans.), S*elected Scientific Works of Hans Christian Ørsted*, Princeton University Press, Princeton, New Jersey, NJ, 1998, pp. 413–416.

[21] A.-M. Ampère, "Analyse des mémoires lus par M. Ampere à l'Académie des Sciences," dans les séances des 18 et 25 septembre, des 9 et 30 octobre 1820, *Annales Generales des Sciences Physiques, vol.* 6, pp. 238–257, 1820.

[22] A.-M. Ampère, "Conclusions d'un Mémoire sur l'action mutuelle de deux courants électriques, sur celle qui existe entre un courant électrique et un aimant et celle de deux aimants l'un sur l'autre," lu à l'Académie royale des sciences le 25 septembre 1820, *Journal de Physique, de Chimie, d'Histoire Naturelle et des Arts, vol.* 91, pp. 76–78, 1820.

[23] A.-M. Ampère, "Mémoire sur la détermination de la formule qui représente l'action mutuelle de deux portions infiniment petites de conducteurs voltaïques," lu à l'Académie des sciences le10 juin 1822*, Annales de Chimie et de Physique*, vol. 20, pp. 398–419, 1822.

[24] A.-M. Ampère, "Mémoire sur la théorie mathématique des phénomènes électrodynamiques," uniquement déduite de l'expérience, dans lequel se

trouvent réunis les Mémoires que M. Ampère a communiqués à l'Académie royale des sciences, dans les séances des 4 et 26 décembre 1820, 10 juin 1822, 22 décembre 1823, 12 septembre et 23 novembre 1825, *Mémoires de l'Académie des Sciences de l'Institut de France*, vol. 6, pp. 175–388, 1823.

[25] J.C. Maxwell, "On Faraday's Lines of Force," *Transactions of the Cambridge Philosophical Society*, Part I, vol. 10, pp. 27–83, 1858.

[26] J.C. Maxwell, "On Physical Lines of Force, Part I," *The London, Edinburgh, and Dublin Philosophical Magazine and Journal of Science*, 4th Series, vol. XXI, no. CXXXIX, pp. 161–175, 1861.

[27] J.C. Maxwell, "On Physical Lines of Force, Part II," *The London, Edinburgh, and Dublin Philosophical Magazine and Journal of Science*, 4th Series, vol. XXI, no. CXL, pp. 281–291, 1861.

[28] J.C. Maxwell, "On Physical Lines of Force, Part III," *The London, Edinburgh, and Dublin Philosophical Magazine and Journal of Science*, 4th Series, vol. XXIII, no. CLI, pp. 12–24, 1862.

[29] J.C. Maxwell, "On Physical Lines of Force, Part IV," *The London, Edinburgh, and Dublin Philosophical Magazine and Journal of Science*, 4th Series, vol. XXIII, no. CLII, pp. 85–95, 1862.

[30] M. Faraday, "Experimental Researches in Electricity," *Philosophical Transactions of the Royal Society*, vol. 122, pp. 125–162, 1832.

[31] M. Faraday, "The Bakerian Lecture - Experimental Researches in Electricity – Second Series," *Philosophical Transactions of the Royal Society*, vol. 122, pp. 163–194, 1832.

[32] M. Faraday, "A Speculation Touching Electric Conduction and the Nature of Matter," *The London, Edinburgh, and Dublin Philosophical Magazine and Journal of Science*, 3rd Series, vol. XXIV, no. CLVII, pp. 136–144, 1844.

[33] M. Faraday, "Thoughts on Ray-vibrations," *The London, Edinburgh, and Dublin Philosophical Magazine and Journal of Science*, 3rd Series, vol. XXIV, no. CLXXXVIII, pp. 345–349, 1846.

[34] https://en.wikipedia.org/wiki/Speed_of_light

[35] J.H. Poynting, "On the Transfer of Energy in the Electromagnetic Field," *Philosophical Transactions of the Royal Society*, vol. 175, Part II, pp. 343–361, 1884.

[36] O. Heaviside, "Electromagnetic Induction and Its Propagation," *The Electrician*, vol. XIV, pp. 306–307, 1885.

[37] O. Heaviside, "Electromagnetic Induction and Its Propagation," In *Electrical Papers of Oliver Heaviside*, vol. I, pp. 449–450, Macmillan and Co., London, 1892.

[38] J.C. Maxwell, *A Treatise on Electricity and Magnetism*, vol. II, Oxford at the Clarendon Press, London, 1873, pp. 291–292.

[39] H.H. Lamb, "On Electrical Motions in a Spherical Conductor," *Philosophical Transactions of the Royal Society*, vol. 174, pp. 519–549, 1883.

[40] J.H. Poynting, "On the Connexion Between Electric Current and the Electric and Magnetic Inductions in the Surrounding Field," *Philosophical Transactions of the Royal Society*, vol. 176, Part II, pp. 277–306, 1885.

[41] D.E. Hughes, "Inaugural Address: The Self-Induction of an Electric Current in Relation to the Nature and Form of its Conductor," *Journal of the Society of Telegraph-Engineers and Electricians*, vol. 15, no. 60, pp. 6–25, 1886.

[42] C.T. Bright, J. Hopkinson, D.E. Hughes, F.L. Pope, and J.W. Strutt (Lord Rayleigh), "Discussion on the President's Address," *Journal of the Society of Telegraph-Engineers and Electricians*, vol. 15, no. 60, pp. 27–52, 1886.

[43] W.E. Ayrton, D.G. Fitzgerald, J.A. Fleming, W.H. Preece, J.W. Strutt (Lord Rayleigh), and S.P. Thomson, "Continuation of the Discussion on the President's Address," *Journal of the Society of Telegraph-Engineers and Electricians*, vol. 15, no. 60, pp. 54–86, 1886.

[44] D.E. Hughes, "The President's Reply to the Discussion on His Inaugural Address, With an Account of Further Researches," *Journal of the Society of Telegraph-Engineers and Electricians*, vol. 15, no. 60, pp. 87–100, 1886.

[45] O. Heaviside, "On the Self-induction of Wires," *The London, Edinburgh, and Dublin Philosophical Magazine and Journal of Science*, 5th Series, vol. XXII, no. XIV, pp. 118–138, 1886.

[46] O. Heaviside, "Electromagnetic Theory – III," *The Electrician*, vol. XXVI, pp. 388–389, 1891.

[47] J.W. Strutt (Lord Rayleigh), "On the Self-induction and Resistance of Straight Conductors," *The London, Edinburgh, and Dublin Philosophical Magazine and Journal of Science*, 5th Series, vol. XXI, pp. 381–394, 1886.

[48] H.F. Weber, "Critical Remarks on the Discoveries of Mr. Hughes Regarding Self-Induction in Metallic Conductors," *The Telegraphic Journal and Electrical Review*, vol. XVIII, no. 437, pp. 321–323, 1886.

[49] J. Stefan, "Ueber electrische Schwingungen in geraden Leitern," *Annalen der Physik und Chemie*, vol. XLI, no. 11, pp. 400–420, 1890.

[50] O.J. Lodge, "On the Theory of Lightning-Conductors," *The London, Edinburgh, and Dublin Philosophical Magazine and Journal of Science*, 5th Series, vol. XXVI, no. XXI, pp. 217–230, 1888.

[51] W. Thomson, "Inaugural Address: Ether, Electricity, and Ponderable Matter," *Journal of the Institution of Electrical Engineers*, vol. 18, no. 77, pp. 4–37, 1889.

[52] H. Hertz, "Ueber die Fortleitung electrischer Wellen durch Drähte," *Annalen der Physik und Chemie*, vol. XXXVII, no. 7, pp. 395–407, 1889.

[53] J.J. Thomson, "Chapter VI. The Distribution of Rapidly Alternating Currents," *Notes on Recent Researches in Electricity and Magnetism*, Oxford at the Clarendon Press, 1893, pp. 512–537.

[54] G.L. Addenbrooke, W.E. Ayrton, G. Kapp, *et al.*, "Continuation of Discussion on Mr. Preece's Paper 'On Some Points Connected with Mains for Electric Lighting,' and Mr. Crompton's Paper 'On the Most Economical Mode of Feeding a Low-pressure Network,'" *Journal of the Institution of Electrical Engineers*, vol. XX, no. 94, pp. 461–491, 1891.

Chapter 4
Electromagnetics of wire structures

In this chapter, we begin to explore in more depth the application of information we have learnt so far, with a primary interest in cylindrical conductors (wires). Subjects addressed include Poynting vector relationships for cylindrical conductors, self and mutual inductance and capacitance, simple transmission line structures fabricated from wires, crosstalk, and skin and proximity effects pertinent to cylindrical conductors. Additional information is provided in an appendix detailing the effects of coatings on the impedance of wires, with emphasis on Sn, Ag, and Ni coatings on copper conductors. Many of the references from the previous two chapters are equally applicable to this chapter, and the reader is encouraged to seek out those references for further investigations.

4.1 Solid homogeneous cylindrical conductors

To begin the study of conductors (wires), we begin with the basic relationships that describe the current density and impedance of a solid, homogeneous, cylindrical conductor. The interest in this chapter is in long cylindrical conductors, so the relationships will be developed in the cylindrical coordinate system for convenience and clarity.

4.1.1 Current density

Consider a solid, homogeneous, infinitely long cylindrical conductor of radius a, as shown in Figure 4.1, whose axis is coincident with the z-axis of a cylindrical coordinate system. The permeability, μ, the permittivity, ε, and the conductivity, σ,

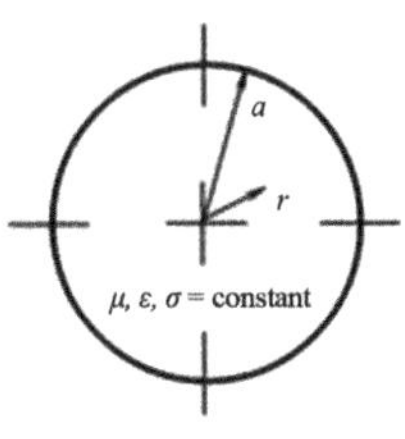

Figure 4.1 Illustration of cylindrical conductor

of the conductor are constant throughout, and the net charge density, ρ, inside the conductor is zero. The permittivity is assumed to be that of free space. The conductor is provided a time-domain current from a source whose output is defined as

$$i(t) = I_0 e^{j\omega t} \tag{4.1}$$

where I_0 is the phasor current.

The current is only a function of time and is steady otherwise. The conductor itself is assumed to have the characteristic

$$\omega\epsilon \ll \sigma \tag{4.2}$$

so that it is possible to ignore any displacement currents.

Everywhere in the conductor, the electromagnetic fields are described by the following relationships:

$$\nabla \times \boldsymbol{E}_z = -j\omega\mu \boldsymbol{H}_\phi \tag{4.3}$$

$$\nabla \times \boldsymbol{H}_\phi = \boldsymbol{J}_z \tag{4.4}$$

$$\nabla \cdot \boldsymbol{E} = 0 \tag{4.5}$$

$$\nabla \cdot \boldsymbol{H} = 0 \tag{4.6}$$

$$\boldsymbol{J}_z = \sigma \boldsymbol{E}_z \tag{4.7}$$

With the electromagnetic fields thus described, it is now possible to derive a differential relationship that describes the current density as a function of position inside the conductor. Taking the curl of (4.3),

$$\nabla \times \nabla \times \boldsymbol{E}_z = -j\omega\mu(\nabla \times \boldsymbol{H}_\phi) \tag{4.8}$$

then substituting (4.4) and (4.7) yields

$$\nabla \times \nabla \times \boldsymbol{J}_z = -j\omega\mu\sigma(\boldsymbol{J}_z) \tag{4.9}$$

$$\equiv \nabla(\nabla \cdot \boldsymbol{J}_z) - \nabla^2 \boldsymbol{J}_z \tag{4.10}$$

The divergence of $\boldsymbol{J}_z$ in the second expression is identically equal to zero, so that

$$\nabla^2 \boldsymbol{J}_z = j\omega\mu\sigma(J_z) \tag{4.11}$$

where $J_z = |\boldsymbol{J}_z|$

Recognizing the current density is not a function of z or ϕ, the left-hand side of (4.11) can be expressed in cylindrical components as a partial differential.

$$\nabla^2 \boldsymbol{J}_z = \frac{1}{r}\frac{\partial}{\partial r}\left(r\frac{\partial J_z}{\partial r}\right) \tag{4.12}$$

so that we have

$$\frac{1}{r}\frac{\partial}{\partial r}\left(r\frac{\partial J_z}{\partial r}\right) - j\omega\mu\sigma(J_z) = 0$$

$$\frac{\partial^2 J_z}{\partial r^2} + \frac{1}{r}\frac{\partial J_z}{\partial r} + (-jk^2)J_z = 0 \tag{4.13}$$

where $k^2 = \omega\mu\sigma$.

Equation (4.13) is in the form of a modified Bessel differential equation of order $\nu = 0$. Equation (4.13) can be reduced by using a change of variable.

Let $\xi = kr\sqrt{j}$, such that $\frac{\partial\xi}{\partial r} = k\sqrt{j}$

Now

$$\frac{\partial J_z}{\partial r} = \frac{\partial J_z}{\partial \xi}\frac{\partial \xi}{\partial r} = \frac{\partial J_z}{\partial \xi}\left(k\sqrt{j}\right) \tag{4.14}$$

from which we obtain the operator

$$\frac{\partial}{\partial r} = \left(k\sqrt{j}\right)\frac{\partial}{\partial \xi} \tag{4.15}$$

Also,

$$\frac{\partial^2 J_z}{\partial r^2} = \frac{\partial}{\partial r}\left(\frac{\partial J_z}{\partial r}\right) = \frac{\partial}{\partial r}\left[\left(k\sqrt{j}\right)\frac{\partial J_z}{\partial \xi}\right] = \left(k\sqrt{j}\right)\frac{\partial}{\partial \xi}\left[\left(k\sqrt{j}\right)\frac{\partial J_z}{\partial \xi}\right]$$

$$\left(k\sqrt{j}\right)\left[\left(k\sqrt{j}\right)\frac{\partial^2 J_z}{\partial \xi^2} + \frac{\partial J_z}{\partial \xi}\right] = \left(k\sqrt{j}\right)^2\frac{\partial^2 J_z}{\partial \xi^2} + \left(k\sqrt{j}\right)\frac{\partial J_z}{\partial \xi} \tag{4.16}$$

Substituting (4.16) into (4.13) yields

$$\left(k\sqrt{j}\right)^2\frac{\partial^2 J_z}{\partial \xi^2} + \frac{(k\sqrt{j})}{r}\frac{\partial J_z}{\partial \xi} + (-jk^2)J_z = 0$$

$$\frac{\partial^2 J_z}{\partial \xi^2} + \frac{1}{kr\sqrt{j}}\frac{\partial J_z}{\partial \xi} + \left(\frac{-jk^2}{jk^2}\right)J_z = \frac{\partial^2 J_z}{\partial \xi^2} + \frac{1}{\xi}\frac{\partial J_z}{\partial \xi} - J_z = 0 \tag{4.17}$$

In general, (4.17) has two solutions, expressed in general form as

$$J_z(\xi) = C_1 I_\nu(\xi) + D_1 K_\nu(\xi) \tag{4.18}$$

where C_1 and D_1 are arbitrary constants, and $I_\nu(\xi)$ and $K_\nu(\xi)$ are modified Bessel functions of the first and second kind, respectively. The variable ν represents the order of the Bessel functions.

The modified Bessel function of the first kind is equal to one when the argument equals zero for all real values of ξ. The modified Bessel function of the second kind is discontinuous at the origin. Since J_z in this case is finite at $r = 0$, D_1 must be equal to zero.

The general series expression for a modified Bessel function of the first kind of any order is given by

$$I_\nu(z) = \left(\frac{z}{2}\right)^\nu \sum\nolimits_{k=0}^{\infty} \left(\frac{1}{k!}\right)\left[\frac{1}{(k+\nu)!}\right]\left(\frac{z^2}{4}\right)^k \tag{4.19}$$

For zeroth order, the expression becomes

$$I_0(z) = \left(\frac{z}{2}\right)^0 \sum\nolimits_{k=0}^{\infty} \left(\frac{1}{k!}\right)\left[\frac{1}{(k+0)!}\right]\left(\frac{z^2}{4}\right)^k = \sum\nolimits_{k=0}^{\infty} \left(\frac{1}{k!}\right)^2 \left(\frac{z^2}{4}\right)^k \tag{4.20}$$

The solution for $J_z(\xi)$ that is finite at $r = 0$ may be thus expressed as an infinite series

$$\begin{aligned} J_z(\xi) &= C_1 I_0(\xi) \\ &= C_1\left[1 + \left(\frac{\xi^2}{4}\right)\left(\frac{1}{1!}\right)^2 + \left(\frac{\xi^2}{4}\right)^2\left(\frac{1}{2!}\right)^2 + \left(\frac{\xi^2}{4}\right)^3\left(\frac{1}{3!}\right)^2 + \dots\right] \end{aligned} \tag{4.21}$$

Equation (4.21) describes a modified Bessel function of the first kind, zeroth order, in terms of powers of ξ.

The constant C_1 in (4.21) can be evaluated simply by setting the current density equal to that at the surface of the cylindrical conductor, $r = a$.

$$J_z\left(ka\sqrt{j}\right) = C_1 I_0\left(ka\sqrt{j}\right) = \sigma E_0 \xrightarrow{yields} C_1 = \frac{\sigma E_0}{I_0(ka\sqrt{j})} \tag{4.22}$$

so that the current density everywhere in the solid, homogeneous, cylindrical conductor is then given by

$$J_z(\xi) = C_1 I_0(\xi) = \frac{\sigma E_0 I_0(kr\sqrt{j})}{I_0(ka\sqrt{j})} \tag{4.23}$$

In general, the current density is not known a priori. The total current is typically known, however. It is therefore useful to express (4.23) in terms of the total current.

$$I_z = \iint J_z \cdot ds = \int_0^{2\pi}\int_0^a \frac{\sigma E_0 I_0(kr\sqrt{j})}{I_0(ka\sqrt{j})} \cdot r dr d\varphi \tag{4.24}$$

$$= \frac{\sigma E_0}{I_0(ka\sqrt{j})}\int_0^{2\pi}\int_0^a I_0\left(kr\sqrt{j}\right) \cdot r dr d\varphi \tag{4.25}$$

$$= \frac{2\pi\sigma E_0}{I_0(ka\sqrt{j})}\int_0^a I_0\left(kr\sqrt{j}\right) \cdot r dr \tag{4.26}$$

Let $kr\sqrt{j} = x$. Then if $r = 0, x = 0$, and when $r = a, x = ka\sqrt{j}$, and $dr = \frac{dx}{k\sqrt{j}}$. With this change of variable, (4.26) becomes

$$I_z = \frac{2\pi\sigma E_0}{I_0(ka\sqrt{j})}\left(\frac{ka\sqrt{j}}{k^2 j}\right)\int_0^{ka\sqrt{j}} xI_0(x)dx \tag{4.27}$$

Using the relationship

$$\int z^\nu I_{\nu-1}(z)dz = z^\nu I_\nu(z) \tag{4.28}$$

Equation (4.27) becomes

$$I_z = \frac{2\pi\sigma a E_0}{k\sqrt{j}}\left(\frac{I_1(ka\sqrt{j})}{I_0(ka\sqrt{j})}\right) \tag{4.29}$$

4.1.2 Impedance

Building on the previous section, the impedance of cylindrical conductors can be determined. The voltage per length on the cylindrical conductor is found as

$$V = -\int_a^b E \cdot dl = -\int_l^0 E_0 \cdot dz = E_0 l \tag{4.30}$$

Using (4.29), the internal impedance is then given by

$$Z_i = \frac{E_0 l}{I} = \frac{E_0 l}{\frac{2\pi\sigma a E_0}{k\sqrt{j}}\frac{I_1(ka\sqrt{j})}{I_0(ka\sqrt{j})}} = \frac{l(k\sqrt{j})}{2\pi\sigma a}\frac{I_0(ka\sqrt{j})}{I_1(ka\sqrt{j})} \tag{4.31}$$

And the internal impedance per unit length is then

$$Z_l = \frac{Z_i}{l} = \frac{k\sqrt{j}\, I_0(ka\sqrt{j})}{2\pi\sigma a\, I_1(ka\sqrt{j})} \tag{4.32}$$

Using the definitions

$$I_0\left(ka\sqrt{j}\right) = ber\ ka + j\ bei\ ka$$

$$I_1\left(ka\sqrt{j}\right) = I_0'\left(ka\sqrt{j}\right) = (j)^{-1/2}(ber'\ ka + j\ bei'\ ka) \tag{4.33}$$

Equation (4.32) becomes

$$Z_l = \frac{k\sqrt{j}\, I_0(ka\sqrt{j})}{2\pi\sigma a\, I_1(ka\sqrt{j})} = \frac{k}{2\pi\sigma a}\frac{(-bei\ ka + j\ ber\ ka)}{(ber'\ ka + j\ bei'\ ka)} \tag{4.34}$$

Approximations of the ber and bei functions can be made for both small and large values of (ka). For small values of (ka), the ber, ber′, bei, and bei′ functions may be expressed as series expansions, shown as follows:

$$ber_\nu z = \sum_{p=0}^{\infty} \frac{(-1)^{\nu+p}\left(\frac{z}{2}\right)^{\nu+2p}}{p!(\nu+p)!}\cos\left(\frac{(\nu+2p)\pi}{4}\right)$$

$$= \left[1 - \frac{(z)^4}{(2!)^4} + \frac{(z)^8}{(2!)^8(4!)^2} - \cdots\right] \tag{4.35}$$

$$bei_\nu z = \sum_{p=0}^{\infty} \frac{(-1)^{\nu+p+1}\left(\frac{z}{2}\right)^{\nu+2p}}{p!(\nu+p)!}\sin\left(\frac{(\nu+2p)\pi}{4}\right)$$

$$= \left[\frac{(z)^2}{4} - \frac{(z)^6}{(2!)^6(3!)^2} + \cdots\right] \tag{4.36}$$

$$ber'_\nu z = \sum_{p=0}^{\infty} \frac{(-1)^{\nu+p}\left(\frac{\nu}{p}+p\right)\left(\frac{z}{2}\right)^{\nu+2p-1}}{p!(\nu+p)!}\cos\left(\frac{(\nu+2p)\pi}{4}\right)$$

$$= \left[-\frac{(z)^3}{(2!)^4} + \frac{(z)^7}{(2!)^4 3!4!} - \cdots\right] \tag{4.37}$$

$$bei'_\nu z = \sum_{p=0}^{\infty} \frac{(-1)^{\nu+p+1}\left(\frac{z}{2}\right)^{\nu+2p-1}}{p!(\nu+p)!}\sin\left(\frac{(\nu+2p)\pi}{4}\right) = \left[\frac{z}{2} - \frac{(z)^5}{(2!)^6 3!} + \cdots\right] \tag{4.38}$$

Discarding terms of $(ka)^4$ and greater and substituting (4.35) through (4.38) into (4.34) yields the low-frequency approximation for the impedance of a solid cylindrical conductor

$$Z_{l(LF)} \approx \frac{k}{2\pi\sigma a}\frac{\left(\frac{-(ka)^2}{4}+j\right)}{\left(\frac{-(ka)^3}{16}+j\frac{ka}{2}\right)} = \frac{1}{\pi\sigma a^2}\left[1+j\frac{(ka)^2}{8}\right] = \frac{1}{\pi\sigma a^2} + j\omega\frac{\mu}{8\pi} \tag{4.39}$$

The left-hand term on the right side of (4.39) is the direct current (DC) resistance of the cylindrical conductor. The right-hand term on the right side of (4.39) is referred to as the internal inductance of the cylindrical conductor. Note this term is a constant dependent on the permeability of the conductor material.

For large values of (ka), it becomes necessary to use asymptotic expressions for the ber, ber′, bei, and bei′ functions, given by

$$ber\ z + j\ bei\ z = \frac{1}{\sqrt{2\pi z}}\exp\left[\frac{z}{\sqrt{2}} + j\left(\frac{z}{\sqrt{2}} - \frac{\pi}{8}\right)\right] \tag{4.40}$$

$$bei'\ z + j\ ber'\ z = \frac{1}{\sqrt{2\pi z}}\exp\left[\frac{z}{\sqrt{2}} + j\left(\frac{z}{\sqrt{2}} + \frac{\pi}{8}\right)\right] \tag{4.41}$$

Starting from (4.34), as was done previously for the small argument case, and substituting (4.40) and (4.41) yields

$$Z_{l(HF)} \sim \frac{k}{2\pi\sigma a} \frac{\frac{1}{\sqrt{2\pi z}} \exp\left[\frac{z}{\sqrt{2}} + j\left(\frac{z}{\sqrt{2}} + \frac{\pi}{8}\right)\right]}{\frac{1}{\sqrt{2\pi z}} \exp\left[\frac{z}{\sqrt{2}} + j\left(\frac{z}{\sqrt{2}} - \frac{\pi}{8}\right)\right]} = \frac{k}{2\pi\sigma a} e^{j\pi/4} = \frac{k\sqrt{j}}{2\pi\sigma a} \tag{4.42}$$

Thus, the high-frequency approximation for the impedance per unit length of a solid cylindrical conductor is

$$Z_{l(HF)} \sim \frac{(1+j)}{2\pi a} \sqrt{\frac{\omega\mu}{2\sigma}} \tag{4.43}$$

It is possible to relate the low-frequency and high-frequency approximations for the impedance per unit length using the skin depth, as follows.

$$r_{(HF)} = \frac{1}{\pi\sigma a^2}\left(\frac{\pi\sigma a^2}{2\pi a}\sqrt{\frac{\omega\mu}{2\sigma}}\right) = r_{(LF)}\left(\frac{\sigma a}{2}\sqrt{\frac{\omega\mu}{2\sigma}}\right) = r_{(LF)}\frac{a}{2}\sqrt{\pi f \mu\sigma} = r_{(LF)}\frac{a}{2\delta} \tag{4.44}$$

$$l_{(HF)} = \frac{\mu}{8\pi}\left(\frac{8\pi}{2\pi a\mu}\sqrt{\frac{\mu}{2\omega\sigma}}\right) = l_{(LF)}\frac{4}{a\mu}\sqrt{\frac{\mu}{2\omega\sigma}} = l_{(LF)}\frac{2}{a}\frac{1}{\sqrt{\pi f\mu\sigma}} = l_{(LF)}\frac{2\delta}{a} \tag{4.45}$$

We now examine how an electromagnetic wave propagates down the length of cylindrical conductors.

4.2 Propagation of electromagnetic waves along cylindrical conductors

Electromagnetic waves may propagate along a single cylindrical conductor or along various transmission line structures, including a single wire above a return plane, an open pair of wires, and a twisted pair of wires. Single wires are used to interconnect multiple electrical and electronic devices. These wires do have a corresponding return conductor, although it may be removed spatially, since current flows in a complete circuit from the source to the load and back to the source. The energy provided by such a circuit flows in only one direction, from the source to the load. For electrostatic (DC) applications, or electroquasistatic (low-frequency alternating current (AC)) applications roughly below 2 MHz, it can be shown that the Poynting vector associated with a single conductor has two components, one that points inwards to the conductor and a second that points in parallel to the axis of the conductor.

Figure 4.2 illustrates a long uniform wire of length L and radius b collinear with the z-axis. The wire is surrounded by a thin hollow open cylinder with infinite conductivity and inner radius b. The large cylinder is closed at the lower end by a thin circular plate with infinite conductivity. The lower end of the wire is connected

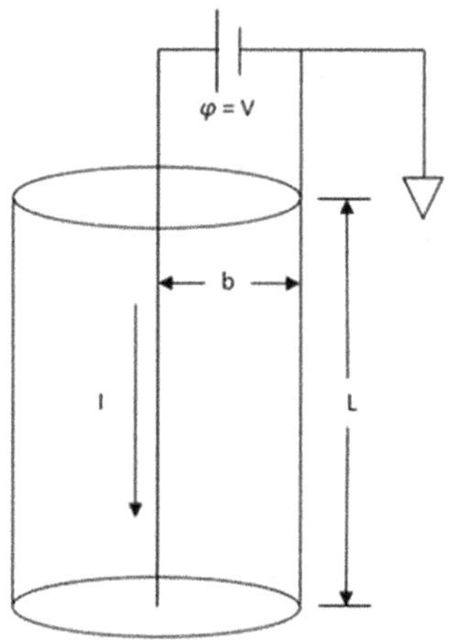

Figure 4.2 Construction for derivation of Poynting vector for single cylindrical conductor

to the centre of the circular plate. The upper end of the wire, at $z = z_1$, is connected to the positive terminal of a battery. The negative terminal of the battery is common with the large surrounding cylinder. The wire is assumed to be long enough that fringe effects at the upper end of the wire may be ignored. The field at any point in the volume between the wire and the large surrounding cylinder as a function of the potential between the wire and the cylinder is given by the Laplace equation.

$$\frac{\partial^2 V}{\partial r^2} + \frac{1}{r}\frac{\partial V}{\partial r} + \frac{\partial^2 V}{\partial z^2} = 0 \tag{4.46}$$

Equation (4.46) can be solved using the technique known as separation of variables. Assuming a solution in the form

$$V = R(r)Z(z) \tag{4.47}$$

let

$$\frac{V}{V_0} = RZ \tag{4.48}$$

where

$R =$ some function of r only
$Z =$ some function of z only
$V_0 =$ some constant

Substituting (4.48) into (4.46) and dividing by RZ, we can write

$$\frac{1}{R}\frac{\partial^2 R}{\partial r^2} + \frac{1}{rR}\frac{R}{\partial r} + \frac{1}{Z}\frac{\partial^2 Z}{\partial z^2} = 0$$

$$\frac{1}{R}\frac{\partial^2 R}{\partial r^2} + \frac{1}{rR}\frac{R}{\partial r} = -\frac{1}{Z}\frac{\partial^2 Z}{\partial z^2} \tag{4.49}$$

The terms on the left-hand side are independent of z and the term on the right-hand side is independent of r. Each of the sides must be equal to some constant determined by the applicable boundary conditions.

The applicable boundary conditions are the following: (1) V is continuous at the surface of the conductor; (2) the tangential components of the electric field on either side of the surface of the conductor are equal; and (3) the difference between the normal components of the electric flux density, $\boldsymbol{D}$, on either side of the wire surface is equal to the surface charge density, ρ_s.

Restating the boundary conditions in terms of the potential, we have

$$V_{z=0} = 0; \quad a \leq r \leq b \tag{4.50}$$

$$V_{r=b} = 0; \quad 0 \leq z \leq z_1 \tag{4.51}$$

$$V_{z=z_1} = V_1; \quad r = a \tag{4.52}$$

Also, at any distance z from the lower end of the wire/cylinder structure, the potential along the surface of the wire, $r = a$, is given by

$$\frac{\partial V}{\partial z} = \frac{V_1}{z_1} \tag{4.53}$$

Integrating (4.53) yields

$$V = \frac{V_1}{z_1} z \tag{4.54}$$

When $z = z_1$ in (4.54), we have (4.52)

Because the term on the right-hand side of (4.49) is solely a function of z, we can write

$$\frac{\partial^2 Z}{\partial z^2} = a_z^2 Z \tag{4.55}$$

Given that (4.53) is a constant, (4.55) must equal zero, and thus $a_z = 0$. A corresponding solution for (4.55) is then given by

$$Z = Az + B \tag{4.56}$$

Since (4.55) is equal to zero, we can write (4.49) as

$$\frac{1}{R}\frac{\partial^2 R}{\partial r^2} + \frac{1}{rR}\frac{R}{\partial r} = 0 \tag{4.57}$$

which has solution

$$R = C \ln r + D \tag{4.58}$$

Substituting (4.56) and (4.58) into (4.48) yields

$$\frac{V}{V_0} = (C \ln r + D)(Az + B) \tag{4.59}$$

$$\frac{V}{V_0} = (ACz \ln r + ADz + BC \ln r + BD) \tag{4.60}$$

Applying (4.50)–(4.52) yields

$$0 = (ACz \ln b + ADz) \tag{4.61}$$

$$\frac{V_1}{V_0} = (ACz_1 ln\, a + ADz_1) \tag{4.62}$$

Solving (4.61) and (4.62) yields

$$AC = \frac{V_1}{V_0 z_1 \ln\left(\frac{a}{b}\right)} \text{ and; } AD = \frac{-V_1 \ln(b)}{V_0 z_1 \ln\left(\frac{a}{b}\right)} \tag{4.63}$$

which lead to

$$V = V_1 \frac{z}{z_1} \frac{\ln\left(\frac{b}{r}\right)}{\ln\left(\frac{b}{a}\right)} = E_1 z \frac{\ln\left(\frac{b}{r}\right)}{\ln\left(\frac{b}{a}\right)} \tag{4.64}$$

Finally, using the boundary condition for the flux density, we can express the surface charge distribution as

$$\rho_s = \frac{\partial}{\partial r}[E_1 z] - \frac{\partial}{\partial r}\left[E_1 z \frac{\ln\left(\frac{b}{r}\right)}{\ln\left(\frac{b}{a}\right)}\right] = \frac{E_1 z}{b \ln\left(\frac{b}{a}\right)} \tag{4.65}$$

Using (4.64), we can write expressions for the longitudinal and radial components of the electric field as

$$\frac{\partial}{\partial z}\left[E_1 z \frac{\ln\left(\frac{b}{r}\right)}{\ln\left(\frac{b}{a}\right)}\right] = E_1 \frac{\ln\left(\frac{b}{r}\right)}{\ln\left(\frac{b}{a}\right)} \tag{4.66}$$

$$\frac{\partial}{\partial r}\left[E_1 z \frac{\ln\left(\frac{b}{r}\right)}{\ln\left(\frac{b}{a}\right)}\right] = -E_1 \frac{z}{r} \frac{1}{\ln\left(\frac{b}{a}\right)} \tag{4.67}$$

so that the complete electric field is given by

$$\boldsymbol{E}_{tot} = E_1 \frac{\ln\left(\frac{b}{r}\right)}{\ln\left(\frac{b}{a}\right)} \boldsymbol{z} - E_1 \frac{z}{r} \frac{1}{\ln\left(\frac{b}{a}\right)} \boldsymbol{r} \tag{4.68}$$

The magnetic field is given by

$$\boldsymbol{B}_{tot} = \frac{-I\mu_0}{2\pi r} \boldsymbol{\phi} = \frac{-\mu_0}{2\pi} E_1 \frac{\pi \sigma a^2}{r} \boldsymbol{\phi} \tag{4.69}$$

The Poynting vector is then given by

$$S = \frac{1}{\mu_0} E_{tot} \times B_{tot} = \frac{1}{\mu_0} \left[\left(E_1 \frac{\ln\left(\frac{b}{r}\right)}{\ln\left(\frac{b}{a}\right)} z - E_1 \frac{z}{r} \frac{1}{\ln\left(\frac{b}{a}\right)} r \right) \times \left(\frac{-\mu_0}{2\pi} E_1 \frac{\pi \sigma a^2}{r} \phi \right) \right] \tag{4.70}$$

$$= -\frac{1}{2} E_1^2 \frac{\ln\left(\frac{b}{r}\right)}{\ln\left(\frac{b}{a}\right)} \frac{\sigma a^2}{r} r - \frac{1}{2} E_1^2 \frac{1}{\ln\left(\frac{b}{a}\right)} \frac{z \sigma a^2}{r^2} z \tag{4.71}$$

In (4.71), we see two components of the Poynting vector, one radial directed inwards to the conductor and the second longitudinal directed parallel to the current along the surface of the conductor. The radial term represents the Joule heating associated with Ohm's law inside the conductor. The longitudinal term represents the energy propagated along the conductor surface. The two energies are the same as can be shown by replacing r with a in the first term and multiplying by the surface area of the wire from $(0 \leq z \leq z_1) = (2\pi a z)$, and then comparing that result to the integration of the second term with respect to r.

$$\frac{1}{2} E_1^2 \frac{\ln\left(\frac{b}{a}\right)}{\ln\left(\frac{b}{a}\right)} \frac{\sigma a^2}{a} (2\pi a z) = E_1^2 \sigma \pi a^2 z \tag{4.72}$$

$$\int_a^b \frac{1}{2} E_1^2 \frac{1}{\ln\left(\frac{b}{a}\right)} \frac{z \sigma a^2}{r^2} 2\pi r dr = \frac{1}{2} E_1^2 \frac{2\pi z \sigma a^2}{\ln\left(\frac{b}{a}\right)} \int_a^b \frac{dr}{r} \tag{4.73}$$

$$\int_a^b \frac{1}{2} E_1^2 \frac{1}{\ln\left(\frac{b}{a}\right)} \frac{z \sigma a^2}{r^2} 2\pi r dr = \frac{1}{2} E_1^2 \frac{2\pi z \sigma a^2}{\ln\left(\frac{b}{a}\right)} \int_a^b \frac{dr}{r} \tag{4.74}$$

$$= E_1^2 \sigma \pi a^2 z \tag{4.75}$$

It is clear the energy propagated is equal to the energy dissipated in the conductor by Joule heating.

4.3 Fundamental characteristics of selected wire structure transmission lines

In most applications, a minimum of two wires are employed as transmission line structures. A single wire above a common reference plane conceptually fits neatly between the single wire case and a pair of wires. For some time, the single wire above the earth configuration was the preferred solution for cross-country telegraph and early telephone systems operating at audio frequencies into the low kilohertz range. Open two-wire systems competed with the single wire above the earth configuration, and these were sufficient to the task until the introduction of voice signals. The introduction of voice signals brought with it the spectre of electromagnetic interference. In fact, this problem was originally addressed by none other than Alexander Graham Bell with his patented metallic lines, essentially a twisted pair of wires with the signal and its return physically intertwined.

Today the single wire above a common reference plane topology remains one of many popular types of transmission line structures and is often employed for applications such as DC and low-frequency AC power distribution systems and discrete (on/off) signals. Most wires and cables in modern vehicles, regardless of end use, are installed in configurations exhibiting a topology much like that of a single wire above a common reference plane, with the wires or cables mechanically supported above local structure that may be metallic, composite, or some combination thereof. The characteristics of a single wire above a common reference plane are frequently employed for initial calculations for field-to-wire coupling.

Examples of applications in which a simple pair of wires is employed as a transmission line include 300Ω TwinAx or Twin Lead, and 450Ω and 600Ω ladder line. TwinAx was used for lead lines for early VHF and UHF television antennas. Ladder line was and is used extensively by Ham radio operators. An identical topology is exhibited by both TwinAx and ladder line, with variations in wire gauge, spacing, and dielectric material.

A great many applications today operate at frequencies well above 2 MHz (think gigabit Ethernet), and accepted engineering practice has shown, depending on operational frequencies, power levels, and required system efficiencies and mechanical constraints, that using a configuration comprising a signal line wire combined with its corresponding return line wire generally offers high operational efficiency and improved electromagnetic interference performance for a comparative economic benefit in terms of acquisition, fabrication, installation, and maintenance. Of course, wires and cables have mass and volume for which incorporation into a given design must be afforded due consideration. While it is true that in many high-speed applications optical fibre is being considered as a replacement for copper, fibre implementation brings mass and volume impacts to a given design in terms of the need for robust electrical interface electronics, adding additional complexity to equipment designs. Recent research and development efforts have demonstrated that under certain circumstances fibre is able to propagate optical power that can then be converted into electrical power with an acceptable level of efficiency [1], although this has yet to be widely implemented as common practice. Because of the foregoing, fibre remains less economical than copper. Given the advantages, and despite the disadvantages of mass and volume impact, copper is still the solution of choice for many, various configurations employed in a wide variety of applications over a rather wide range of frequencies. These configurations include the ubiquitous single wire above a common reference plane, two-wire balanced lines, and multiple-wire lines with one or two conductors connected to a common reference plane.

Extensive literature is available for transmission line theory and application. From the perspective of this text, we are only concerned with the fundamental behaviour of simple wire structure lines. Our interest in these lines has primarily to do with their inductance, capacitance, resistance, conductance, and impedance, referred to as the parameters or constants of these transmission line structures. The following sections will briefly examine the derivation of the Telegrapher's Equations and determination of the parameters for some of the configurations mentioned previously. Inductive and capacitive crosstalk will be discussed. Characteristics associated with wire structures as radiating sources will be discussed in Chapter 5. The result of exposure to external stimuli will be examined in a later chapter.

4.4 Telegrapher's equations

As discussed at some length in Chapter 2, the design of the Trans-Atlantic cable was a lengthy and complex process. Many of the best and brightest, including William Thompson (Lord Kelvin), were engaged in analysis and design of the cables. Oliver Heaviside was able to develop a great deal of the applicable theory of operation, incorporating the series inductance of the line into work by Lord Kelvin [2]. Referencing previous work by Gustav Robert Kirchoff [3] and Sir Joseph John (JJ) Thomson [4] in an August 1886 paper published in *The Electrician* [5], Heaviside provided the mathematical basis for what are now known as the Telegrapher's Equations. Equation (4.76) provides Heaviside's equation (numbered 72b in the cited paper).

$$\frac{d^2V}{dx^2} = RS\frac{dV}{dt} + LS\frac{d^2V}{dt^2} \tag{4.76}$$

where V = the distributed voltage on the transmission line
R = the distributed resistance on the transmission line
L = the distributed inductance on the transmission line
S = the distributed capacitance on the transmission line

Regardless of the physical configuration of the line, the modern versions of this equation are based on a general simple four-element circuit as shown in Figure 4.3 and can be derived as follows.

Note that in the schematic shown in Figure 4.3, the parameters R, L, G, and C, are all functions of length.

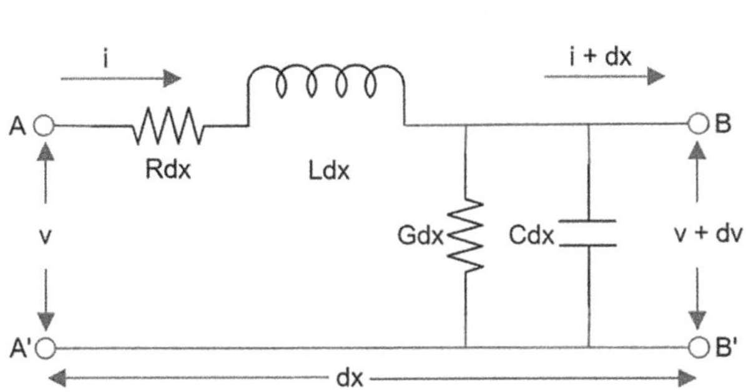

Figure 4.3 Simple four-element circuit for derivation of the Telegrapher's equations

1. The resistance $R\Delta x$ represents the combined ohmic resistance of both conductors of the transmission line in series. R is resistance per unit length, Ω/m.
2. The conductance $G\Delta x$ represents the leakage of current directly from one conductor to the other through the media between the conductors. When $G\Delta x > 0$, current may flow from one conductor to the other through the space between the conductors. Any such current flow represents a loss of power separate from the loss associated with R above. G has units of S/m. Note that G and R are completely different physical mechanisms and are not related.
3. The capacitance $C\Delta x$ represents the capacitance of the transmission line structure. Transmission line capacitance may be defined between the conductors or between each conductor and a common reference plane, or both, depending on the cross-sectional geometry and any media separating the conductors. C has units of F/m.
4. The inductance $L\Delta x$ represents the inductance of the transmission line structure, including both self-inductance of the conductors as well as the mutual inductance between the two conductors. Like transmission line capacitance, transmission line inductance may be defined between the conductors or between each conductor and a common reference plane, or both, depending on the cross-sectional geometry and any media separating the conductors. L has units of H/m.

The voltage change from A to B in Figure 4.3 is given by

$$\Delta v = v_B - v_A = -(R\Delta x)i(x,t) - (L\Delta x)\frac{\partial i(x,t)}{\partial t} \tag{4.77}$$

The current flowing into B from B′ is given by

$$\Delta i = -(G\Delta x)(v + \Delta v) - (C\Delta x)\frac{\partial}{\partial t}(v + \Delta v) \tag{4.78}$$

$$= -G\Delta x v - C\Delta x\frac{\partial v}{\partial t} + G\Delta x^2\left(Ri(x,t) + L\Delta x\frac{\partial i(x,t)}{\partial t}\right)$$
$$+ C\Delta x^2\left(R\frac{\partial i(x,t)}{\partial t} + L\frac{\partial^2 i(x,t)}{\partial t^2}\right) \tag{4.79}$$

Dividing each term in (4.77) by Δx and allowing Δx to approach zero yields the first fundamental equation

$$\lim_{\Delta x \to 0}\left(\frac{\Delta v(x,t)}{\Delta x}\right) = \frac{\partial v(x,t)}{\partial x} = -Ri(x,t) - L\frac{\partial i(x,t)}{\partial t} \tag{4.80}$$

Repeat this dividing each term in (4.79) by Δx, and allow Δx to approach zero yields the second fundamental equation

$$\lim_{\Delta x \to 0}\left(\frac{\Delta i(x,t)}{\Delta x}\right) = \frac{\partial i(x,t)}{\partial x} = -Gv(x,t) - C\frac{\partial v(x,t)}{\partial t} \tag{4.81}$$

Differentiating (4.80) with respect to x yields

$$\frac{\partial^2 v(x,t)}{\partial x^2} = -\left(R\frac{\partial i(x,t)}{\partial x} + L\frac{\partial}{\partial x}\frac{\partial i(x,t)}{\partial t}\right) = -\left(R\frac{\partial i(x,t)}{\partial x} + L\frac{\partial}{\partial t}\frac{\partial i(x,t)}{\partial x}\right) \tag{4.82}$$

Replace $\frac{\partial i(x,t)}{\partial x}$ in (4.82) with (4.81) to yield

$$\frac{\partial^2 v(x,t)}{\partial x^2} = -R\left(-Gv(x,t) - C\frac{\partial v(x,t)}{\partial t}\right) + L\frac{\partial}{\partial t}\left(-Gv(x,t) - C\frac{\partial v(x,t)}{\partial t}\right) \tag{4.83}$$

$$= RGv(x,t) + RC\frac{\partial v(x,t)}{\partial t} + LG\frac{\partial v(x,t)}{\partial t} + LC\frac{\partial^2 v(x,t)}{\partial t^2} \tag{4.84}$$

$$= RGv(x,t) + (RC + LG)\frac{\partial v(x,t)}{\partial t} + LC\frac{\partial^2 v(x,t)}{\partial t^2} \tag{4.85}$$

Repeat the previous steps by differentiating (4.81) with respect to x, then replace $\frac{\partial v}{\partial x}$ in the resulting equation with (4.80) to yield

$$\frac{\partial^2 i(x,t)}{\partial x^2} = RGi(x,t) + (RC + LG)\frac{\partial i(x,t)}{\partial t} + LC\frac{\partial^2 i(x,t)}{\partial t^2} \tag{4.86}$$

Equations (4.85) and (4.86) are modern versions of the Heaviside equation (4.76) accounting for voltage and current behaviour, and any loss in the media separating the two conductors and are a form of the fundamental Telegrapher's equations expressed in (4.80) and (4.81).

Using phasor notation, (4.80) and (4.81) may be expressed in the frequency domain as

$$\frac{\partial \hat{\boldsymbol{V}}_\omega(x)}{\partial x} = -(R_\omega + j\omega L_\omega)\hat{\boldsymbol{I}}_\omega(x) \tag{4.87}$$

$$\frac{\partial \hat{\boldsymbol{I}}_\omega(x)}{\partial x} = -(G_\omega + j\omega C_\omega)\hat{\boldsymbol{V}}_\omega(x) \tag{4.88}$$

Equations (4.80) and (4.81) may be combined to form two single variable partial differential equations, as was done in (4.85) and (4.86).

$$\frac{\partial^2 \hat{\boldsymbol{V}}_\omega(x)}{\partial x^2} = \gamma^2 \hat{\boldsymbol{V}}_\omega(x) \tag{4.89}$$

$$\frac{\partial^2 \hat{\boldsymbol{I}}_\omega(x)}{\partial x^2} = \gamma^2 \hat{\boldsymbol{I}}_\omega(x) \tag{4.90}$$

where $\gamma \equiv \sqrt{(R_\omega + j\omega L_\omega)(G_\omega + j\omega C_\omega)} = \alpha + j\beta$, and is known as the propagation constant for the line (see (3.137) in Chapter 3). α is the attenuation constant and β is the phase constant.

The square root of the ratio between α and β is defined as the characteristic impedance of the line.

$$Z_c \equiv \sqrt{\frac{(R_\omega + j\omega L_\omega)}{(G_\omega + j\omega C_\omega)}} \tag{4.91}$$

If R_ω and G_ω are small but not negligible, (4.91) may also be expressed as

$$Z_c \cong \sqrt{\frac{L_\omega}{C_\omega}}\left[1 + j\left(\frac{G_\omega}{2\omega C_\omega} - \frac{R_\omega}{2\omega L_\omega}\right)\right] \tag{4.92}$$

Equations (4.91) and (4.92) are generally complex valued.
When $R_\omega \ll j\omega L_\omega$ and $G_\omega \ll j\omega C_\omega$, (4.91) reduces to

$$Z_c = \sqrt{\frac{L_\omega}{C_\omega}} \tag{4.93}$$

Equation (4.93) is real valued and is sometimes referred to as the characteristic resistance of the line. Note that when the Heaviside Condition ((2.2) in Chapter 2) prevails, (4.92) reduces to (4.93), and the line is considered distortionless.

To determine the impedance of the line, it is advantageous to determine the parameters of the line as a function of length. Derivations of these parameters have been determined for different types of wire structure transmission lines and are available in multiple references. Table 4.1 lists approximations of the parameters for basic single-wire transmission lines. Table 4.2 lists approximations of the parameters for basic two-wire transmission lines. The transmission line structures listed in Tables 4.1 and 4.2 are elementary and form essential building blocks towards more elaborate configurations.

4.4.1 Cable-to-cable crosstalk

Two or more individual circuits are said to be coupled when signals in individual circuits cause unintended, potentially undesirable, responses in other individual circuits. Coupling in this text is defined as changes in the voltages and/or currents in a transmission line, wire, or cable, referred to as the source, as compared to corresponding changes in the voltages and/or currents in another, nearby, transmission line, wire, or cable, referred to as the victim. This sort of coupling is also known as cable-to-cable crosstalk. Note hereinafter the term "lines" may be used to refer collectively to transmission lines, wires, and cables.

Interpretation of the previous definition includes the reciprocal effects of changes in the voltages and/or currents in the victim, as compared to corresponding changes in the voltages and/or currents in the source. The nature of this effect is that with each successive response the effect is diminished by the losses of the victim and of the source circuits. The magnitude of the initial response of the victim to changes in the source is often considered in approximate analytical estimation to be sufficient to envelop any reciprocal responses.

Table 4.1 Approximate parameters of basic single-wire transmission lines

Line type	Characteristic impedance (Z) in (Ω)	Resistance (R) in (Ω/m)	Capacitance (C) in (F/m)	Conductance (G) in (S/m)	Inductance (L) in (H/m)
Single wire isolated in free space $f \leq \frac{4}{\pi\mu a^2\sigma}$	(internal) $R + j\omega L$	(internal) $\frac{1}{\pi a^2 \sigma_c}$	$\frac{2\pi\varepsilon_d}{\ln\left(\frac{2len}{a}\right)-1}$	$C\left(\frac{\sigma_d}{\varepsilon_d}\right)$	(internal) $\frac{\mu}{2\pi}\left[\ln\left(\frac{2len}{a}\right)+\frac{a}{2len}-1\right]+\frac{\mu_c}{8\pi}$
Single wire isolated in free space $f > \frac{4}{\pi\mu a^2\sigma}$	(internal) $\frac{1}{2\pi a}(1+j)\sqrt{\frac{\pi f\mu_c}{\sigma_c}}$	(internal) $\frac{1}{2\pi a}\sqrt{\frac{\pi f\mu_c}{\sigma_c}}$	$\frac{2\pi\varepsilon_d}{\ln\left(\frac{2len}{a}\right)-1}$	$C\left(\frac{\sigma_d}{\varepsilon_d}\right)$	(internal) $\frac{\mu}{2\pi}\left[\ln\left(\frac{2len}{a}\right)+\frac{a}{2len}-1\right]+\frac{\mu_c}{8\pi}\frac{2\delta}{a}$
Single wire above reference plane in free space $f \leq \frac{4}{\pi\mu a^2\sigma}$	$\frac{60}{\sqrt{\varepsilon_r}}\ln\left[\frac{h}{a}+\sqrt{\left(\frac{h}{a}\right)^2-1}\right]$	(internal) $\frac{1}{\pi a^2\sigma_c}$	$\frac{2\pi\varepsilon_d}{\ln\left(\frac{h}{a}+\sqrt{\left(\frac{h}{a}\right)^2-1}\right)}$	$C\left(\frac{\sigma_d}{\varepsilon_d}\right)$	$\frac{\mu}{2\pi}\ln\left(\frac{h}{a}+\sqrt{\left(\frac{h}{a}\right)^2-1}\right)$
Single wire above reference plane in free space $f > \frac{4}{\pi\mu a^2\sigma}$	$\sqrt{L/C}$ $\frac{60}{\sqrt{\varepsilon_r}}\ln\left[\frac{h}{a}+\sqrt{\left(\frac{h}{a}\right)^2-1}\right]$	(internal) $\frac{1}{2\pi a}\sqrt{\frac{\pi f\mu_c}{\sigma_c}}$	$\frac{2\pi\varepsilon_d}{\ln\left(\frac{h}{a}+\sqrt{\left(\frac{h}{a}\right)^2-1}\right)}$	$C\left(\frac{\sigma_d}{\varepsilon_d}\right)$	$\frac{\mu}{2\pi}\ln\left(\frac{h}{a}+\sqrt{\left(\frac{h}{a}\right)^2-1}\right)$

Notes:

1. All calculated values are to be considered as external to the conductor(s) unless otherwise indicated.
2. Variables list:
   (a) $\varepsilon_d = \varepsilon_0\varepsilon_r$, where $\varepsilon_0 \cong 1/(36\pi \times 10^9)$, and ε_r represents the relative permittivity of the dielectric material surrounding the wire. Conservatively, it is assumed if a wire has a dielectric coating the characteristics of that material dominate any calculations as compared to those for a bare wire in free space.
   (b) $\mu = \mu_0\mu_r$, where $\mu_0 \cong 4\pi \times 10^{-7}$, and μ_r represents the relative permeability of the dielectric material surrounding the wire. Most, not all, dielectrics exhibit little to no magnetic permeability.
   (c) μ_c, σ_c represent the permeability and the conductivity of the wire.
   (d) σ_d represents the electrical conductivity of the dielectric material surrounding the wire. It is generally preferred for a given dielectric this characteristic possess a very low numerical value.
   (e) For single wire calculations, *len* represents the length of the wire.
   (f) $\delta = 1/\sqrt{\pi f \mu_c \sigma_c}$ represents the skin depth in a wire.
   (g) h, a, and d represent the height above a common reference, the radius of the wire, and the distance of separation between two wires.

Table 4.2 Approximate parameters of basic two-wire transmission lines

Line type	**Characteristic impedance (Z) in (Ω)**	**Resistance (R) in (Ω/m)**	**Capacitance (C) in (F/m)**	**Conductance (G) in (S/m)**	**Inductance (L) in (H/m)**
Wire Pair in Free Space, (wires same diameter) $f > \frac{4}{\pi\mu a^2\sigma}$	$\sqrt{L/C}$ $\frac{120}{\sqrt{\varepsilon_r}}\cosh^{-1}\left(\frac{d}{2a}\right)$	(internal) $\frac{1}{\pi a}\sqrt{\frac{\pi f\mu_c}{\sigma_c}}$	$\frac{\pi\varepsilon_d}{\cosh^{-1}\left(\frac{d}{2a}\right)}$	$C\left(\frac{\sigma_d}{\varepsilon_d}\right)$	$\frac{\mu}{\pi}\cosh^{-1}\left(\frac{d}{2a}\right)$
Wire Pair Above Common Reference Plane (wires same diameter; differential mode)	$\frac{60}{\sqrt{\varepsilon_r}}\ln\left(\frac{2h}{a\sqrt{1+\left(\frac{2h}{d}\right)^2}}\right)$	(internal) $\frac{1}{\pi a}\sqrt{\frac{\pi f\mu_c}{\sigma_c}}$	$\frac{4\pi\varepsilon_d\ln\left[1+\left(\frac{2h}{d}\right)^2\right]}{4\left[\ln\left(\frac{2h}{a}\right)\right]^2-\left[\ln\left[1+\left(\frac{2h}{d}\right)^2\right]\right]^2}$	$C\left(\frac{\sigma_d}{\varepsilon_d}\right)$	$\frac{\mu}{2\pi}\ln\left[1+\left(\frac{2h}{d}\right)^2\right]$
Wire Pair Above Common Reference Plane (wires same diameter, common mode)	$\frac{60}{\sqrt{\varepsilon_r}}\ln\left(\frac{2h}{a}\sqrt{1+\left(\frac{2h}{d}\right)^2}\right)$	(internal) $\frac{1}{\pi a}\sqrt{\frac{\pi f\mu_c}{\sigma_c}}$	$\frac{4\pi\varepsilon_d}{\ln\left[\left(\frac{2h}{a}\right)\sqrt{1+\left(\frac{2h}{d}\right)^2}\right]}$	$C\left(\frac{\sigma_d}{\varepsilon_d}\right)$	$\frac{\mu}{2\pi}\ln\sqrt{1+\left(\frac{2h}{d}\right)^2}$
Coaxial Cable	$\frac{60}{\sqrt{\varepsilon_r}}\ln\left(\frac{r2}{r1}\right)$	$\frac{1}{2\pi}\sqrt{\frac{\pi f\mu_c}{\sigma_c}}\left(\frac{1}{r1}+\frac{1}{r2}\right)$	$\frac{2\pi\varepsilon_d}{\ln\left(\frac{r2}{r1}\right)}$	$C\left(\frac{\sigma_d}{\varepsilon_d}\right)$	$\frac{\mu}{2\pi}\ln\left(\frac{r2}{r1}\right)$

Notes: See *Notes* of Table 4.1.

The term "cable-to-cable crosstalk" might be interpreted as a descriptive phrase used to describe the exchange of energy between lines in the electric and magnetic fields generated by the charges and currents on the individual lines via mutual capacitance and/or mutual inductance. The total amount of crosstalk between lines is likely to be the sum of mutual capacitive and mutual inductive coupling, combined with secondary effects attributable to self-inductance and self-capacitance of the various circuit/line components. The determination of coupling is in general an estimation process, as the true coupling is a function of materials, geometry, and multiple other characteristics of which most tend to change as frequency changes.

The goal of any estimation process is to establish bounding values for the amount of coupling that may conveniently be expressed as a ratio between the source voltage and the voltage appearing across one or more resistances (impedances) in the victim circuit. This approach is useful in that it can provide information to inform any possible steps that might be taken initially to confirm if crosstalk is indeed the cause of any observed interference and to mitigate the observed/calculated coupling. It may be necessary to also estimate the so-called near- and far-end voltages and/or currents, depending on the application. For example, differential data bus architectures often impose limits on these near- and far-end voltages and/or currents. Any estimation method selected for these sorts of applications will need to provide more than a general sense of the amount of coupling. With the additional information specific to near- and/or far-end interference levels, possible mitigation steps to be considered may be more readily focused. Ultimately, for the most critical circuits, the accuracy of a numerical solution may be the only acceptable path to take.

Various authors over time (several are listed here: [6–15]) have employed models with varying degrees of complexity while still allowing for ease of manual computation of both capacitive and inductive crosstalk using approaches wherein the mutual capacitance and mutual inductance terms of the coupling are treated separately, and results from these separate efforts are then superimposed to form a solution. An approach of this ilk typically employs multiple assumptions, primary among them (1) the coupled length of the lines is $\leq \frac{\lambda}{10}$, where λ is the operating wavelength of interest; (2) circuit terminating impedances are all purely resistive; and (3) the lines are weakly coupled, that is, the effects of the victim on the source are sufficiently small to ignore. Additionally, somewhat less restrictive assumptions may be made to simplify the estimation process, including (1) the medium in which the lines are located is homogeneous, linear, isotropic, and lossless; (2) the lines are uniform and exhibit no variation in cross section; (3) the lines are parallel to each other; and (4) all conductors are lossless. These less restrictive assumptions apply in many examples of practical transmission line structures.

Simple models often are focused on the effects of mutual capacitance and inductance and as a result fail to fully account for the effects of self-capacitive and self-inductive terms or those of the terminating impedances on the overall coupling. Such models that depend on the primary assumptions previously listed tend to fail as the operating frequency increases, with the magnitude of prediction errors proportional to frequency.

Clayton Paul [16,17] was able to develop generally accurate equations that avoid the constraints of coupling length $\leq \frac{\lambda}{10}$, purely resistive terminations, and notably weak coupling. Paul's more accurate results embody the additional less restrictive assumptions listed and account for the impact of mutual capacitance, mutual inductance, self-capacitance, and self-inductance terms as well as the ratios between terminating and line characteristic impedances. Despite the somewhat more complicated appearance of the equations, Paul's equations still allow for simple hand calculations and serve as a solid baseline for yet more detailed numerical investigation if crosstalk is believed to be the source of interference and the criticality of the involved circuits is felt to warrant the additional effort.

For those who seek an even more accurate approach, Paul also developed a strong matrix approach [18,19] to the crosstalk problem suitable for efficient numerical processing. In this text, methods of computation suitable for hand calculation are provided, with references to provide additional information for readers interested in pursuing a numerical solution process or pursuing more information regarding the relationships shown herein.

4.4.2 Simple model for capacitive crosstalk

Figure 4.4 is a simple schematic representation of two transmission lines assembled from two identical cylindrical conductors run in parallel to one another, with each line at the same height above a common reference plane.

We want to model this schematic in a somewhat generic fashion to bring focus on the coupling mechanism. Accordingly, the following assumptions are made: (1) the coupled length of the lines is $\leq \frac{\lambda}{10}$; (2) circuit impedances are all purely resistive; (3) the two capacitances to the common reference, C_{1r} and C_{2r}, are assumed to be equal in magnitude as well; (4) the lines are weakly coupled; and (5) the source $V_2 = 0$.

Based on the foregoing assumptions a simple nodal analysis of Figure 4.4 to ascertain the ratio of the interference voltage V_N across $RL2$ to the source voltage

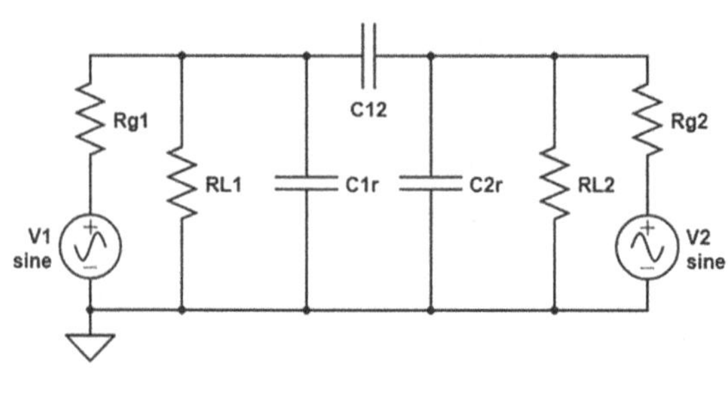

Figure 4.4 Simple circuit to illustrate capacitive coupling

V_1 yields the following relationship:

$$\frac{V_N}{V_1} = \frac{j\omega RC_m}{[2 + j\omega R(C_m + C)]^2 - (j\omega RC_m)^2}$$
$$= \frac{j\omega RC_m}{4[1 + j\omega R(C_m + C)] - (\omega RC)^2(1 + 2C_m/C)} \tag{4.94}$$

where $C_m = C_{12}$, $C = C_{1r} = C_{2r}$, and $R = R_{g1} = RL1 = R_{g2} = RL2$.

If the circuit impedances are less than the sum of the impedances of the mutual capacitance C_m and the capacitance between the lines and the common reference plane, C,

$$R \ll \frac{1}{j\omega(C_m + C)} \tag{4.95}$$

the magnitude of (4.94) becomes

$$\left|\frac{V_N}{V_1}\right| \cong \left|\frac{j\omega RC_m}{4}\right| \tag{4.96}$$

If the circuit impedances are greater than the sum of the impedances of the mutual capacitance C_m and the capacitance between the lines and the common reference plane, C,

$$R \gg \frac{1}{j\omega(C_m + C)} \tag{4.97}$$

the magnitude of (4.94) becomes

$$\left|\frac{V_N}{V_1}\right| \cong \left|\frac{C_m}{4(C_m + C)}\right| \tag{4.98}$$

Equation (4.98) represents the maximum coupled noise predicted by the model shown in Figure 4.4. The capacitances in (4.94) can be computed using expressions in Table 4.1 for a single wire above a common reference plane and Table 4.2 for the mutual capacitance between the two wires.

Figure 4.5 graphically illustrates the behaviour of the coupling model as captured in (4.94). The lines are supported 5 cm above the common reference plane, are 20 AWG (0.465 mm $< r <$ 0.50 mm), are separated by 4 times the wire radius, and have polytetrafluoroethylene (PTFE, also known as Teflon) insulation. Examination of Figure 4.5 highlights the high-pass filter characteristic of capacitive coupling and illustrates the impact of circuit resistances (impedances) on the degree of coupling. As circuit resistances (impedances) increase, the coupling shifts to the left, lower in frequency, and increases. Figure 4.5 suggests that reducing circuit resistances (impedances) will shift the frequency response to the right, higher in

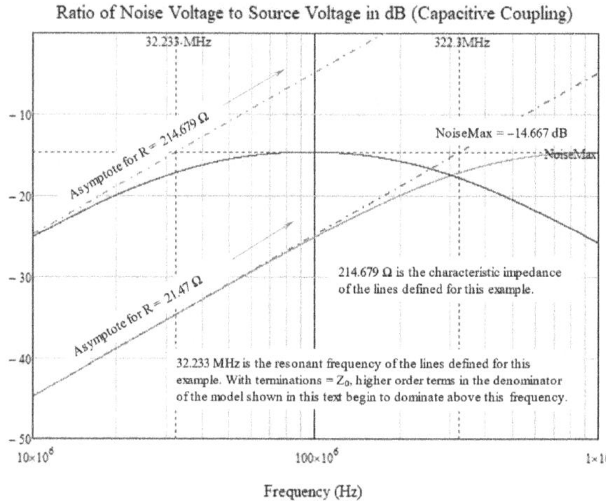

Figure 4.5 Ratio of coupled noise voltage to source voltage for capacitive coupled circuits

frequency, thus reducing the coupling. A very important aspect of Figure 4.5 is to illustrate the frequency range over which the asymptotic estimate of the model in (4.94) is strictly valid. In this example, the coupling length has an electrical length of 1 metre, corresponding to a frequency of 30 MHz. Close examination of the darker coloured lines and their relationship to the lighter coloured asymptotic lines shows that with a constant mutual capacitance and operational frequency, lowering the circuit resistances (impedances) may reduce crosstalk behaviour. The takeaway message here is the baseline asymptotic expression for this example is valid as a low-frequency estimation and is not strictly valid much beyond 30 MHz.

Note also the straight red dashed line indicating the expected maximum coupling predicted by this model regardless of frequency or terminating impedances. This value is directly proportional to the mutual capacitance between the circuits. The most effective and direct means of reducing this mutual capacitance is to physically separate the two coupled lines.

Routing wiring as close to a common reference plane as possible acts to reduce the magnitude of mutual capacitance, although it will increase the capacitance of the line to the common reference. Reducing the coupling length can reduce the impact of self-capacitances and self-inductances in the circuit that contribute to the coupling. Routing lines near to right angles when they cross effectively reduces any coupling length. These are all layout and routing topics.

Lowering circuit resistances (impedances) may reduce capacitive crosstalk behaviour. Altering layout and routing or lowering circuit resistances (impedances) are usually not available options unless they are implemented in early design efforts. If such options prove to be difficult or not possible, cable shielding may be employed. The use of cable shielding as a mitigation of capacitive coupling will be addressed in the shielding chapter.

4.4.3 Simple model for inductive crosstalk

Figure 4.6 is a simple schematic representation of two transmission lines assembled from two identical cylindrical conductors run in parallel to one another, with each line at the same height above a common reference plane.

As we did for capacitive coupling, we want to model this schematic in a somewhat generic fashion to bring focus on the coupling mechanism. Accordingly, the following assumptions are made: (1) the coupled length of the lines is $\leq \frac{\lambda}{10}$; (2) circuit impedances are all purely resistive; (3) the two inductances, L_1 and L_2, are assumed to be equal in magnitude as well; (4) the lines are weakly coupled; and (5) the source $V_2 = 0$.

Based on the foregoing assumptions, a simple Kirchoff Voltage Law (KVL) analysis of Figure 4.6 to ascertain the ratio of the interference voltage V_N across $RL2$ to the source voltage V_1 yields the following relationship:

$$\frac{V_N}{V_1} = \frac{-j\omega RL_m}{4R^2 + j\omega RL - \omega^2\left(L^2 - L_m{}^2\right)} \tag{4.99}$$

where $L = L1 = L2$, and $R = R_{g1} = RL1 = R_{g2} = RL2$.

If the circuit impedances are less than the sum of the impedances of the mutual inductance L_m and the inductance between the lines and the common reference plane, L,

$$R \ll j\omega(L_m + L) \tag{4.100}$$

the magnitude of (4.99) becomes

$$\left|\frac{V_N}{V_1}\right| \cong \left|\frac{-j\omega RL_m}{4R}\right| \tag{4.101}$$

If the circuit impedances are greater than the sum of the impedances of the mutual inductance L_m and the inductance between the lines and the common reference plane, L,

$$R \gg j\omega(L_m + L) \tag{4.102}$$

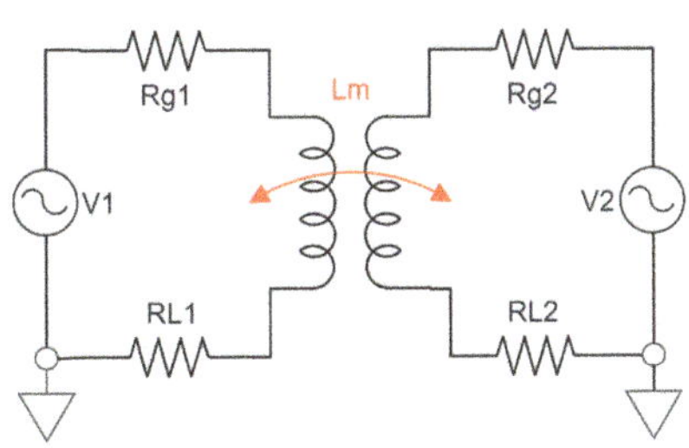

Figure 4.6 Simple circuit to illustrate inductive coupling

the magnitude of (4.99) becomes

$$\left|\frac{V_N}{V_1}\right| \cong \left|\frac{-L_m}{4(L)}\right| \tag{4.103}$$

Equation (4.103) represents the maximum coupled noise predicted by the model shown in Figure 4.6. The inductances in (4.99) can be computed using expressions in Table 4.1 for a single wire above a common reference plane and Table 4.2 for the mutual inductance between the two wires.

Figure 4.7 graphically illustrates the behaviour of the coupling model as captured in (4.99). The lines are supported 5 cm above the common reference plane, are 20 AWG (0.465 mm $< r <$ 0.50 mm), are separated by 4 times the wire radius, and have PTFE insulation. Examination of Figure 4.7 highlights the high-pass filter characteristic of inductive coupling, and it illustrates the impact of circuit resistances (impedances) on the degree of coupling. As circuit resistances (impedances) decrease, the coupling shifts to the left, lower in frequency, and increases. Figure 4.7 suggests that increasing circuit resistances (impedances) will shift the frequency response to the right, higher in frequency, thus reducing the coupling. A very important aspect of Figure 4.7 is to illustrate the frequency range over which the asymptotic estimate of the model in (4.99) is strictly valid. In this example, the coupling length has an electrical length of 1 metre, corresponding to a frequency of 30 MHz. Close examination of the darker coloured lines and their relationship to the lighter coloured asymptotic lines shows that with a constant mutual inductance and operational frequency, raising the circuit resistances (impedances) may reduce crosstalk behaviour. The takeaway message here is, like the previous model for capacitive coupling, the asymptotic

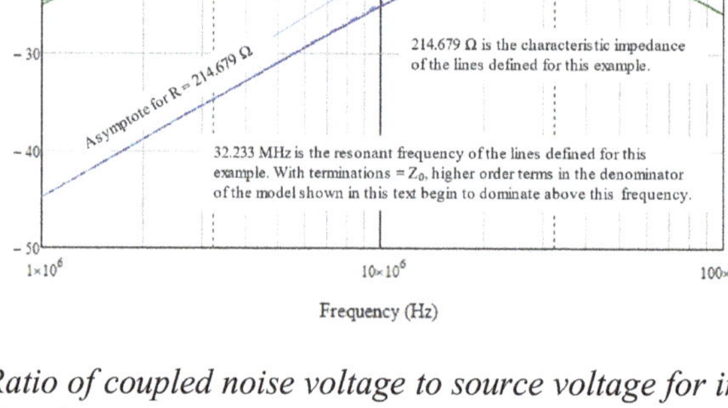

Figure 4.7 Ratio of coupled noise voltage to source voltage for inductive coupled circuits

expression for this example is valid as a low-frequency estimation and is not strictly valid much beyond 30 MHz.

Note also the straight red dashed line indicating the expected maximum coupling predicted by this model regardless of frequency or terminating impedances. This value is directly proportional to the mutual inductance between the circuits. The most effective and direct means of reducing this mutual inductance is to physically separate the two coupled lines. Minimizing adjacent loop areas can be very effective as well.

Routing wiring as close to a common reference plane as possible is an excellent means of minimizing loop areas and acts to reduce the magnitude of mutual inductance. Reducing the coupling length can reduce the impact of self-capacitances and self-inductances in the circuit that contribute to the coupling. Routing lines near to right angles when they cross effectively reduces any coupling length. These are all layout and routing topics.

Raising circuit resistances (impedances) may reduce inductive crosstalk behaviour. Altering layout and routing or raising circuit resistances (impedances) are usually not available options unless they are implemented in early design efforts. If such options prove to be difficult or not possible, cable shielding may be employed. The use of cable shielding as a mitigation of inductive coupling will be addressed in the shielding chapter.

4.4.4 Simple test to identify the low-frequency dominant mode of coupling

In general, noise voltage resulting from inductive coupling is in series with the victim line, whereas noise voltage resulting from capacitive coupling is in parallel with the victim line. In low-frequency applications, it may be possible to perform a simple test to discern if either of the two modes is dominant. Select one end of the victim line and measure the voltage across the termination. While observing the measured voltage, decrease the termination impedance at the opposite end of the victim line. If the measured voltage decreases, the dominant mode of coupling is most likely to be capacitive. If the measured voltage increases, the dominant mode of coupling is most likely to be inductive.

4.4.5 Simple models for near- and far-end effects

Figure 4.8 is a schematic representation of two transmission lines assembled from two identical cylindrical conductors run in parallel to one another, with each line at the same height above a common reference plane.

Once again, we want to model this schematic in a somewhat generic fashion to bring focus on the coupling mechanism. Accordingly, the following assumptions are made: (1) the coupled length of the lines is $\leq \frac{\lambda}{10}$; (2) circuit impedances are all purely resistive; (3) the lines are weakly coupled; (4) the source $V_2 = 0$; and (5) only the mutual capacitance and the mutual inductance are shown, no self-terms.

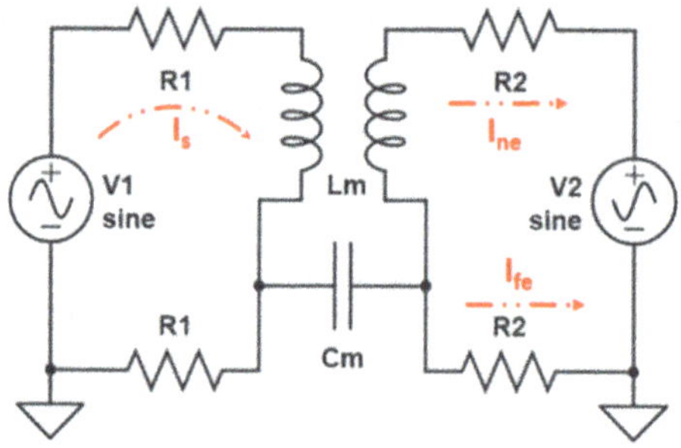

Figure 4.8 Simple circuit to illustrate combined capacitive and inductive coupling

In Figure 4.8, we have a source voltage V_s measured from the junction of $R1$ and L_m to the common reference. The corresponding near-end voltage appears across $R2$, shown at the upper right-hand of Figure 4.8, and the far-end voltage appears across $R2$, shown at the lower right-hand of Figure 4.8. It is assumed that $(R1 = R2 = R)$. Proceeding on the basis that we can separate the capacitive and inductive coupling for calculation, we begin by ignoring C_m and write

$$V_{ne} - V_{fe} = I_s(j\omega L_m) \tag{4.104}$$

Now ignoring L_m we can write

$$I_{ne} + I_{fe} = V_s(j\omega C_m) \tag{4.105}$$

Since $V_s \cong I_s R1$, and $R1 = R$, we can write (4.104) as

$$V_{ne} - V_{fe} = V_s\left(\frac{j\omega L_m}{R}\right) \tag{4.106}$$

Also, since $V_{ne} = I_{ne}R2$ and $V_{fe} = I_{fe}R2$, and $R2 = R$, (4.105) can be written as

$$V_{ne} + V_{fe} = V_s(j\omega C_m R) \tag{4.107}$$

Combining (4.106) and (4.107), we have

$$V_{ne} = \frac{V_s}{2}(j\omega C_m R) + \frac{V_s}{2}\left(\frac{j\omega L_m}{R}\right) \tag{4.108}$$

$$V_{fe} = \frac{V_s}{2}(j\omega C_m R) - \frac{V_s}{2}\left(\frac{j\omega L_m}{R}\right) \tag{4.109}$$

Examination of (4.108) and (4.109) indicates that as circuit resistances (impedances) increase, capacitive coupling increases and inductive coupling decreases, and as circuit resistances (impedances) decrease, capacitive coupling decreases and inductive coupling increases. V_{ne} and V_{fe} are in-phase at the near end, and 180° out of phase at the far end and thus will tend to cancel each other out.

The simple capacitive and simple inductive models converge with terminating resistances equal to the characteristic impedance, Z_0, of each of the two lines. For the sake of comparison, the same line characteristics, with the terminating resistance values equal to Z_0, as were used for the simple capacitive and inductive models, are substituted into (4.108) and (4.109). Figure 4.9 illustrates the results and the comparison. As might be expected, the simple capacitive and inductive models and the simple near- and far-end models generate very similar behaviour.

Note in Figure 4.9 the near- and far-end calculations appear to be more conservative by roughly 6 dB, predicting somewhat greater crosstalk levels than those predicted by the sums and differences of the simple coupling networks previously discussed. It bears repeating these are low-frequency estimations, with the greatest fidelity near or below 30 MHz for these examples. The curving dashed orange lines capture the sum and difference of the two simple coupling results. Note the maximum predicted noise, −8.568 dB, based on the sum of the two simple coupling equations, is nearly 16 dB less than the V_{ne} asymptote at 100 MHz. The delta is comparable at the same frequency between the difference of the two simple coupling equations and the V_{fe} asymptote by just over 13 dB. These errors diminish as frequency decreases towards 30 MHz, emphasizing the low frequency and approximate nature of these asymptotic relationships.

Figure 4.10 illustrates the changes associated with either reducing or increasing the terminating impedances by a factor of 10. The asymptotic behaviour exhibited by the simple capacitive and inductive models, as well as that of the simple near-/far-end model, is essentially identical. However, the delta between the asymptotic behaviour and the sum and difference of the simple capacitive and inductive models is much greater than illustrated in Figure 4.9 above 10 MHz and

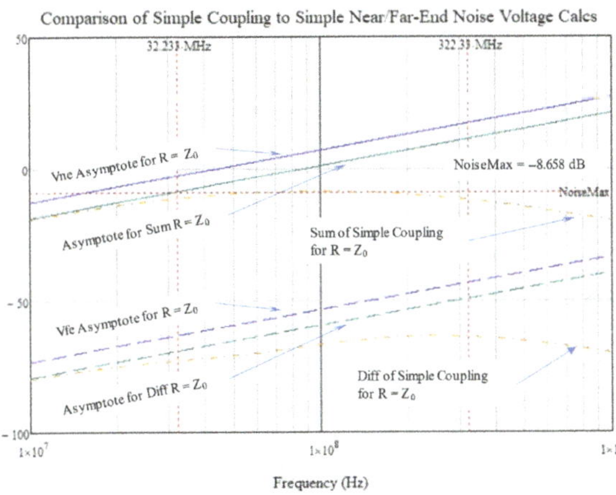

Figure 4.9 Comparison of simple coupling to simple near-/far-end voltage calculations for R = Z_0

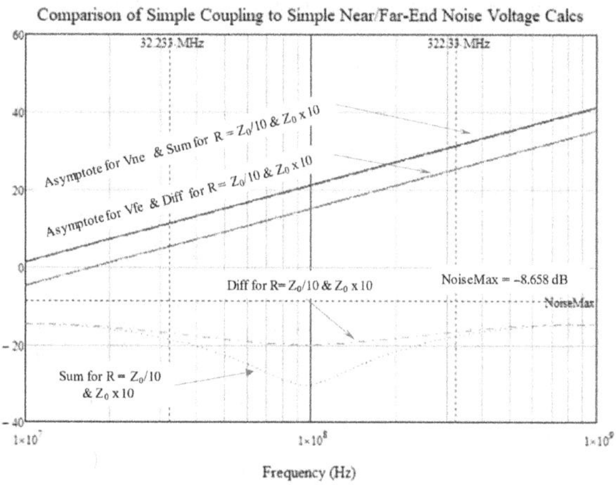

Figure 4.10 Comparison of simple coupling to simple near-/far-end voltage calculations for $R = Z_{10}/10$ and $Z_0 \times 10$

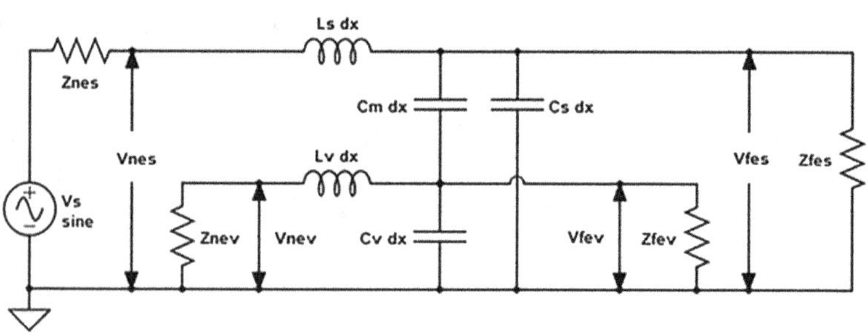

Figure 4.11 Fundamental transmission line structure used for exact equations

continues to increase with frequency. This sort of behaviour cannot be physically realizable and, as illustrated in Figure 4.9, emphasizes the "low-frequency" applicability of the asymptotic behaviour of the simple models.

4.4.6 Exact equations for crosstalk estimation

The following are equations developed by Clayton Paul, as discussed previously. These equations assume the conductance, *G*, is negligible and are provided without derivation. The reader who is interested in detail is encouraged to access references [16,17]. Explanatory remarks are included herein to aid in understanding the variables. Figure 4.11 illustrates the fundamental transmission line structure upon which the equation set is based.

The basic coupling equations are given as expressions for the coupled voltage at the near and far ends of the victim line in (4.110) and (4.111):

$$V_{NE} = \frac{S'}{DEN}\left[\left(\frac{Z_{NEV}}{Z_{NEV}+Z_{FEV}}\right)(j\omega l_m L)F_L I_{sdc} + \left(\frac{Z_{NEV}Z_{FEV}}{Z_{NEV}+Z_{FEV}}\right)(j\omega c_m L)F_C V_{sdc}\right] \tag{4.110}$$

$$V_{FE} = \frac{S'}{DEN}\left[-\left(\frac{Z_{FEV}}{Z_{NEV}+Z_{FEV}}\right)(j\omega l_m L)I_{sdc} + \left(\frac{Z_{NEV}Z_{FEV}}{Z_{NEV}+Z_{FEV}}\right)(j\omega c_m L)V_{sdc}\right] \tag{4.111}$$

The DC terms in (4.110) and (4.111) are given by

$$V_{sdc} = \frac{Z_{FES}}{Z_{NES}+Z_{FES}}V_S;\, I_{sdc} = \frac{V_S}{Z_{NES}+Z_{FES}} \tag{4.112}$$

where
V_s = applied source voltage
Z_{NEV}, Z_{FEV} = impedances at the near and far end of the victim circuit
Z_{NES}, Z_{FES} = impedances at the near and far end of the source circuit
L = the coupled length between the two lines
l_m = mutual inductance between the two lines
c_m = mutual capacitance between the two lines

Auxiliary terms for (4.110) and (4.111):

$$F_L = \left(C + \frac{j2\pi\left(\frac{L}{\lambda}\right)S'}{\sqrt{1-k^2}}\alpha_{FES}\right);\, F_C = \left(C + \frac{j2\pi\left(\frac{L}{\lambda}\right)S'}{\sqrt{1-k^2}}\frac{1}{\alpha_{FES}}\right) \tag{4.113a, b}$$

$$DEN = C^2 - (S'\omega)^2\tau_v\tau_s\left[1 - k^2\frac{(1-\alpha_{NES}\alpha_{FEV})(1-\alpha_{FES}\alpha_{NEV})}{(1+\alpha_{NEV}\alpha_{FEV})(1+\alpha_{NES}\alpha_{FES})}\right] + j\omega CS'(\tau_v+\tau_s) \tag{4.114}$$

$$C = \cos(\beta L);\, S' = \frac{\sin(\beta L)}{(\beta L)} \tag{4.115a, b}$$

$$\beta = \frac{2\pi}{\lambda} = \text{phase constant};\, \lambda = \text{operating wavelength} \tag{4.116a, b}$$

$$\omega = 2\pi f = \text{operating frequency of interest} \tag{4.117}$$

The "alpha" terms in (4.113) and (4.114) are given by

$$\alpha_{NES} = \frac{Z_{NES}}{Z_{cs}};\, \alpha_{FES} = \frac{Z_{FES}}{Z_{cs}};\, \alpha_{NEV} = \frac{Z_{NEV}}{Z_{cv}};\, \alpha_{FEV} = \frac{Z_{FEV}}{Z_{cv}} \tag{4.118a–d}$$

The equivalent time constants for the source and victim circuits in (4.114) are given by

$$\tau_s = \frac{l_S L}{Z_{NES} + Z_{FES}}(1 + \alpha_{NES}\alpha_{FES}) \tag{4.119}$$

$$\tau_v = \frac{l_v L}{Z_{NEV} + Z_{FEV}}(1 + \alpha_{NEV}\alpha_{FEV}) \tag{4.120}$$

The denominator terms in (4.118) represent the respective characteristic impedances for the source and victim circuits and are given by

$$Z_{cs} = \sqrt{\frac{l_s}{(c_s + c_m)}} = v l_s \sqrt{1 - k^2} = \frac{1}{\sqrt{\mu\varepsilon}} l_s \sqrt{1 - k^2} \tag{4.121}$$

$$Z_{cv} = \sqrt{\frac{l_v}{(c_v + c_m)}} = v l_v \sqrt{1 - k^2} = \frac{1}{\sqrt{\mu\varepsilon}} l_v \sqrt{1 - k^2} \tag{4.122}$$

The coupling coefficient in (4.113), (4.114), (4.121), and (4.122) is given by

$$k = \frac{l_m}{\sqrt{l_s l_v}} = \frac{c_m}{\sqrt{(c_s + c_m)(c_v + c_m)}} \tag{4.123}$$

where l_s = inductance of the source circuit
l_v = inductance of the victim circuit
c_s = capacitance of the source circuit
c_v = capacitance of the victim circuit

Substitution of the characteristics of the previous two lines analysed in the simple capacitive, simple inductive, and simple near- and far-end calculations into the foregoing equation set produces the following graphs. Note that Table 4.1 provides typical expressions for source and victim capacitance and inductance terms in (4.119) through (4.123). Table 4.2 provides expressions for typical mutual inductance and capacitance terms in (4.110), (4.111), (4.119), (4.120), and (4.123).

Several comments about these results are in order. The close agreement in Figure 4.12 between the summed results of the simple capacitive and inductive coupling models and the exact equation set not only at and below 30 MHz but also across frequency might suggest that the simple models are "good enough", and indeed they are in this particular case, primarily because all the line reactive elements are the same, and all the terminating impedances are real, are of the same magnitude, and are equal to the characteristic impedances of the lines. This was done intentionally in this text for a fair basis of comparison. Figure 4.12 should not be interpreted as evidence the simple expressions are "good enough" under all circumstances. This is made clear by examination of Figures 4.13 and 4.14. In Figure 4.13, the envelope of the near-end coupling predicted by the exact equations falls well below the sum and difference of the simple capacitive and

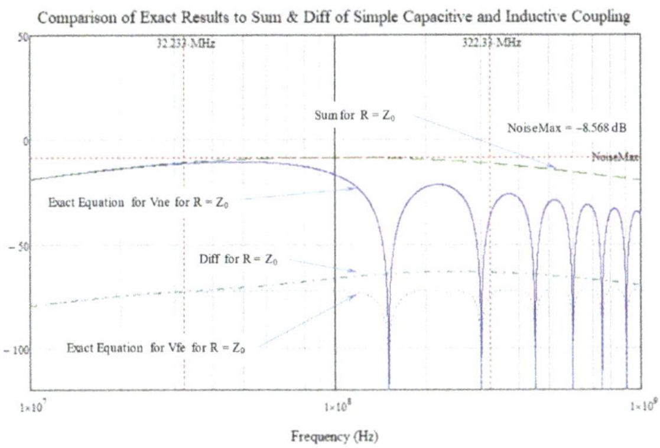

Figure 4.12 Comparison of exact results to Sum and Diff of simple capacitive and inductive coupling for $R = Z_0$

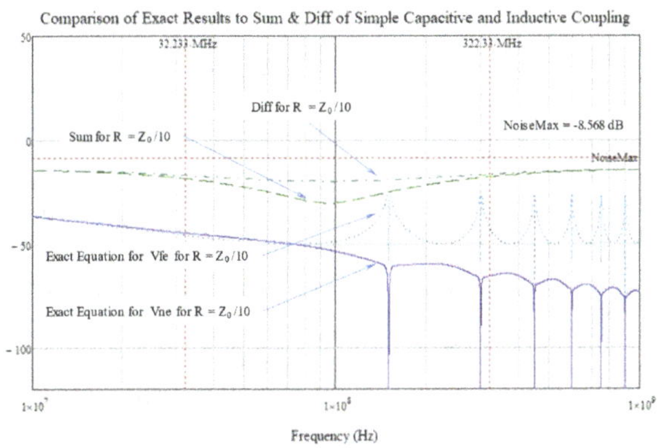

Figure 4.13 Comparison of exact results to Sum and Diff of simple capacitive & inductive coupling for $R = Z_0/10$

inductive models, while the envelope of the far-end coupling not only falls well below the sum and difference of the simple capacitive and inductive models, but it has also increased above that of the near end. Similar behaviour is observed in Figure 4.14, with noticeable differences. In Figure 4.14, the envelope of the near-end coupling predicted by the exact equations falls well below the sum and difference of the simple capacitive and inductive models, while the envelope of the far-end coupling not only falls well below the sum and difference of the simple capacitive and inductive models, but it is also below that of the near end by a large delta.

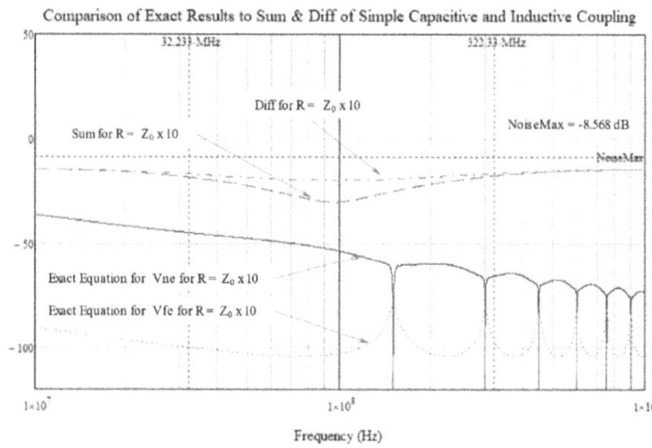

Figure 4.14 Comparison of exact results to Sum and Diff of simple capacitive and inductive coupling for $R = Z_0 \times 10$

The strength of Paul's exact equation set lies not only in the flexibility to avoid the restrictions of a coupling length $\leq \frac{\lambda}{10}$, purely resistive terminations, and notably weak coupling, but also in the fact that Paul's equations fully account for the impact of self-capacitance and self-inductance terms as well as properly account for the ratio between the various terminating impedances and the characteristic impedances of the lines in the presence of one another. A review of the cited references will quickly show that, under circumstances wherein coupling is strong, and/or circuit impedances are complex and tend not to be equal in magnitude as a function of frequency, the simple models fail when compared to measured data.

4.4.7 Use of twisted-pair wiring for mitigation of crosstalk

The following discussion is most applicable in the context of the simple capacitive and inductive models previously discussed, wherein multiple assumptions were made, namely (1) the coupled length of the lines is $\leq \frac{\lambda}{10}$, where λ is the operating wavelength of interest; (2) circuit terminating impedances are all purely resistive; and (3) the lines are weakly coupled.

Examination of Figures 4.5 and 4.7 illustrates the fundamental responses for crosstalk of either type based on circuit impedances. Later in discussions of the exact equations, it was further illustrated that the terminating impedances can play a significant role in crosstalk behaviour. Twisted pairs exhibit similar behaviour, with some interesting differences.

Twisting the wires of a transmission line structure where the return side is twisted with the corresponding high side can be highly effective in the reduction of inductive coupling. By twisting the wires together, the loop area between the two wires is effectively minimized to the greatest extent. Secondly, when a pair of wires is twisted together, small loops are formed adjacent to one another. In accordance with Faraday's and Lenz's laws, when an external magnetic field penetrates these loops, an electromotive force (emf) is induced in each loop. Each loop is 180° out of phase with its neighbour, and thus the induced emfs are opposite in sign and tend to cancel each other out.

Theoretically, if the number of loops is even and each loop is physically the same as the next, the inductive coupling can be reduced to zero. If the number of loops is odd, and each loop is physically the same as the next, the inductive coupling can be reduced to that associated with a single loop.

Figure 4.15 illustrates a model of a short section of a twisted pair showing a source line, an idealized twisted pair, and a common reference.

The model illustrated in Figure 4.15 incorporates approximately rectangular segments or loops and abrupt transitions between each segment. This is because a twisted pair is not a uniform line, as is an untwisted pair. In an actual twisted pair, the segments are smoothly curved and the transitions between segments are correspondingly smooth as well, such that the mutual inductance and capacitance vary continuously along the length of the line. By using approximately rectangular segments and abrupt transitions between each segment, the model is made simpler, and the corresponding mathematics are made simpler as well. Assuming the transitions are vanishingly small, any variation of the mutual terms in the transitions may be ignored, enabling employment of the simpler expressions used for the mutual terms as in the previous section. Note that each segment represents one half of a complete single twist, so that two segments or loops make up a single twist. Figure 4.16 illustrates an equally simple circuit model that can be applied for each half-twist of the twisted pair.

In each segment or loop, a voltage will be developed by the line current as it interacts with the line mutual inductance of the segment. A corresponding

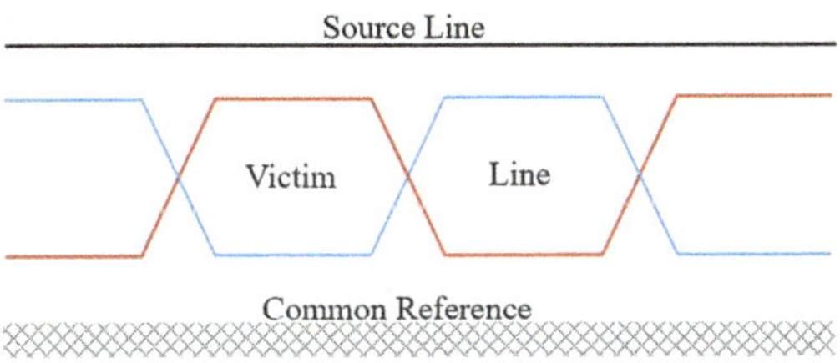

Figure 4.15 Simple model of twisted pair

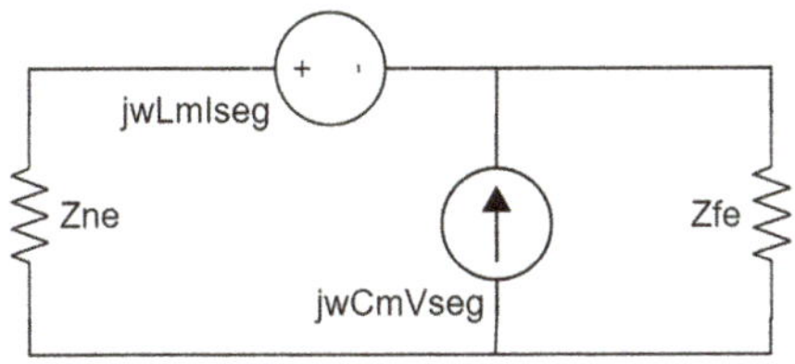

Figure 4.16 Simple circuit model of each half-twist of a twisted pair

current source develops for each of the two conductors in each segment or loop as the voltage of the segment interacts with the mutual capacitance of each of the two lines between the two lines of the segment and the common reference.

That said, for an unbalanced line, the total crosstalk may still be dominated by capacitive coupling if the terminating impedances are large, and the total crosstalk may be dominated by inductive coupling if the terminating impedances are small. Review Figures 4.5 and 4.7 to see these effects illustrated.

Oftentimes, an engineer may be confronted by a wiring scheme using untwisted wires in which crosstalk has been identified as the likely culprit and assumes that twisting the wires will resolve the crosstalk interference. A typical example of this is the replacement of a non-twisted pair with a twisted pair to mitigate coupling between circuits, with the frustrating result that the coupling is relatively unaffected. In further review of the circuits involved, it is ultimately revealed the terminating impedances are large (on the order of a few hundred Ω or greater), indicating the dominant coupling mode is most likely to be capacitive and therefore relatively unimpacted by use of a twisted pair.

If the transmission line pair is balanced, meaning the impedances between each of the two lines making up the pair are equal in magnitude, capacitive coupling may be greatly reduced, regardless of twisting. Finally, twisting the two wires of a line, balanced or not, will have little impact on any common mode currents that may be flowing along the pair.

It is important to ensure that if twisting is employed, the twist rates of wires that share a common cable envelope are different. If the pairs involved have the same twist rate, the coupling between the wires or cables will exhibit a systematic aspect and will increase as a function of length. Note the efficacy of using twisting as a mitigation is also a function of the wavelength of the signals involved. Coupling will increase as the twist length approaches an appreciable fraction of a wavelength. Regardless of the use of twisted pair topology, the use of electromagnetic shielding as a means of field cancellation may be employed to further mitigate any undesirable coupling.

Twisting the wires in a transmission line will increase the length of the conductors, thereby increasing the series resistance. Twisting will also increase the line's capacitance as "seen" by the terminating circuits.

Twist rate (aka pitch) in a twisted wire pair has a significant impact on the capacitance of the pair, and thus on the transmission line characteristic impedance of the pair. This is true for either unshielded or shielded wire pairs. In both cases, an equivalent dielectric constant can be determined that is a function of the relative dielectric constants of the surrounding environment and the wire insulation, and the pitch or twist rate of the pair.

When two dielectrics are combined, as in a dielectric-covered wire in free space, the resultant effective dielectric constant is a combination of the two individual dielectric constants. This is often expressed as a volumetric relationship such as

$$\epsilon_{r_{eq}} = \alpha\epsilon_{r_1} + \beta\epsilon_{r_2} \tag{4.124}$$

In this expression, α and β represent the proportions of two dielectric constants involved in a bi-phase relationship such as exists when an insulated wire is immersed in free space or some other medium. In this relationship, $\alpha + \beta = 1$.

Simplification of (4.124) yields an expression capturing this effect for unshielded twisted pair is given here [20]:

$$\epsilon_{r_{eq}} = \epsilon_{r_1} + \beta(\epsilon_{r_2} - \epsilon_{r_1}) \tag{4.125}$$

where

$\epsilon_{r_{eq}}$ = the equivalent relative dielectric constant of the combination of the insulation surrounding the conductor(s) (it is assumed the insulation is the same on each wire in the twisted pair) and the environment surrounding the conductor pair

ϵ_{r_1} = the relative dielectric constant of the surrounding environment ($\epsilon_{r_1} \cong 1$ for air)

ϵ_{r_2} = the relative dielectric constant of the conductor insulation

β = the twist angle correction factor: this term captures the volumetric relationship between the two dielectrics, recognizing the volume occupied by each of the two dielectrics in the mutually shared space of the twisted pair is a function of the twist rate

The twist angle correction factor for a typical pair of 24 AWG wires with PTFE insulation is approximately given by an empirical relationship based on measured data:

$$\beta = 0.25 + 0.001\theta^2 \tag{4.126}$$

where θ = is the twist angle in degrees.

The twist angle can be found from

$$\tan\theta = \pi TD \tag{4.127}$$

where T = the number of twists/metre

D = the outer diameter of each conductor with insulation, metres

The wires in a typical Ethernet pair are equivalent to a 24 AWG PTFE-insulated stranded silver-coated copper conductor IAW Mil-W-16878/5. Per multiple vendor sources, T (the nominal twist rate) for CAT 5, CAT 5E, and CAT 6 falls between 60 and 200 twists/metre (approx. 1.50 twists/inch (TPI) up to 5.00 TPI). Assuming each wire in the pair has an overall diameter (conductor plus insulation) of approximately 1.37 mm, (4.127) yields a θ between 14.5° and 40.7°.

Experimental data suggest that a twist angle between 20° and 45° provides optimal performance. At a twist angle $< 20°$, the wires in the pair become increasingly subject to loss of line geometrical uniformity caused by flexure of the pair and may introduce a randomness that can dominate the characteristic impedance of the line. At a twist angle $>45°$, a typical copper conductor may be approaching a yield point leading to mechanical failure.

To increase the twist angle from 14.5° to equal the minimum recommended 20°, the number of twists/metre must be increased to 85 twists/metre (approx. 2.16 TPI). Note that a twist angle of 45° is equivalent to a twist rate of 232.34 twists/metre (approx. 5.90 TPI). Substituting twist angles between 20° and 45° in (4.126), β falls between 0.650 and 2.275.

Now, assuming we have air surrounding the pair, $\epsilon_{r1} = 1$, and each wire in the wire pair is covered with PTFE insulation, $\epsilon_{r2} = 2.1$, using the values determined for β using (4.126) and the bounding twist angles 20° and 45°, using (4.125) we find

$$\epsilon_{req} = 1 + 0.650(2.1 - 1) = 1.715 \tag{4.128}$$

or

$$\epsilon_{req} = 1 + 2.275(2.1 - 1) = 3.503 \tag{4.129}$$

Now, the capacitance for two parallel conductors having the same diameter from Table 4.2 is given by

$$C = \frac{\pi\varepsilon_d}{\cosh^{-1}\left(\frac{d}{2a}\right)} \tag{4.130}$$

where $d =$ the separation distance between the conductor centres; $d = 1.37$ mm
$2a =$ the diameter of each 24 AWG conductor w/o insulation;
$2a = 0.635$ mm
$\varepsilon_d =$ the product of the permittivity of free space and the relative dielectric constant of the insulation covering the conductors

For comparison, C is determined assuming the insulation has a relative dielectric constant of PTFE, or the equivalent relative dielectric constant as calculated by (4.128) or (4.129). Note the calculation for PTFE assumes no twisting and thus serves as a baseline for comparison.

$$\begin{aligned} C_{PTFE} &= \frac{\pi\varepsilon_d}{\cosh^{-1}\left(\frac{d}{2a}\right)} \\ &= \frac{\pi\varepsilon_0\varepsilon_{PTFE}}{\cosh^{-1}\left(\frac{d}{2a}\right)} \\ &= 41.56\ \text{pF/m} \end{aligned} \tag{4.131}$$

$$\begin{aligned} C_{128} &= \frac{\pi\varepsilon_d}{\cosh^{-1}\left(\frac{d}{2a}\right)} \\ &= \frac{\pi\varepsilon_0\varepsilon_{128}}{\cosh^{-1}\left(\frac{d}{2a}\right)} \\ &= 33.94\ \frac{\text{pF}}{\text{m}} \end{aligned} \tag{4.132}$$

$$
\begin{aligned}
C_{129} &= \frac{\pi\varepsilon_d}{\cosh^{-1}\left(\frac{d}{2a}\right)} \\
&= \frac{\pi\varepsilon_0\varepsilon_{129}}{\cosh^{-1}\left(\frac{d}{2a}\right)} \\
&= 69.33 \text{ pF/m}
\end{aligned} \tag{4.133}
$$

Assuming the conductors and their insulation coverings are non-magnetic, the inductance of the line can be determined from Table 4.2 as

$$
\begin{aligned}
L &= \frac{\mu}{\pi}\cosh^{-1}\left(\frac{d}{2a}\right) \\
&= 0.561\mu H/m
\end{aligned} \tag{4.134}
$$

The resultant line impedance can then be determined (from Table 4.2) as

$$
Z = \sqrt{L/C}
$$

$$
Z_{PTFE} = \sqrt{L/C_{PTFE}} = 116.1833\ \Omega \tag{4.135}
$$

$$
Z_{125} = \sqrt{L/C_{125}} = 128.5279\ \Omega \tag{4.136}
$$

$$
Z_{126} = \sqrt{L/C_{126}} = 89.9541\ \Omega \tag{4.137}
$$

These results may also be obtained by using (from Table 4.2)

$$
Z = \frac{120}{\sqrt{\varepsilon_r}}\cosh^{-1}\left(\frac{d}{2a}\right)
$$

$$
Z_{PTFE} = \frac{120}{\sqrt{2.1}}\cosh^{-1}\left(\frac{d}{2a}\right) = 116.2167\ \Omega \tag{4.138}
$$

$$
Z_{125} = \frac{120}{\sqrt{1.715}}\cosh^{-1}\left(\frac{d}{2a}\right) = 128.6015\ \Omega \tag{4.139}
$$

$$
Z_{126} = \frac{120}{\sqrt{3.503}}\cosh^{-1}\left(\frac{d}{2a}\right) = 89.9825\ \Omega \tag{4.140}
$$

The adjusted dielectric constant accounting for the twist rate has yielded corresponding changes in the line capacitance and the line impedance compared to simply using the original, unadjusted PTFE dielectric constant. The lower twist rate yields a higher impedance, and vice versa.

CAT 6 shielded ethernet cables using 24 AWG conductors "off the shelf" exhibit a typical capacitance on the order of 56 pF/m and a characteristic impedance near 100 Ω, ±15 Ω. These data should be compared to the results in (4.131)–(4.140).

The impact of using the same twist rate for multiple pairs is most easily understood by considering the placement of two helices next to each other with the same wire diameters and the same twist rate. This will result in the same two wires of each of the two pairs coming into proximity every other twist, consequently increasing the inductive and capacitive coupling between the wire pairs. By using different twist rates, this coupling, and thus the crosstalk between the pairs, can be dramatically reduced.

Appendix A

A.1 Plots of impedance of coated conductors

"Pure" metals follow basic relationships for impedance, and this behaviour is captured in Figure A.1. The impedance of coated metals also follows basic relationships, but the behaviour is clearly affected by the presence of the coating material, as can be seen in the following plots. In particular, when copper is coated with nickel, the behaviour is markedly different. Careful consideration of the effects of nickel coated on copper may lead to the ability to specify a nickel coating thickness to tailor the wiring response and introduce a level of control using the wire itself in addition to any behavioural alterations associated with embedded ferrite in the dielectric or distributed filtering in sensitive circuits (Figures A.2–A.5).

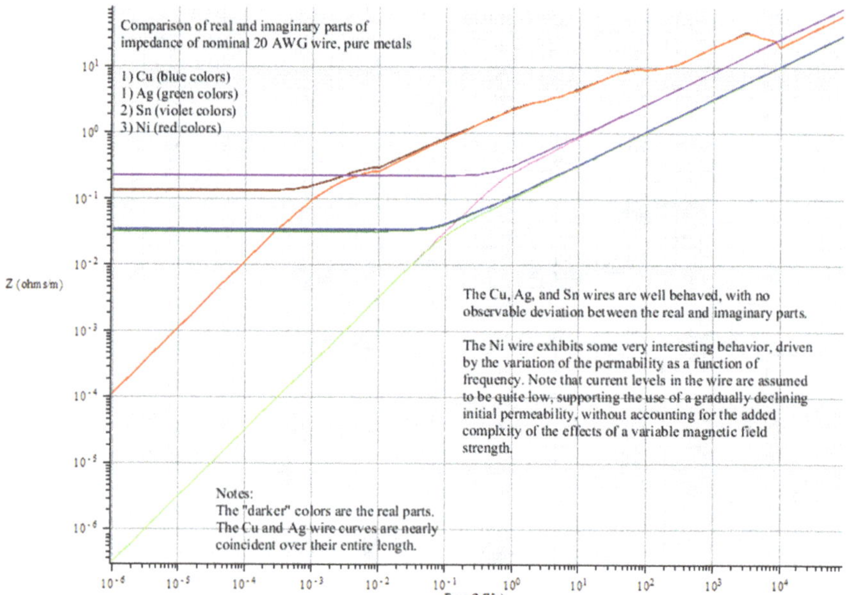

Figure A.1 Impedance behaviour of "pure" metals

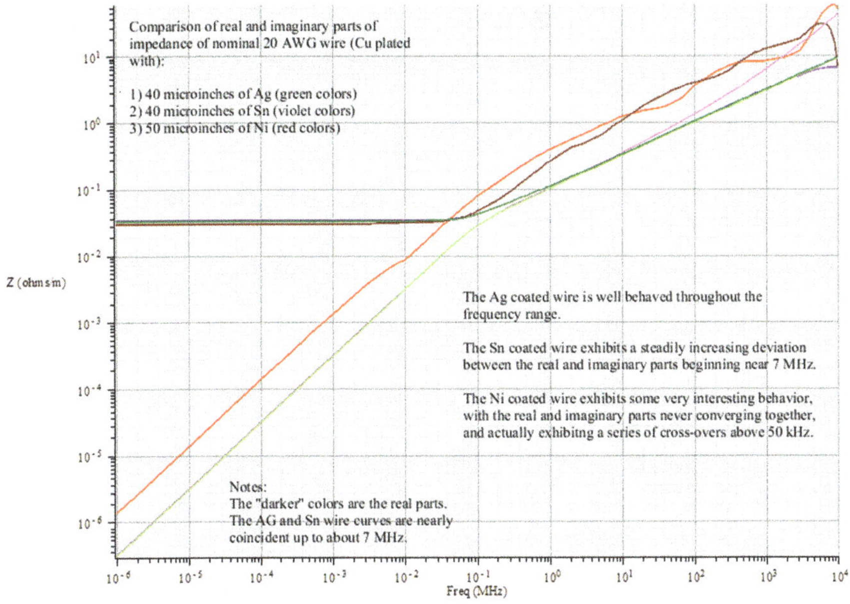

Figure A.2 Impedance behaviour of coated metals

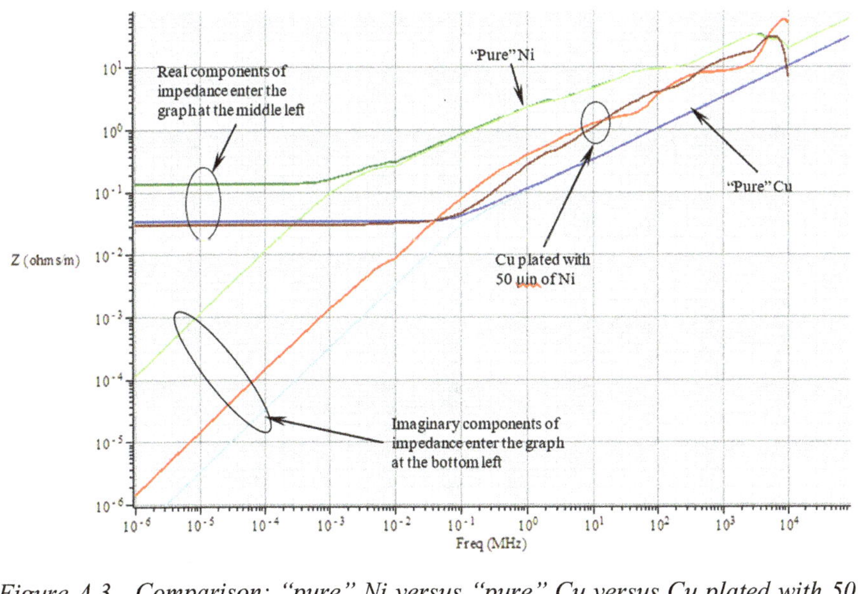

Figure A.3 Comparison: "pure" Ni versus "pure" Cu versus Cu plated with 50 μin of Ni

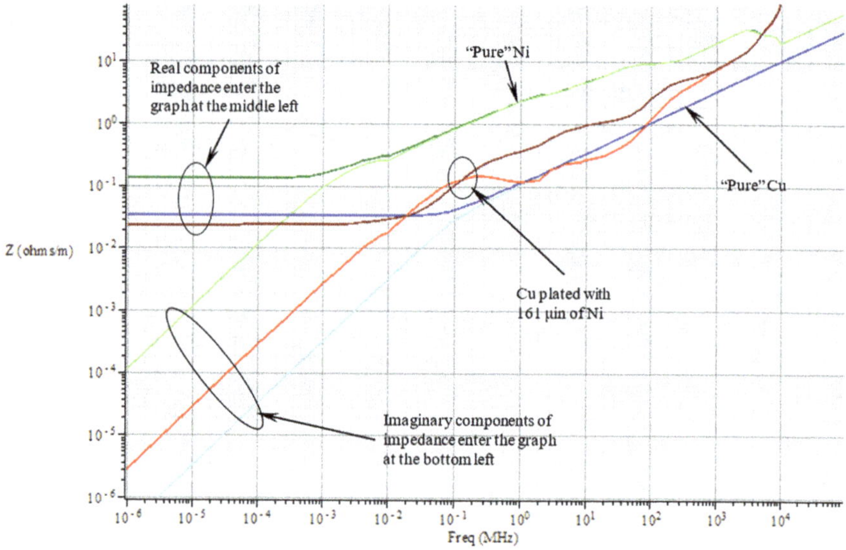

Figure A.4 Comparison: "pure" Ni versus "pure" Cu versus Cu plated with 161 µin of Ni

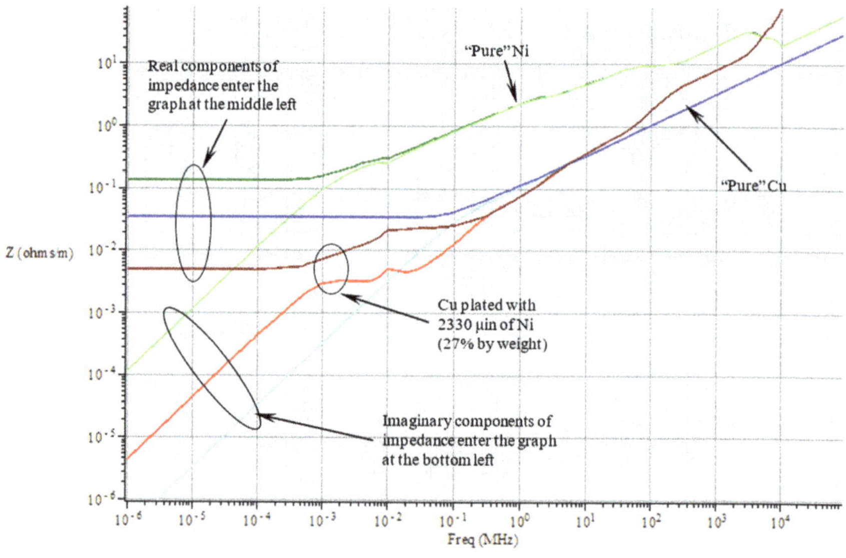

Figure A.5 Comparison: "pure" Ni versus "pure" Cu versus Cu plated with 2330 µin of Ni

References

[1] M. Matsuura, "Recent Advancement in Power-over-Fiber Technologies," *Photonics*, vol. 8, no. 8, p. 335, 2021. DOI:10.3390/photonics8080335.

[2] W. Thomson, "On the Theory of the Electric Telegraph," *Proceedings of the Royal Society of London (1854–1855)*, vol. 7, pp. 382–399, 1855, DOI: 10.1098/rspl.1854.0093.

[3] G. Kirchhoff, "On the Motion of Electricity in Wires," *London, Edinburgh and Dublin Philosophical Magazine and Journal of Science, 4th series*, vol. 13, no. 88, pp. 393–412, 1857. DOI:10.1080/14786445708642318

[4] J.J. Thomson, "Electrical Theories," *The Electrician*, vol. 27, pp. 138–139, 1886.

[5] O. Heaviside, "Electromagnetic Induction and Its Propagation – XXXIII," *The Electrician*, vol. 28, pp. 10–11, 1886.

[6] L.J. Greenstein and H.G. Tobin, "Analysis of Cable-coupled Interference," *IEEE Transactions on Radio Frequency Interference*, vol. 5, no. 1, pp. 43–55, 1963. DOI:10.1109/TRIF1.1963.4607266.

[7] R.J. Mohr, "Coupling between Open and Shielded Wires over a Ground Plane," *IEEE Transactions on Electromagnetic Compatibility*, vol. EMC-9, no. 2, pp. 34–45, 1967.

[8] R.J. Mohr, "Coupling between Lines at High Frequencies," *IEEE Transactions on Electromagnetic Compatibility*, vol. EMC-9, no. 3, pp. 127–129, 1967.

[9] R.J. Mohr, "Coupling between Open Wires over a Ground Plane," *IEEE Electromagnetic Compatibility Symposium Record*, Seattle, WA, USA, 1968, pp. 404–413, DOI:10.1109/TEMC.1968.4307156.

[10] J.R. Barnes, *Electronic System Design Interference and Noise Control Techniques*, Prentice Hall, Englewood Cliffs, NJ, 1987.

[11] J. Goedbloed, *Electromagnetic Compatibility*, Prentice Hall International (UK) Ltd, Hemel Hempstead, Hertfordshire, 1992.

[12] M.D. Heerema, *HP11949B Designing for Electromagnetic Compatibility*, Hewlett Packard, 1997.

[13] K.L. Kaiser, *Electromagnetic Compatibility Handbook*, CRC Press, Boca Raton, FL, 2005.

[14] H. Ott, *Electromagnetic Compatibility Engineering*, John Wiley & Sons, Hoboken, NJ, 2009.

[15] D.A. Weston, *Electromagnetic Compatibility Methods, Analysis, Circuits, and Measurement*, Taylor and Francis Group, Boca Raton, FL, 2017.

[16] C.R. Paul, "Solution of the Transmission-Line Equations for Three-Conductor Lines in Homogeneous Media," *IEEE Transactions on Electromagnetic Compatibility*, vol. EMC-20, no. 1, pp. 216–222, 1978.

[17] C.R. Paul, "On the Superposition of Inductive and Capacitive Coupling in Crosstalk-Prediction Models," *IEEE Transactions on Electromagnetic Compatibility*, vol. EMC-24, no. 3, pp. 335–343, 1982.

[18] C.R. Paul and J.W. McKnight, "Prediction of Crosstalk Involving Twisted-Pairs of Wires – Part I: A Transmission-Line Model for Twisted-Wire pairs," *IEEE Transactions on Electromagnetic Compatibility*, vol. EMC-21, no. 2, pp. 92–105, 1979.

[19] C.R. Paul and J.W. McKnight, "Prediction of Crosstalk Involving Twisted-Pairs of Wires – Part II: A Simplified Low-Frequency Prediction Model," *IEEE Transactions on Electromagnetic Compatibility*, vol. EMC-21, no. 2, pp. 105–114, 1979.

[20] P. Lefferson, "Twisted Magnet Wire Transmission Line," *IEEE Transactions on Parts, Hybrids, and Packaging*, vol. PHP-7, no. 4, pp. 148–154, 1971.

Chapter 5
Basic antenna theory

Antennas are devices that transform currents and voltages into radiated electromagnetic fields, or conversely transform electromagnetic fields into voltages and currents. This general description includes both intentional and unintentional antennas. Intentional antennas are those specifically designed to perform the task as described. Unintentional antennas are not so designed and include isolated metallic structures, and gaps and seams in electrical and system enclosures. Although these sorts of mechanical features can be troublesome, in comparison, interconnecting wires and wire harnesses that route electrical signals between subsystem components are more often the largest offenders. These wires and wire harnesses may act as monopole and dipole antennas to radiate undesirable emissions into sensitive antenna-connected receivers and couple external electromagnetic signals into sensitive circuitry. The reader is encouraged to consult an antenna design text (several are listed here [1–12]) for a more in-depth treatment of wire antennas.

It is useful to discuss how charges and currents radiate electromagnetic energy as a starting point. Discussion begins with a brief review of electrostatic and magnetostatic fields as defined by Maxwell's equations and as described by Coulomb's law and the Biot-Savart law. Table 5.1 shows Maxwell's equations for electrostatic and magnetostatic cases. For simple geometries, these relationships are relatively easy to solve. As the geometry for the configurations under study becomes more complex, these relationships become equally more complex, and the associated mathematical complexity makes solutions at best tedious. Use of potential functions facilitates solutions in such cases, and these are also briefly discussed as they pertain to electrostatic and magnetostatic fields. From there, discussion moves on to encompass changes to the relationships for electrostatic and magnetostatic fields once time variation is introduced, leading to a macroscopic

Table 5.1 Maxwell's equations for electrostatic and magnetostatic cases

Popular title	Integral form	Differential (point) form
Ampere's law	$\oint \boldsymbol{H} \cdot \boldsymbol{dl} = \sigma \int \boldsymbol{E} \cdot \boldsymbol{ds}$	curl $\boldsymbol{H} = \sigma \boldsymbol{E}$
Faraday's law	$\oint \boldsymbol{E} \cdot \boldsymbol{dl} = 0$	curl $\boldsymbol{E} = 0$
Gauss' law for magnetic fields	$\oint \boldsymbol{H} \cdot \boldsymbol{ds} = \boldsymbol{0}$	div $\boldsymbol{H} = 0$
Gauss' law for electric fields	$\oint \boldsymbol{E} \cdot \boldsymbol{ds} = \frac{1}{\varepsilon} \int \rho dv$	div $\boldsymbol{E} = \frac{\rho}{\varepsilon}$

description of how accelerated charges and time-varying currents create electromagnetic fields. After this, Hertzian and Fitzgerald dipoles are discussed, followed by a discussion of field regions surrounding these structures.

5.1 Review of electrostatic and magnetostatic fields

For the electrostatic case, the electric field associated with a charge density distribution can be solved directly from Coulomb's law, and is given by

$$\boldsymbol{E}(\boldsymbol{r}) = \frac{1}{4\pi\varepsilon_0}\int \rho(\boldsymbol{r}')\frac{\mathbf{1}}{|\boldsymbol{r}-\boldsymbol{r}'|^2}\boldsymbol{u}_r d^3r'; \quad \boldsymbol{u}_r = \frac{\boldsymbol{r}-\boldsymbol{r}'}{|\boldsymbol{r}-\boldsymbol{r}'|} \tag{5.1}$$

where

$\boldsymbol{u}_r =$ a unit vector directed towards the observation point from the charge distribution
$d^3r' = dx'dy'dz'$ is a three-dimensional volume element located at r'
$\boldsymbol{r} =$ the distance from the origin to the observation point, m
$\boldsymbol{r}' =$ the distance from the origin to the charge distribution, m

Equation (5.1) expresses that the electric field at some distance from the origin, denoted by $\boldsymbol{r}-\boldsymbol{r}'$, is associated with the sum total of the charges, q, in the charge distribution located at a distance from the origin, denoted by $\boldsymbol{r}'$.

Recognizing the product of the second and third terms in the integrand can be expressed as

$$\frac{\boldsymbol{r}-\boldsymbol{r}'}{|\boldsymbol{r}-\boldsymbol{r}'|^3} = -\nabla\left(\frac{\mathbf{1}}{|\boldsymbol{r}-\boldsymbol{r}'|}\right) \tag{5.2}$$

Equation (5.1) may also be expressed as

$$\boldsymbol{E}(\boldsymbol{r}) = -\nabla\left(\frac{1}{4\pi\varepsilon_0}\int \rho(\boldsymbol{r}')\frac{\mathbf{1}}{|\boldsymbol{r}-\boldsymbol{r}'|}d^3r'\right) \tag{5.3}$$

Based on the vector identity

$$\nabla \times \nabla\psi = 0 \tag{5.4}$$

substitution of (5.3) into Faraday's law for the electrostatic case

$$\nabla \times \boldsymbol{E} = \nabla \times \left[-\nabla\left(\frac{1}{4\pi\varepsilon_0}\int \rho(\boldsymbol{r}')\frac{\mathbf{1}}{|\boldsymbol{r}-\boldsymbol{r}'|}d^3r'\right)\right] = 0 \tag{5.5}$$

leads to

$$\boldsymbol{E} = -\nabla\left(\frac{1}{4\pi\varepsilon_0}\int \rho(\boldsymbol{r}')\frac{\mathbf{1}}{|\boldsymbol{r}-\boldsymbol{r}'|}d^3r'\right) = -\nabla\psi \tag{5.6}$$

Combining (5.6) with Gauss' law for electric fields yields

$$\nabla \cdot \boldsymbol{E} = \nabla \cdot (-\nabla \psi) = -\nabla^2 \psi = \frac{\rho}{\varepsilon} \tag{5.7}$$

$$= \nabla^2 \psi = -\frac{\rho}{\varepsilon} \tag{5.8}$$

Equation (5.8) is Poisson's equation for electrostatics. A solution to (5.8) using Green's theorem is given by

$$\psi(\boldsymbol{r}) = \frac{1}{4\pi\varepsilon_0} \int \frac{\rho(\boldsymbol{r}')}{|\boldsymbol{r} - \boldsymbol{r}'|} d^3 r' \tag{5.9}$$

which fully agrees with (5.6).

Following on the work of J.L. Lagrange, A. Legendre, and P.S. Laplace, G. Green's first published work, in 1828, was titled "An Essay on the Application of Mathematical Analysis to the Theories of Electricity and Magnetism". This major work contains the derivation of Green's theorem and applies the theorem, in conjunction with Green functions, to electrostatic problems. Green did not publish this paper in either of the mainstream journals of the time and it consequently failed to garner much response in the scientific community. William Thomson (Lord Kelvin) rediscovered Green's paper in 1845, and subsequently had it republished in *Crelle's Journal* in 1852 [13]. *Crelle's Journal*, or just *Crelle*, was the common name for the mathematics journal *Journal für die reine und angewandte Mathematik* (in English, *Journal for Pure and Applied Mathematics*). The journal is the oldest mathematics periodical still in existence and was established by August Leopold Crelle in 1826. Crelle served as the editor of the journal until his death in 1855. Green's work has proven invaluable in multiple areas of physics since then.

For the magnetostatic case, the magnetic field associated with a steady-current volume density can be solved directly from the Biot-Savart law, and is given by

$$\boldsymbol{B}(\boldsymbol{r}) = \frac{\mu_0}{4\pi} \int \frac{\boldsymbol{J}(\boldsymbol{r}') \times \boldsymbol{u}_r}{|\boldsymbol{r} - \boldsymbol{r}'|^2} d^3 r' \tag{5.10}$$

Equation (5.10) expresses that the magnetic field at some distance, denoted by $\boldsymbol{r}-\boldsymbol{r}'$, from a current volume density, denoted by $\boldsymbol{J}(\boldsymbol{r}')$, is associated with the sum total of the effects of all the infinitesimal current density elements within the volume containing the individual current densities.

Based on Gauss' law for magnetic fields, $\boldsymbol{B}$ must be equal to the curl of some other vector, expressed as

$$\boldsymbol{B} = \nabla \times \boldsymbol{A} \tag{5.11}$$

Combining (5.11) with Ampere's law for magnetostatic fields yields

$$\nabla \times \boldsymbol{B} = \nabla \times (\nabla \times \boldsymbol{A}) = \mu \boldsymbol{J} \tag{5.12}$$

Using the vector identity

$$\nabla \times (\nabla \times \boldsymbol{A}) = \nabla(\nabla \cdot \boldsymbol{A}) - (\nabla \cdot \nabla)\boldsymbol{A} = \nabla(\nabla \cdot \boldsymbol{A}) - \nabla^2 \boldsymbol{A}$$

Equation (5.12) becomes

$$\nabla(\nabla \cdot \boldsymbol{A}) - \nabla^2 \boldsymbol{A} = \mu \boldsymbol{J} \tag{5.13}$$

Imposing the Coulomb Gauge,

$$\nabla \cdot \boldsymbol{A} = 0 \tag{5.14}$$

yields

$$\nabla^2 \boldsymbol{A} = -\mu \boldsymbol{J} \tag{5.15}$$

Equation (5.15) is Poisson's vector equation for magnetostatics. Like (5.8), a solution to (5.15) using Green's theorem is given by

$$\boldsymbol{A}(\boldsymbol{r}) = \frac{\mu_0}{4\pi} \int \frac{\boldsymbol{J}(\boldsymbol{r}')}{|\boldsymbol{r} - \boldsymbol{r}'|} d^3 r' \tag{5.16}$$

Equations (5.1), (5.9), (5.10), and (5.16) facilitate the computation of static electric and magnetic fields associated with concentrations of charge and with the contributions of a constant flow of charge (steady, non-time variant current) around a closed loop. Maxwell's equations for these cases indicate that the electric and magnetic fields are not coupled and can be determined as independent solutions. Both Coulomb's law and the Biot-Savart law define fields that are proportional to the inverse of r^2 or greater. These static fields extend into space surrounding a region of stationary charge or a steady-state current that follows a closed loop, but do not radiate into space.

5.2 Effects of time variation on electrostatic and magnetostatic fields

With time variation, the static solutions no longer hold. In the static case, electric and magnetic fields can be determined as if the effects of all the charges or current elements are instantaneous. This is not true when the positions of charges and infinitesimal current elements are allowed to vary with time. It requires a finite amount of time for the effects of positional changes in charge or current elements to be felt at the point of observation.

Thus, it becomes necessary to account for the positions of these charges and current elements at earlier times. Table 5.2 illustrates time-varying Maxwell's equations. In this table, Faraday's law and Gauss' law for magnetic fields are source-free and are sometimes referred to as the homogeneous equations. In contrast, Ampere's law and Gauss' law for electric fields have charges and currents as sources and are sometimes referred to as the inhomogeneous equations.

Table 5.2 Time-varying versions of Maxwell's equations

Popular title	Integral form	Differential (point) form
Ampere's law	$\oint \boldsymbol{H} \cdot d\boldsymbol{l} = \int \boldsymbol{J} \cdot d\boldsymbol{s} + \frac{d}{dt}\int \boldsymbol{D} \cdot d\boldsymbol{s}$	$\nabla \times \boldsymbol{H} = \boldsymbol{J} + \frac{\partial \boldsymbol{D}}{\partial t}$
Faraday's law	$\oint \boldsymbol{E} \cdot d\boldsymbol{l} = -\frac{d}{dt}\int \boldsymbol{B} \cdot d\boldsymbol{s}$	$\nabla \times \boldsymbol{E} + \frac{\partial \boldsymbol{B}}{\partial t} = 0$
Gauss' law for magnetic fields	$\oint \boldsymbol{B} \cdot d\boldsymbol{s} = \boldsymbol{0}$	$\nabla \cdot \boldsymbol{B} = 0$
Gauss' law for electric fields	$\oint \boldsymbol{E} \cdot d\boldsymbol{s} = \frac{1}{\varepsilon}\int \rho dv$	$\nabla \cdot \boldsymbol{E} = \frac{\varrho}{\varepsilon}$

Note: The Del operator is used here for convenience.

Using (5.11) with Faraday's law from Table 5.2 for time-dependent cases yields

$$\nabla \times \boldsymbol{E} + \frac{\partial \boldsymbol{B}}{\partial t} = \nabla \times \boldsymbol{E} + \frac{\partial (\nabla \times \boldsymbol{A})}{\partial t} \tag{5.17}$$

which can be written as

$$\nabla \times \left(\boldsymbol{E} + \frac{\partial \boldsymbol{A}}{\partial t} \right) = 0 \tag{5.18}$$

Equation (5.18) defines a curl-free vector field and as such can be expressed as the gradient of a scalar potential as was shown in (5.6).

$$\boldsymbol{E} + \frac{\partial \boldsymbol{A}}{\partial t} = -\nabla \Phi \tag{5.19}$$

Equation (5.19) can be rearranged as

$$\boldsymbol{E} = -\nabla \Phi - \frac{\partial \boldsymbol{A}}{\partial t} \tag{5.20}$$

indicating that in a time-varying case, the electric field comprises a Coulombic field attributable to the presence of electric charge and a dynamic field that arises from time-varying currents. Equations (5.11) and (5.20) identically satisfy the two homogeneous equations in Table 5.2. Assuming we are in free space vacuum conditions, the two inhomogeneous equations can be expressed in terms of the A, Φ potentials as well.

Substituting (5.20) into Gauss' law for electric fields produces

$$\nabla \cdot \left(-\nabla \Phi - \frac{\partial \boldsymbol{A}}{\partial t} \right) = -\nabla^2 \Phi - \frac{\partial (\nabla \cdot \boldsymbol{A})}{\partial t} = \frac{\rho}{\varepsilon} \tag{5.21}$$

$$\nabla^2 \Phi + \frac{\partial (\nabla \cdot \boldsymbol{A})}{\partial t} = -\frac{\rho}{\varepsilon} \tag{5.22}$$

which reduces to (5.8) for the electrostatic case using (5.14).

Replacing $\boldsymbol{H}$ with $\boldsymbol{B}$ in Ampere's law

$$\nabla \times \boldsymbol{H} = \boldsymbol{J} + \frac{\partial \boldsymbol{D}}{\partial t} = \nabla \times \boldsymbol{B} = \mu \boldsymbol{J} + \mu\varepsilon \frac{\partial \boldsymbol{E}}{\partial t} \tag{5.23}$$

and then substituting (5.11) into (5.23) yields

$$\nabla \times \boldsymbol{B} = \nabla \times \nabla \times \boldsymbol{A} = \mu \boldsymbol{J} + \mu\varepsilon \frac{\partial \boldsymbol{E}}{\partial t} \tag{5.24}$$

Using the vector identity

$$\nabla \times (\nabla \times \boldsymbol{A}) = \nabla(\nabla \cdot \boldsymbol{A}) - (\nabla \cdot \nabla)\boldsymbol{A} = \nabla(\nabla \cdot \boldsymbol{A}) - \nabla^2 \boldsymbol{A}$$

Equation (5.24) becomes

$$\nabla(\nabla \cdot \boldsymbol{A}) - \nabla^2 \boldsymbol{A} = \mu \boldsymbol{J} + \mu\varepsilon \frac{\partial \boldsymbol{E}}{\partial t} \tag{5.25}$$

Substituting (5.20) into (5.25) then yields

$$\nabla(\nabla \cdot \boldsymbol{A}) - \nabla^2 \boldsymbol{A} = \mu \boldsymbol{J} + \mu\varepsilon \frac{\partial}{\partial t}\left(-\nabla \Phi - \frac{\partial \boldsymbol{A}}{\partial t}\right) \tag{5.26}$$

which after some additional manipulation becomes

$$\nabla^2 \boldsymbol{A} - \mu\varepsilon \frac{\partial^2 \boldsymbol{A}}{\partial t^2} - \nabla\left[(\nabla \cdot \boldsymbol{A}) + \mu\varepsilon \frac{\partial \Phi}{\partial t}\right] = -\mu \boldsymbol{J} \tag{5.27}$$

Equations (5.22) and (5.27) are identically equal to the four of Maxwell's equations, but they are still coupled. As was shown before, the imposition of a gauge choice can be used to simplify the equations. The use of gauges in these equations is known as gauge transformation. The fields as defined in Maxwell's equations can be produced by many different potentials. Thus, the definition of the potentials as related to the choice of gauge should be chosen carefully to simplify the equations to the greatest extent possible. The Coulomb gauge provides the advantage that the scalar potential is quite easy to calculate, having the same form as that found for the electrostatic case. The primary disadvantage is that the magnetic vector potential, given by (5.27), is much more difficult to calculate.

A different choice that makes the potentials symmetrical and completely uncoupled is the Lorenz gauge, given by

$$\nabla \cdot \boldsymbol{A} = -\mu\varepsilon \frac{\partial \Phi}{\partial t} \tag{5.28}$$

Substituting (5.28) into (5.22) and (5.27) yields the following two wave equations:

$$\nabla^2 \Phi - \mu\varepsilon \frac{\partial^2 \Phi}{\partial t^2} = -\frac{\rho}{\varepsilon} \tag{5.29}$$

$$\nabla^2 \boldsymbol{A} - \mu\varepsilon \frac{\partial^2 \boldsymbol{A}}{\partial t^2} = -\mu \boldsymbol{J} \tag{5.30}$$

As discussed previously, the effects of moving charges and currents are not observed instantaneously over some distance, but rather travel with a finite speed. According to Maxwell's predictions, in free space, that speed is equal to that of light. Thus, it becomes necessary to know the positions of charges and currents at a time previous to that at which it is desired to know the effects of those charges and currents. This is known as time retardation. The solutions of (5.29) and (5.30) using this time retardation approach may be written as

$$\Phi(\boldsymbol{r}, t) = \frac{1}{4\pi\varepsilon_0} \int \frac{\rho(\boldsymbol{r}', t - |\boldsymbol{r} - \boldsymbol{r}'|/c)}{|\boldsymbol{r} - \boldsymbol{r}'|} d^3 r' \tag{5.31}$$

$$\boldsymbol{A}(\boldsymbol{r}, t) = \frac{\mu_0}{4\pi} \int \frac{\boldsymbol{J}(\boldsymbol{r}', t - |\boldsymbol{r} - \boldsymbol{r}'|/c)}{|\boldsymbol{r} - \boldsymbol{r}'|} d^3 r' \tag{5.32}$$

Using (5.11), (5.20), (5.29), and (5.30), it is possible to arrive at wave equations for the electric and magnetic fields in free space with given charge and current densities:

$$\nabla^2 \boldsymbol{E} - \mu\varepsilon \frac{\partial^2 \boldsymbol{E}}{\partial t^2} = -\frac{1}{\boldsymbol{\varepsilon}}\left(-\nabla \rho - \mu\varepsilon \frac{\partial \boldsymbol{J}}{\partial t}\right) \tag{5.33}$$

$$\nabla^2 \boldsymbol{B} - \mu\varepsilon \frac{\partial^2 \boldsymbol{B}}{\partial t^2} = -\mu \nabla \times \boldsymbol{J} \tag{5.34}$$

Using (5.31) and (5.32), time-retarded solutions for (5.33) and (5.34) are given by

$$\boldsymbol{E}(\boldsymbol{r}, t) = \frac{1}{4\pi\varepsilon_0} \int \frac{\mathbf{1}}{|\boldsymbol{r} - \boldsymbol{r}'|} \left(-\nabla' \rho - \mu\varepsilon \frac{\partial \boldsymbol{J}}{\partial t'}\right)_{\text{ret}} d^3 r' \tag{5.35}$$

$$\boldsymbol{B}(\boldsymbol{r}) = \frac{\mu_0}{4\pi} \int \frac{\mathbf{1}}{|\boldsymbol{r} - \boldsymbol{r}'|} (\nabla' \times \boldsymbol{J})_{\text{ret}} d^3 r' \tag{5.36}$$

With some additional manipulations, (5.35) and (5.36) become

$$\boldsymbol{E}(\boldsymbol{r}, t) = \frac{1}{4\pi\varepsilon_0} \int \left\{ \frac{\frac{\widehat{\boldsymbol{R}}}{R^2} [\rho(\boldsymbol{r}', t')]_{\text{ret}} + \widehat{\boldsymbol{R}}}{cR} \left[\frac{\partial \rho(\boldsymbol{r}', t')}{\partial t'}\right]_{\text{ret}} - \frac{\mu\varepsilon}{R} \left[\frac{\partial \mathbf{J}(\boldsymbol{r}', t')}{\partial t'}\right]_{\text{ret}} \right\} d^3 r' \tag{5.37}$$

$$B(r) = \frac{\mu_0}{4\pi} \int \left\{ [\mathbf{J}(r', t')]_{\text{ret}} \times \frac{\widehat{R}}{R^2} + \left[\frac{\partial \mathbf{J}(r', t')}{\partial t'} \right]_{\text{ret}} \times \frac{\widehat{R}}{cR} \right\} d^3 r' \tag{5.38}$$

where

$$\boldsymbol{R} = \boldsymbol{r} - \boldsymbol{r}'$$

$$R = |\boldsymbol{r} - \boldsymbol{r}'|$$

$$\widehat{\boldsymbol{R}} = \boldsymbol{R}/R$$

Equations (5.37) and (5.38) are sometimes referred to as Jefimenko's generalizations [14,15] of the Coulomb and Biot-Savart laws. The important takeaways here are (1) that the electric field has three causative sources: the charge density, the time derivative of the charge density, and the time derivative of the electric current density; (2) the magnetic field has two causative sources: the electric current density and the time derivative of the electric current density; (3) both fields are dominated by the acceleration of charge represented by the 3rd term in in (5.37) and the 2nd term in (5.38); and (4) the 2nd and 3rd terms in (5.37) and the 3rd term in (5.38) represent the radiated portions of the fields as a function of $1/r$. In short, an electric field and a magnetic field are always simultaneously created by the same time-variable electric current, the same accelerated charge.

It can be shown that the propagating electric and magnetic fields of an accelerated charge can be expressed as

$$E_\theta = \frac{qa\sin\theta}{4\pi\epsilon_0 c^2 r} \tag{5.39}$$

$$H_\phi = \frac{qa\sin\theta}{4\pi cr} \tag{5.40}$$

where

$q =$ the charge, C

$a =$ the acceleration of the charge, m/s^2

$\theta =$ the angle of propagation, radians

$r =$ the distance to the point of measurement, m

The radiated power in the direction of r from the charge can be found using Poynting's theorem, to wit:

$$\boldsymbol{S} = E_\theta \times H_\phi = \left(\frac{qa\sin\theta}{4\pi\epsilon_0 c^2 r}\right)\left(\frac{qa\sin\theta}{4\pi cr}\right)\boldsymbol{u}_r \tag{5.41}$$

$$= \left(\frac{qa\sin\theta}{4\pi r}\right)^2 \left(\frac{\mu_0}{c}\right)\boldsymbol{u}_r \tag{5.42}$$

The total power can be determined using spherical coordinates and integrating $\boldsymbol{S}$ over a complete sphere at radius r:

$$P_T = \int_0^{\pi}\int_0^{2\pi}\left[\left(\frac{qa\sin\theta}{4\pi r}\right)^2\left(\frac{\mu_0}{c}\right)\boldsymbol{u}_r\right]\cdot\left(r^2\sin\theta d\phi d\theta\right)\boldsymbol{u}_r \tag{5.43}$$

$$= \frac{(qa)^2}{6\pi\epsilon_0 c^3} \tag{5.44}$$

which is the Larmor formula. Equation (5.44) clearly shows that the total power radiated by an accelerated charge depends on the magnitude of the charge and the charge's acceleration, plus universal constants. Note this radiation is not uniformly distributed but is rather strongest in the direction perpendicular to the direction of acceleration, decreasing to zero in the axis of acceleration.

The Larmor formula provides a simple mathematical description of the radiated electromagnetic power of an accelerated charge and was derived by J.J. Larmor in 1897 in his paper "On the Theory of the Magnetic Influence on Spectra; and on the Radiation from Moving Ions" [16].

Similarly, a steady-state harmonic time-varying current produces an acceleration of charge as the magnitude varies and the polarity reverses. The following processes are primarily responsible for electromagnetic radiation from wires:

1. the acceleration (deceleration) of charge caused by the imposition of an exciting electric field or a steady-state harmonic voltage;
2. the acceleration (deceleration) of charge caused by reflection from impedance discontinuities in the wire, such as a terminating load, a lack of termination, or an abrupt bend in the wire;
3. the acceleration (deceleration) of charge caused by smooth curves;
4. the acceleration (deceleration) of charge caused by dispersion along the length of the wire.

5.3 Hertzian dipoles

These same fundamental mechanisms of electromagnetic radiation for a single wire also work for a pair of wires that are not terminated at one end. Assume we have a pair of cylindrical conductors across which we have imposed a steady-state harmonic voltage source at one end, with the opposite end left unterminated. Application of the voltage source will create time-varying electric field lines tangential to the two wires of the wire pair. These tangent electric field lines of force accelerate charges towards the open end of the pair. The motion of the charges constitutes a current, and a magnetic field is created by this current. The electric and magnetic field lines form a wavefront that moves along the wire pair away from the source. The current density that is varying in time is creating a propagating electric and magnetic field between the wire pair. As the wavefront reaches the end of the wire pair, the charges are decelerated and reflected back towards the source. The radiation caused by this combines with the wave structure moving along the wire pair, and the resulting ensemble of waves moves out of the open end of the pair of conductors. The travelling electric and magnetic field lines will join in loops, electric lines to electric lines, and magnetic lines to magnetic lines. These loops will form a new detached wavefront that then propagates away from the end of the conductors into free space.

It is only a small step to move from an unterminated pair of conductors to a classic electric dipole structure, in which the ends of the dipole are not connected to each other, and the mechanism of radiation is the same as described for the unterminated pair of conductors. The simplest configuration is the elementary doublet, also known as a Hertzian dipole. The Hertzian dipole is a theoretical construct, comprising an infinitesimal length of conductor with an RF current exhibiting a constant amplitude along its length. The Hertzian dipole is often used in superposition to model longer, finite-length dipole structures. To find the solution for $\boldsymbol{E}$ and $\boldsymbol{H}$, it is easiest to solve first for the magnetic vector potential $\boldsymbol{A}$, then for $\boldsymbol{H}$, and then finally for $\boldsymbol{E}$.

The solution begins with stating the retarded time magnetic vector potential (5.32) in time harmonic form, allowing the use of phasors moving forward to simplify some of the mathematics. The magnetic vector potential in time harmonic form for a z-oriented infinitesimal linear source is given by

$$\tilde{\boldsymbol{A}}(\boldsymbol{r}) = \int \frac{\mu_0 I}{4\pi} \frac{e^{-jk|\boldsymbol{r}-\boldsymbol{r}'|}}{|\boldsymbol{r}-\boldsymbol{r}'|} dl = \frac{\mu_0 I}{4\pi} \boldsymbol{z} \int \frac{e^{-jk|\boldsymbol{r}-\boldsymbol{r}'|}}{|\boldsymbol{r}-\boldsymbol{r}'|} dl \tag{5.45}$$

where $\boldsymbol{z}$ = unit vector in the z-direction
I = the current on the dipole, A
l = the length of the dipole, m
k = the wavenumber, $\omega\sqrt{\mu_0 \varepsilon_0}$

If the incremental length $\Delta l \ll |\boldsymbol{r} - \boldsymbol{r}'|$, (5.45) can be simplified further to

$$\boldsymbol{A}(\boldsymbol{r}) = \frac{\mu_0 I \Delta l}{4\pi} \frac{e^{-jkr}}{r} \boldsymbol{z} \tag{5.46}$$

Using the transformation matrix from rectangular to spherical components

$$\begin{bmatrix} \tilde{A}_r \\ \tilde{A}_\theta \\ \tilde{A}_\phi \end{bmatrix} = \begin{bmatrix} \sin\theta\cos\phi & \sin\theta\sin\phi & \cos\theta \\ \cos\theta\cos\phi & \cos\theta\sin\phi & -\sin\theta \\ -\sin\phi & \cos\phi & 0 \end{bmatrix} \begin{bmatrix} \tilde{A}_x \\ \tilde{A}_y \\ \tilde{A}_z \end{bmatrix} \tag{5.47}$$

and recognizing that since the current element is z-directed,

$$\tilde{A}_x = \tilde{A}_y = 0 \tag{5.48}$$

Equation (5.46) can be expressed in spherical coordinates as

$$\tilde{A}_r = \tilde{A}_z \cos\theta = \frac{\mu_0 I \Delta l}{4\pi} \frac{e^{-jkr}}{r} \cos\theta \tag{5.49}$$

$$\tilde{A}_\theta = -\tilde{A}_z \sin\theta = \frac{-\mu_0 I \Delta l}{4\pi} \frac{e^{-jkr}}{r} \sin\theta \tag{5.50}$$

$$\tilde{A}_\phi = 0 \tag{5.51}$$

Using (5.11), the components of the $\boldsymbol{H}$ field are determined as

$$\tilde{H}_r = \tilde{H}_\theta = 0 \tag{5.52a}$$

$$\tilde{H}_\phi = \frac{k^2 I \Delta l}{4\pi} \sin\theta \left(\frac{j}{kr} + \frac{1}{(kr)^2} \right) e^{-jkr} \tag{5.52b}$$

and then applying Faraday's law, the components of the E field are determined as

$$\tilde{E}_r = \frac{2\eta_0 k^2 I \Delta l}{4\pi} \cos\theta \left(\frac{1}{(kr)^2} - \frac{j}{(kr)^3} \right) e^{-jkr} \tag{5.53a}$$

$$\tilde{E}_\theta = \frac{\eta_0 k^2 I \Delta l}{4\pi} \sin\theta \left(\frac{j}{kr} + \frac{1}{(kr)^2} - \frac{j}{(kr)^3} \right) e^{-jkr} \tag{5.53b}$$

$$\tilde{E}_\phi = 0 \tag{5.53c}$$

where $\eta_0 = \sqrt{\frac{\mu_0}{\varepsilon_0}}$ is the impedance of free space, Ω

5.4 Fitzgerald loops

Equations (5.52) and (5.53) express the electric and magnetic field components for an infinitesimal electric dipole. An infinitesimal magnetic loop, also known as a Fitzgerald loop or dipole, exhibits behaviour very similar to a Hertzian dipole. The name derives from analysis performed and published by George Francis Fitzgerald, who calculated the power radiated by a circulating current in two separate papers in 1883 [17,18]. The corresponding electric and magnetic fields for the infinitesimal magnetic loop can be most easily found using the principle of duality.

Duality in electromagnetics arises from the fact that EM fields may be described using two sets of quantities that correspond to each other. This correspondence allows substitution of a quantities in one set with their counterparts in the other set to generate valid equations. Tables 5.3 and 5.4 illustrate the

Table 5.3 Duality in electromagnetics

Electric sources ($J \neq 0$, $M = 0$)	Magnetic sources ($J = 0$, $M \neq 0$)
$\nabla \times \boldsymbol{E}_A = -j\omega\mu\boldsymbol{H}_A$	$\nabla \times \boldsymbol{H}_F = j\omega\varepsilon\boldsymbol{E}_F$
$\nabla \times \boldsymbol{H}_A = j\omega\varepsilon\boldsymbol{E}_A + \boldsymbol{J}$	$\nabla \times \boldsymbol{E}_F = -j\omega\mu\boldsymbol{H}_F - \boldsymbol{M}$
$\nabla \cdot \boldsymbol{D} = \rho$	$\nabla \cdot \boldsymbol{B} = \rho_m$
$\nabla \cdot \boldsymbol{B} = 0$	$\nabla \cdot \boldsymbol{D} = 0$
$\nabla \cdot \boldsymbol{J} = -j\omega\rho$	$\nabla \cdot \boldsymbol{M} = -j\omega\rho_m$
$\nabla^2\boldsymbol{A} + k^2\boldsymbol{A} = -\mu\boldsymbol{J}$	$\nabla^2\boldsymbol{F} + k^2\boldsymbol{F} = -\varepsilon\boldsymbol{M}$
$\boldsymbol{A} = \iiint \mu\boldsymbol{J}\frac{e^{-jkR}}{4\pi R}dv$	$\boldsymbol{F} = \iiint \varepsilon\boldsymbol{M}\frac{e^{-jkR}}{4\pi R}dv$
$\boldsymbol{H}_A = \frac{\nabla}{\mu} \times \boldsymbol{A}$	$\boldsymbol{E}_F = -\frac{\nabla}{\varepsilon} \times \boldsymbol{F}$
$\boldsymbol{E}_A = -j\omega\boldsymbol{A} + \frac{1}{j\omega\mu\varepsilon}\nabla(\nabla \cdot \boldsymbol{A})$	$\boldsymbol{H}_F = -j\omega\boldsymbol{F} + \frac{1}{j\omega\mu\varepsilon}\nabla(\nabla \cdot \boldsymbol{F})$

Table 5.4 Dual quantities for electric ($\boldsymbol{J}$) and magnetic ($\boldsymbol{M}$) sources

Electric sources ($J \neq 0, M = 0$)	Magnetic sources ($J = 0, M \neq 0$)
$\boldsymbol{E}_A$	$\boldsymbol{H}_F$
$\boldsymbol{H}_A$	$-\boldsymbol{E}_F$
$\boldsymbol{J}$	$\boldsymbol{M}$
$\boldsymbol{M}$	$-\boldsymbol{J}$
$\boldsymbol{A}$	$\boldsymbol{F}$
$\boldsymbol{F}$	$-\boldsymbol{A}$
ε	μ
μ	ε
η	$1/\eta$
$1/\eta$	η
k	k

correspondence between the two sets of equations. Note that in these tables, a fictitious magnetic current density denoted as $\boldsymbol{M}$ is the counterpart to the electric current density $\boldsymbol{J}$. Use of $\boldsymbol{M}$ facilitates symmetry in Maxwell's equations. This is used to simplify computations, as will become evident.

Applying duality to (5.52) and (5.53), the components of the $\boldsymbol{H}$ field are determined as

$$\tilde{E}_r = \tilde{E}_\theta = 0 \tag{5.54a}$$

$$\tilde{E}_\phi = -\frac{k^2 I_m \Delta l}{4\pi} \sin\theta \left(\frac{j}{kr} + \frac{1}{(kr)^2} \right) e^{-jkr} \tag{5.54b}$$

and

$$\tilde{H}_r = \frac{2k^2 I_m \Delta l}{4\pi\eta_0} \cos\theta \left(\frac{1}{(kr)^2} - \frac{j}{(kr)^3} \right) e^{-jkr} \tag{5.55a}$$

$$\tilde{H}_\theta = \frac{k^2 I_m \Delta l}{4\pi\eta_0} \sin\theta \left(\frac{j}{kr} + \frac{1}{(kr)^2} - \frac{j}{(kr)^3} \right) e^{-jkr} \tag{5.55b}$$

$$\tilde{H}_\phi = 0 \tag{5.55c}$$

The question might arise "What is I_m?" in (5.54) and (5.55). This question can be answered by establishing the equivalency between the fields generated by an electric current density $\boldsymbol{J}$ and the fields generated by a magnetic current density $\boldsymbol{M}$. Table 5.5 illustrates the corresponding relationships.

If the boundary conditions for $\boldsymbol{E}_1$ and $\boldsymbol{E}_2$ in the 3rd equations on each side of Table 5.5 are the same and the excitations for both fields fulfil the relationship

$$j\omega\mu\boldsymbol{J} = \nabla \times \boldsymbol{M} \tag{5.56}$$

then both the $\boldsymbol{E}$ and $\boldsymbol{H}$ fields are identical.

*Table 5.5 Field expressions for electric (**J**) and magnetic (**M**) current densities*

Electric current density	Magnetic current density
$-\nabla \times \boldsymbol{E}_1 = j\omega\mu\boldsymbol{H}_1$	$-\nabla \times \boldsymbol{E}_2 = j\omega\mu\boldsymbol{H}_2 + \boldsymbol{M}$
$\nabla \times \boldsymbol{H}_1 = j\omega\varepsilon\boldsymbol{E}_1 + \boldsymbol{J}$	$\nabla \times \boldsymbol{H}_2 = j\omega\varepsilon\boldsymbol{E}_2$
$\nabla \times \nabla \times \boldsymbol{E}_1 - \omega^2\mu\varepsilon\boldsymbol{E}_1 = -j\omega\mu\boldsymbol{J}$	$\nabla \times \nabla \times \boldsymbol{E}_2 - \omega^2\mu\varepsilon\boldsymbol{E}_2 = -\nabla \times \boldsymbol{M}$

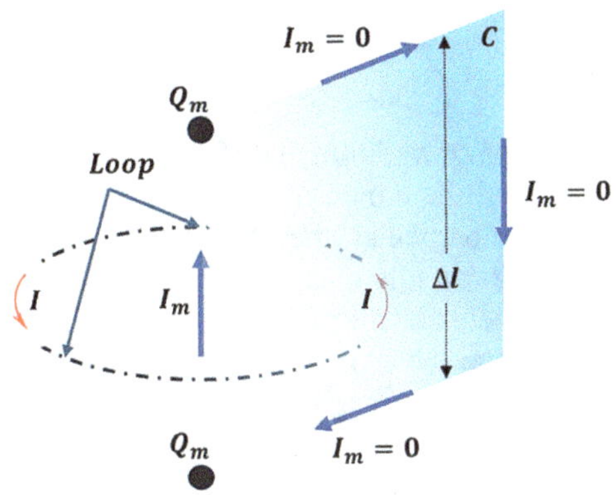

Figure 5.1 Current loop L passing through surface C

Illustrated in Figure 5.1 is a loop L of electric current I. Equation (5.56) can be rewritten for the loop in Figure 5.1 in integral form as

$$j\omega\mu \int \boldsymbol{J} \cdot d\boldsymbol{s} = \oint \boldsymbol{M} \cdot d\boldsymbol{l} \tag{5.57}$$

$$j\omega\mu I = M\Delta l \tag{5.58}$$

$$j\omega\mu IS = MS\Delta l = I_m\Delta l \tag{5.59}$$

where S is the area of C, m^2.

Substituting (5.59) into (5.54) and (5.55) yields

$$\tilde{E}_r = \tilde{E}_\theta = 0 \tag{5.60a}$$

$$\tilde{E}_\phi = \frac{-j\eta_0 k^3 IS}{4\pi} \sin\theta \left(\frac{j}{kr} + \frac{1}{(kr)^2} \right) e^{-jkr} \tag{5.60b}$$

and

$$\tilde{H}_r = \frac{2jk^3 IS}{4\pi} \cos\theta \left(\frac{1}{(kr)^2} + \frac{1}{j(kr)^3} \right) e^{-jkr} \tag{5.61a}$$

$$\tilde{H}_\theta = \frac{jk^2IS}{4\pi}\sin\theta\left(\frac{j}{kr}+\frac{1}{(kr)^2}-\frac{j}{(kr)^3}\right)e^{-jkr} \tag{5.61b}$$

$$\tilde{H}_\phi = 0 \tag{5.61c}$$

Equations (5.60) and (5.61) express the electric and magnetic field components, respectively, for an infinitesimal magnetic loop, also known as a Fitzgerald dipole. These equations assume the loop has a single turn and the radius $a < 0.1\lambda$.

5.5 Field regions

The components in (5.52)–(5.55), (5.60), and (5.61) contain terms that depend on the magnitude of the current, the dimension of the source, the frequency, and the distance from the source. In particular, the distance terms $1/r^2$ and $1/r^3$ illustrate the field magnitude decreases rapidly over distance. These terms define the regions known as the reactive near field and the radiative near field and may be neglected at relatively large distances from the source. The radial components, $\tilde{E}_\theta/\tilde{H}_\phi$ for the electric source and $\tilde{E}_\phi/\tilde{H}_\theta$ for the magnetic source, are orthogonal to each other and their direction of propagation, depend on distance as $1/r$, are related to each other by the impedance of the medium through which they are propagating, and describe a transverse electromagnetic (TEM) wave. These terms define the region known as the far field.

In the reactive near field (very close to the antenna), the relationship between the ***E*** and ***H*** fields is generally quite complicated. In this region, energy slowly changes between electrostatic, magnetostatic, and electric and magnetic inductive effects. Current flowing in the antenna creates a purely magnetic component in the near field, which then collapses as the antenna current reverses direction, transferring the field's magnetic energy back to electrons in the antenna current. This returns energy to the antenna in a regenerative way so that it is not lost. A similar process happens as electric charge accumulates at either end of the dipole structure. The charge accumulation builds up, collapses, and reverses, building, collapsing, and rebuilding an oppositely polarized electrostatic field with each cycle of the applied AC signal. The electric field thus created pushes charge in the opposite direction, also transferring energy back to the antenna current in a regenerative way, so it is not lost. Components of either field may dominate at any one time or spatial position, and the opposite relationship may dominate only a short time or distance away. This makes measuring the true power density over time and spatial location in this region problematic, because not only the magnitudes of ***E*** and ***H*** need to be measured but also their phase relationship as well in every point of space. Locating any sort of probe into the region will distort the field and degrade the measurement results, making determination of the fields more difficult still. Because of this energy storage and replenishment process, if either of the inductive or electrostatic effect in the reactive near field transfers any field energy to charge in a different (nearby) conductor, this energy is lost to the primary antenna.

The radiative near field, or Fresnel zone, is far enough from the antenna that the cycling of electrostatic and magnetostatic energy falls out of phase with that of the original antenna signal and is no longer able to return stored magnetic or electric energy produced by antenna currents or charges. The behaviour of the electric and magnetic fields remains complex in the radiative near field, with inductive effects predominant. Radiant components in this region are not yet orthogonal as they are in the far field. Metal objects in this portion of the near field, such as aluminium supports or structural steel beams, can act as antennas themselves by coupling to the inductive and capacitive elements in the radiative near field and then re-radiating a portion of the energy emanating from the antenna, thereby forming a new radiating surface.

The far field, also known as the radiation zone or Fraunhofer region, is dominated by electric or magnetic fields that are orthogonal to each other and to the direction of propagation. In the far-field region of an antenna, radiated power decreases as the square of distance, and absorption of the radiation does not return any energy back to the antenna. In the far-field region, each part of the EM field (electric and magnetic) is associated with currents and charge densities that are the source of both fields, and the ratio of electric and magnetic field intensities is simply the wave impedance in the medium.

As an illustration of these regions, the power density of the fields for the infinitesimal dipole can be found using the complex Poynting Vector (discussed in section 3.2.9), using all terms of (5.52b) and (5.53a and 5.53b) and substituting into (5.62), as follows:

$$\boldsymbol{S} = \frac{1}{2}\boldsymbol{E} \times \boldsymbol{H}^* \tag{5.62}$$

where $\boldsymbol{E}$ and $\boldsymbol{H}$ are peak phasors, not rms, leading to the factor of 1/2 in power expressions.

$$\begin{aligned}
\boldsymbol{S} &= \frac{1}{2}\boldsymbol{E} \times \boldsymbol{H}^* \\
&= \frac{1}{2}(E_r\boldsymbol{r} + E_\theta\boldsymbol{\theta}) \times \left(H_\phi^*\boldsymbol{\phi}\right) && (5.63)\\
&= \frac{1}{2}\left(E_\theta H_\phi^*\boldsymbol{r} - E_r H_\phi^*\boldsymbol{\theta}\right) && (5.64)\\
&= \frac{\eta_0}{2}\left(\frac{k^2 I\Delta l}{4\pi}\right)^2\left\{\sin^2\theta\left[\frac{1}{(kr)^2} - \frac{j}{(kr)^5}\right]\boldsymbol{r} + \sin 2\theta\left[\frac{j}{(kr)^3} + \frac{j}{(kr)^5}\right]\boldsymbol{\theta}\right\} && (5.65)
\end{aligned}$$

The power density in (5.65) has both real and imaginary components. The real component is radially directed and is a function of $1/r^2$, both characteristics of propagating, radiated power. The imaginary components are representative of both the reactive and radiative regions. These components of power density are wholly imaginary, are dominant for small values of kr, and do not contribute to the real radiated fields. The reactive power density, a function of $1/r^5$, merely changes between outward and inward directions to form a standing wave at a rate of twice per cycle. Note that reactive and radiative power densities do exist in the far field, but their magnitudes are negligible compared to that of the real component of power density. Note also that real power density does exist in the reactive and radiative near fields, but its magnitude is negligible compared to those of the imaginary components in these regions.

Equations (5.52)–(5.55), (5.60), and (5.61) have some very interesting properties, to wit:

(a) The distance $r = \lambda/2\pi$ is known as the radian length. The radian length is the radius of the radian sphere, a hypothetical sphere whose radius is equal to one radian length as measured from the centre of an antenna that is much smaller than the radian sphere. The radian length is directly proportional to wavelength and is also equal to $kr = 1$.
(b) The radian sphere contains the volume occupied mainly by the stored energy of the antenna's electric and magnetic fields. Radiated power density is greater than reactive power density external to the radian sphere and begins to dominate as r increases beyond the radian length. The radian sphere essentially defines the transition between stored energy pulsating primarily in the $\pm\theta$ direction. Similar behaviour, where the power density near the antenna is primarily reactive and far away is primarily real, is exhibited by all antennas, although not exactly at the radian length.
(c) At the radian length, the magnitudes of the two terms within the parentheses in (5.52b) and (5.53a) are identical, indicating that the real and imaginary parts of the fields are of equal magnitude at this distance from the centre of the antenna. At this same distance, all three terms within the parentheses of (5.53b) are identical, and the only term that contributes to the total field is the second, because the first and third terms cancel each other.
(d) At distances greater than the radian length, $r > \lambda/2\pi$, or equivalently $kr > 1$, the first term within the parentheses of (5.52b) and (5.53a) is greater than the magnitude of the second term and begins to dominate as r grows ever larger than $\lambda/2\pi$. In similar fashion, the first term within the parentheses of (5.53b) begins to dominate over the second and third terms as r grows ever larger than $\lambda/2\pi$.
(e) Note that all these properties apply equally to (5.54)–(5.55), (5.60), and (5.61).

Additional insight can be gleaned by examining the behaviour of the wave impedance as a function of distance from an infinitesimal source. Recall the definition of the wave impedance is the ratio of the electric and magnetic fields in a medium. The wave impedance for the infinitesimal dipole can be found by simply dividing the $\boldsymbol{E}$ and $\boldsymbol{H}$ fields, as shown in (5.66) and (5.67).

$$|Z_{Hertzian}| = \left|\frac{\tilde{E}_\theta}{\tilde{H}_\phi}\right| = \left|\frac{\frac{\eta_0 k^2 I\Delta l}{4\pi}\sin\theta\left(\frac{j}{kr} + \frac{1}{(kr)^2} - \frac{j}{(kr)^3}\right)e^{-jkr}}{\frac{k^2 I\Delta l}{4\pi}\sin\theta\left(\frac{j}{kr} + \frac{1}{(kr)^2}\right)e^{-jkr}}\right| \tag{5.66}$$

$$= \left| \frac{\eta_0 \left(\frac{j}{kr} + \frac{1}{(kr)^2} - \frac{j}{(kr)^3} \right)}{\left(\frac{j}{kr} + \frac{1}{(kr)^2} \right)} \right| \tag{5.67}$$

An entirely similar relationship applies to the infinitesimal Fitzgerald loop,

$$|Z_{Fitzgerald}| = \left| \frac{\tilde{E}_\phi}{\tilde{H}_\theta} \right| = \left| \frac{\frac{-j\eta_0 k^3 IS}{4\pi} \sin\theta \left(\frac{j}{kr} + \frac{1}{(kr)^2} \right) e^{-jkr}}{\frac{jk^3 IS}{4\pi} \sin\theta \left(\frac{j}{kr} + \frac{1}{(kr)^2} - \frac{j}{(kr)^3} \right) e^{-jkr}} \right| \tag{5.68}$$

$$= \left| \frac{-\eta_0 \left(\frac{j}{kr} + \frac{1}{(kr)^2} \right)}{\left(\frac{j}{kr} + \frac{1}{(kr)^2} - \frac{j}{(kr)^3} \right)} \right| \tag{5.69}$$

Equations (5.67) and (5.69) are plotted as a function of kr in Figure 5.2.

Note the approximate boundaries expressed as multiples of the wavelength, λ. The boundary between the reactive and the radiative near fields occurs at the distance $r = \lambda/2\pi$, or $r = 0.16\lambda$. At the distance $r = 1.6\lambda$, the impedances of the Hertzian dipole and the Fitzgerald loop converge to the impedance of free space, $\eta_0 = 120\pi$. A more conservative estimate for this convergence is at the distance $r = 5\lambda$, at which point

$$\left| \frac{\tilde{E}_{\theta FarFld} - \tilde{E}_\theta}{\tilde{E}_{\theta FarFld}} \right| = \left| \frac{\left[\frac{\eta_0 k^2 I\Delta l}{4\pi} \sin\theta \left(\frac{j}{kr} \right) e^{-jkr} \right] - \left[\frac{\eta_0 k^2 I\Delta l}{4\pi} \sin\theta \left(\frac{j}{kr} + \frac{1}{(kr)^2} - \frac{j}{(kr)^3} \right) e^{-jkr} \right]}{\left[\frac{\eta_0 k^2 I\Delta l}{4\pi} \sin\theta \left(\frac{j}{kr} \right) e^{-jkr} \right]} \right|$$

$$\leq 1/\sqrt{1000} \tag{5.70}$$

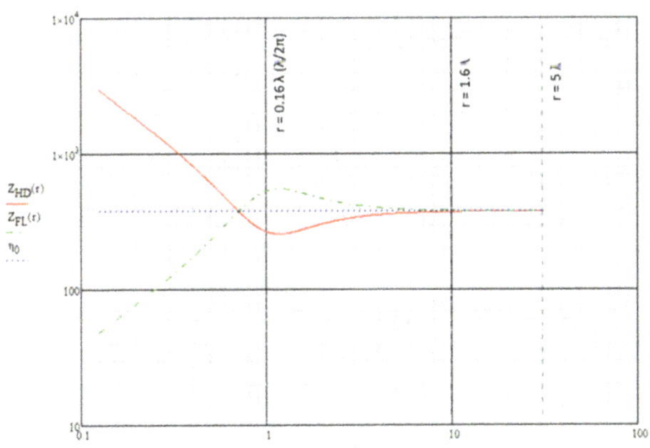

Figure 5.2 Wave impedance curves for Hertzian dipole and Fitzgerald loop antennas

As discussed in the foregoing, the volume of space surrounding an infinitesimal element is divided into three separate regions or zones, the boundaries of which are somewhat arbitrarily determined and are known as the reactive near field, the radiating near field, and the far field. For the infinitesimal elements discussed, the dimensions of the elements and the distances from the different parts of the elements are essentially the same, so that as the distance from the element increases, amplitude and phase errors remain relatively constant. As the dimensions of the source grow larger, the relationships used for the infinitesimal dipole structures alter, most notably in the current distribution. For line currents and general applications, accepted practice suggests the phase error be kept to less than $\pi/8$. This phase error is embodied in the phase factor e^{-jkR} terms in each of the element field expressions in (5.52)–(5.61).

Recall the distance term $R = |\boldsymbol{r}-\boldsymbol{r}'|$ is given in Cartesian coordinates as

$$R = \sqrt{(x-x')^2 + (y-y')^2 + (z-z')^2} \tag{5.71}$$

where the primed coordinates denote the distance from the source to the origin of the coordinate system, and the un-primed coordinate denotes the distance from the origin to the observation point. Assuming the wire used in the antenna elements is infinitesimally thin and thus varies only in the z-direction yields an expression for R:

$$R = \sqrt{x^2 + y^2 + (z-z')^2} \tag{5.72}$$

which becomes, using the law of Cosines and rearranging,

$$R = \sqrt{r^2 + (z'2 - 2rz'\cos\theta)} \tag{5.73}$$

Expanding (5.64) using the binomial expansion yields

$$R = (r^2)^{\frac{1}{2}} + \frac{1}{2}(r^2)^{-\frac{1}{2}}(z' - 2rz'\cos\theta) - \frac{1}{8}(r^2)^{-\frac{3}{2}}(z' - 2rz'\cos\theta)^2 + \cdots \tag{5.74}$$

Equation (5.74) can be simplified to

$$R \approx r - z'^{\cos\theta} + \frac{z'^2\sin^2\theta}{2r} + \frac{z'^3\cos\theta\sin^2\theta}{2r^2} + O^3 \tag{5.75}$$

where O^3 represents terms in $1/r^3$ and greater. These terms may be neglected assuming $r \gg z'$. Equation (5.75) will now be used to define the boundaries of the three different zones, the reactive near field, the radiative near field (aka the Fresnel region), and the far field (aka the Fraunhofer region).

For the far field, only the first two terms of (5.75) are used. The most significant error term neglected by this approach is

$$e(r) = \frac{z'^2\sin^2\theta}{2r} \tag{5.76}$$

The maximum of this term occurs when $\sin\theta = 1$ which occurs when $\theta = \pi/2$, and z' is maximum. (5.76) then becomes

$$e_{\max}(r) = \frac{\left(z'_{\max}\right)^2}{2r} \tag{5.77}$$

From (5.45), the length of the dipole is l. In (5.77) then, $z'_{\max} = l/2$. Substituting this into (5.77), multiplying by the wavenumber $k = 2\pi/\lambda$, and setting the equation equal to or less then the $\pi/8$ criteria yields

$$k\frac{(l/2)^2}{2r} \leq \frac{\pi}{8} \tag{5.78}$$

Solving (5.78) for r yields the far field boundary,

$$r \geq \frac{2l^2}{\lambda} \tag{5.79}$$

Two caveats apply for use of (5.79) in order that the approximation used in the binomial relationship remains valid. First, $(l/r) \ll 1$, and second, $r \gg \lambda$.

The far-field boundary also denotes the outer boundary of the Fresnel region. To determine the inner boundary, it is necessary to consider in the error calculation the third term in (5.75), as shown here

$$R \approx r - z'^{\cos\theta} + \frac{z'^2\sin^2\theta}{2r} \tag{5.80}$$

Now the most significant error neglected is

$$e(r) = \frac{z'^3\cos\theta\sin^2\theta}{2r^2} \tag{5.81}$$

As before, it is necessary to determine the angles of θ for which (5.81) is maximum. This is done by taking the partial derivative of (5.81) with respect to θ and setting the result equal to zero.

$$\frac{\partial e(r)}{\partial\theta} = \frac{z'^3}{2r^2}\sin\theta\left(-\sin^2\theta + 2\cos^2\theta\right) = 0 \tag{5.82}$$

Here, $\theta = 0$ is a trivial solution and represents the minimum error. The maximum error occurs when the term in parentheses, $\left(-\sin^2\theta + 2\cos^2\theta\right) = 0$, which occurs when

$$\theta = \tan^{-1}\left(\pm\sqrt{2}\right) \tag{5.83}$$

The distance r at which the maximum error equals $\pi/8$ can be found from

$$ke(r) = \frac{kz'^3}{2r^2}\cos\theta\sin^2\theta\Bigg|_{\substack{Z' = Z'_{\max} \\ \theta = \tan^{-1}(\pm\sqrt{2})}} \leq \frac{\pi}{8} \tag{5.84}$$

$$= \frac{2\pi(l/2)^3}{\lambda 2r^2} \frac{1}{\sqrt{3}} \frac{2}{3} \leq \frac{\pi}{8} \tag{5.85}$$

or

$$r \geq 0.62\sqrt{\frac{l^3}{\lambda}} \tag{5.86}$$

Equation (5.86) represents the inner boundary of the radiative near-field region. Distances less than (5.86) fall into the reactive near-field region.

For a line source whose length falls between 0.32λ and 2.5λ, the distance $r = 5l$ corresponds to a 10% error in the distance to the far field for a line source. Recall that R is the distance from the source location to the point of observation and r is the distance from the centre of the line source to the point of observation. Figure 5.3 illustrates the geometry.

In Figure 5.3, $R = r - z'\cos\theta$ has a maximum value for $\theta = \pi/2$. If $z' = l/2$ and $\theta = \pi/2$, then $R = r - l/2$. For a 10% error in the distance to the far field for the line source, we have

$$R = r - l/2 = 0.9r \tag{5.87}$$

which leads to $r = 5l$.

Figures 5.4 and 5.5 illustrate graphically the various approximations for the field boundaries for a line source of length l as discussed in the foregoing.

To put the results into some context, the wavelength at 30 MHz is 10 metres. A wire harness with a length of 0.32λ, by definition a Hertzian element, is 3.2 metres long, and the boundary between the reactive and radiating near fields is 1.6 metres from the harness, with the far field for this harness at a minimum distance of 16 metres. At 300 MHz, corresponding lengths are 1 metre, 320 mm, 160 mm, and 1.6 metres. At 3 GHz, corresponding lengths are 0.1 metres, 32 mm, 16 mm, and 160 mm. Any other wire, harness, or structure in near proximity will most likely be in the near radiating or reactive field of the harness of interest, potentially leading

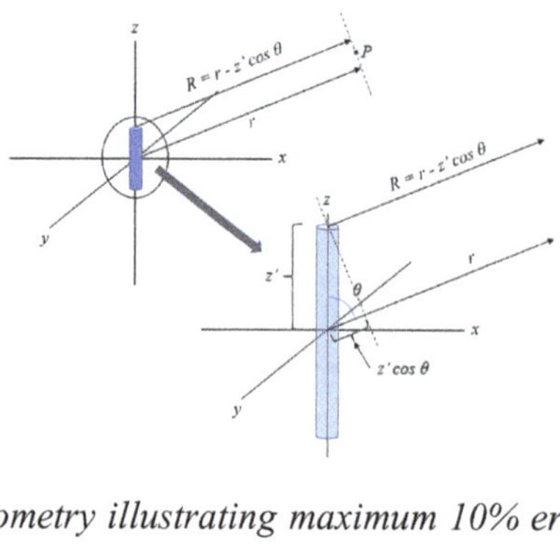

Figure 5.3 Geometry illustrating maximum 10% error construction

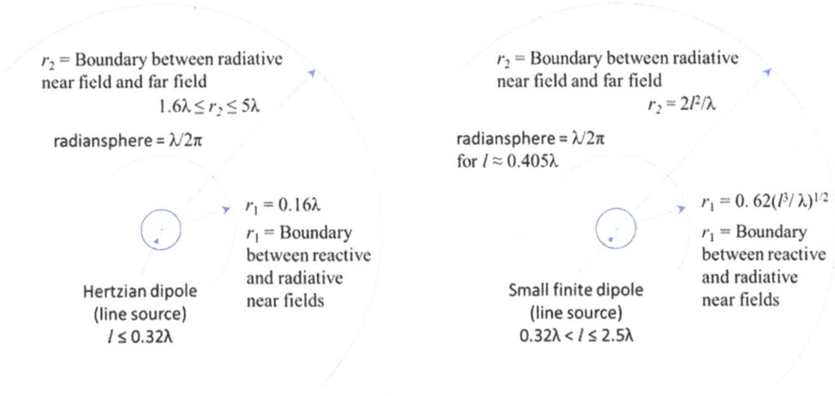

Figure 5.4 Approximate field boundaries for Hertzian dipole and finite length dipole

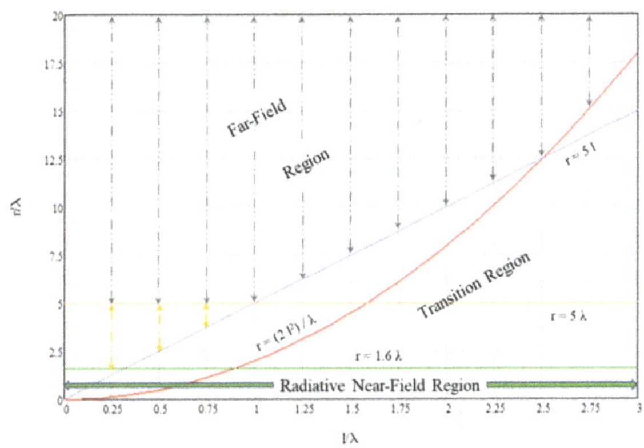

Figure 5.5 Generalized approximate field boundaries as a function of source dimension and radial distance

to capacitive crosstalk coupling, inductive crosstalk coupling, or a combination of both.

5.6 Wires and cables as emitters and receivers of electromagnetic fields

Wires and cables can produce radiated electric and magnetic fields and are also quite capable of acting as receivers of radiated electric and magnetic fields. It might seem intuitive that wires and cables can act as emitters, although many engineers can't explain why other than to say something along the lines of "Well, they're just antennas, right?". While this is in fact basically true, unfortunately just accepting this doesn't generally carry the EMC engineer responsible for correcting the non-compliant behaviour very far towards resolving a conducted or radiated emissions non-compliance.

Hertzian dipoles or Fitzgerald loops are often used to estimate the potential for radiated emissions from wire structures. Before we begin, it may perhaps be a good idea to define what is meant by the terms Hertzian dipole and Fitzgerald loop.

A Hertzian dipole is generally defined as a straight infinitesimally thin filament of current with length which is very small relative to a wavelength, but not precisely zero. In physics, the Hertzian dipole is in fact an elementary doublet rather than a physical antenna design: It is an idealized tiny segment of conductor carrying an RF current with constant amplitude and direction along its entire (short) length. One of the most useful aspects of a Hertzian dipole is that a physically realizable antenna can be modelled as the combination of many Hertzian dipoles laid end-to-end. The name originates with the work of Heinrich Rudolf Hertz, who created what is most likely the first functional dipole structure in his laboratory ca. 1886 [19] in his pursuit of validating the theoretical work of James Clerk Maxwell.

An infinitesimal magnetic loop, also known as a Fitzgerald loop or dipole, exhibits behaviour very similar to a Hertzian dipole. The name derives from analysis performed and published by George Francis Fitzgerald, who calculated the power radiated by a circulating current in two separate papers in 1883 [17,18].

Assuming we have a cable or wire that we wish to treat as a Hertzian dipole, we require the cable or wire to be (1) sufficiently short such that the distance vectors from any point on the cable or wire to some distant measurement point are approximately parallel; (2) the current distribution (phase and magnitude) is constant along the length of the cable or wire; and (3) the distant measurement point is in the far field of the dipole. A small linear dipole exhibits similar behaviour, and we can estimate both an electric and magnetic field radiated by the cable or wire of interest. In this case, the radiated electric field will be polarized along the length of the cable or wire, and the magnetic field will be polarized orthogonally to the electric field. Typically, the fields we are most interested in when we talk about radiated electric emissions from wires and cables are generated by common mode currents. These electric fields are readily modelled by those radiated by a Hertzian dipole.

If, on the other hand, we have some small cable or wire loop that we wish to treat as a Fitzgerald loop, we require the following: (1) the loop has a circumference $< \lambda/4$; (2) the current distribution (phase and magnitude) is constant around the circumference of the cable or wire loop; and (3) the distant measurement point is in the far field of the loop. If these constraints are met, we can estimate both an electric and magnetic field radiated by the cable or wire loop of interest. In this case, the radiated electric field will be polarized along the plane of the cable or wire loop, and the magnetic field will be polarized orthogonally to the electric field. Typically, the fields we are most interested in when we talk about radiated magnetic emissions from wires and cables are generated by differential mode currents. These magnetic fields are readily modelled by those radiated by a Fitzgerald loop.

Table 5.6 Radiated fields of small linear dipole and small loop

Small linear dipole	Small loop
$E_\theta = \frac{j60\pi I_c \sin\theta l}{r\lambda}$	$E_\phi = \frac{120\pi^2 I_D \sin\theta A}{r\lambda^2}$
$H_\phi = \frac{jI_c \sin\theta l}{2r\lambda}$	$H_\theta = \frac{\pi I_D \sin\theta A}{r\lambda^2}$

Table 5.7 Simplified fields of Table 5.6

Small linear dipole	Small loop
$\lvert E_\theta \rvert = 6.28 \times 10^{-7} \frac{I_c l f}{r}$	$\lvert E_\phi \rvert = 131.6 \times 10^{-16} \frac{I_D A f^2}{r}$
$\lvert H_\phi \rvert = 1.67 \times 10^{-9} \frac{I_c l f}{r}$	$\lvert H_\theta \rvert = 3.49 \times 10^{-17} \frac{I_D A f^2}{r}$

The radiated electric and magnetic fields from a small linear dipole and a small loop in free space [3,4] are given here in Table 5.6. These equations are basic expressions for small linear dipoles and small loops in free space. They are intended to be used as a starting point, enabling some simple calculations to inform the situation and provide additional engineering data that can be used for further assessment and evaluation.

In both types, the maxima will occur when $\theta = \pi/2$, allowing some simplification of the equations in Table 5.6 as shown in Table 5.7.

To reduce small linear-dipole-radiated fields, we need to

1. reduce the magnitude of any common mode currents;
2. reduce the frequency or harmonic content of any currents;
3. reduce the cable or wire length.

To reduce small loop radiated fields, we need to

1. reduce the magnitude of any differential mode currents;
2. reduce the frequency or harmonic content of any currents;
3. reduce the cable or wire loop area.

I should mention here that Paul has also developed an approach in which he considers the radiated fields of a pair of wires carrying both differential and common mode currents. For details, I refer the reader to his work.

5.7 Wires and cables as emitters and receivers of electromagnetic fields

Coupled power onto wires and cables from an incident radiated field is often modelled simply as a $\lambda/2$ dipole. The relationship between coupled power and the incident power density on a $\lambda/2$ dipole or loop antenna is given by

$$P_r = A_e P_D = \frac{1.64\lambda^2}{4\pi} P_D = 0.13\lambda^2 P_D \tag{5.88}$$

where

A_e = the effective aperture of the dipole in square metres
P_r = the maximum coupled power in watts
P_D = the incident power density in watts per square metre
λ = the wavelength of the of the incident power density in metres

This approach also assumes the cable or wire is sufficiently far from the structure to eliminate use of a Field to Wire (FTW) approach (aka a single wire to ground (SWAG) plane method, discussed later in this chapter).

As an example, assume we have a radiated field with magnitude of 2 V/m incident on platform cabling. The average length of the cabling is 1 metre, equal to $\lambda/2$ for a 150 MHz signal. We will also assume the cabling is in the far field of the incident radiation such that

$$P_D = E^2/\zeta \tag{5.89}$$

where

P_D = the incident power density in watts per square metre
E = the electric field strength in V/m
$\zeta = 120\pi$, the impedance of free space

Combining all the foregoing yields a maximum received power at 150 MHz

$$P_r = 0.13(2^2)(2^2/120\pi) \cong 5.5\text{mW} \tag{5.90}$$

This will increase or decrease in direct proportion to the received wavelength. If we repeat the former calculation for the same length of cabling at 600 MHz, we find

$$P_r = 0.13(0.5^2)(2^2/120\pi) \cong 345\mu\text{W} \tag{5.91}$$

Further assuming the circuit resistance (impedance) is on the order of 1 Ω, we find the corresponding antenna mode currents to be on the order of 74.2 and 18.6 mA. These are equivalent to common mode induced currents since they will flow in the same direction on both the high and return sides of the circuit, and will typically, but not always, be at their maximum magnitudes near the centre of the length of the cabling [20]. Figures 4–6, from [20], illustrate that for cables and wires whose dimensions are electrically short and are exposed to an external electromagnetic wave stimulus, the resultant common mode currents will dominate a radiated field response to the presence of the incident field. This fact tends to justify the use of (5.88) as a so-called "worst case" assessment of the coupled energy in terms of an antenna-like response.

In [21], discussion focuses on the random aspect angle of an incident electromagnetic wave, described as the relative orientation between the direction of propagation of the incident wave and the geometrical orientation of the cable or wire with respect to the propagation direction, and concludes the power coupled to a cable or wire is a probabilistic function of the aspect angle. The results of multiple experimental setups were used to explore the randomness of the aspect angle and several cable parameters that indicated the measured distribution of coupled power levels was relatively independent of length and load impedance. Reference [21] thus concluded that (5.88) represents a reasonable upper bound for the data gathered from experimentation.

Reference [21] provides three caveats for the use of (5.88), namely,

1. In certain cases, the geometry of the installation will result in cables and wires exhibiting a larger effective aperture, A_e, than is assumed by (5.88). Wires and cables may exhibit large effective apertures caused by specific design and/or fortuitous focusing effects related to nearby structure or apparatus. An example would be a wire or cable located nearby an intentional emitter such as a parabolic dish antenna. Note the definition of effective aperture is the ratio of power delivered by an antenna to a conjugate matched load to the incident power density.
2. If (5.88) is extrapolated to lower frequencies, the calculated effective aperture, A_e, grows without bound, a situation that is counter to experience. This paradox can be resolved by considering that (5.88) is a model for a matched dipole. In other words, the impedance of the driving source and the dipole are equal. The Thevenin driving impedance of a dipole approaches zero as frequency approaches zero, such that the mismatch offsets the effect of the increasing aperture size, and the load continues to see the same amount of power.
3. Finally, because of the statistical nature of the power pickup, it is possible to have a probability density response for an unshielded cable or wire less than that for the same cable or wire in a shielded configuration. The likelihood of this occurring is driven by, and proportional to, the area of overlap between the two probability density functions, as illustrated in Figure 5.6 [21].

The foregoing caveats highlight the guidance that measurements should be performed at the system level to rule out such possibilities.

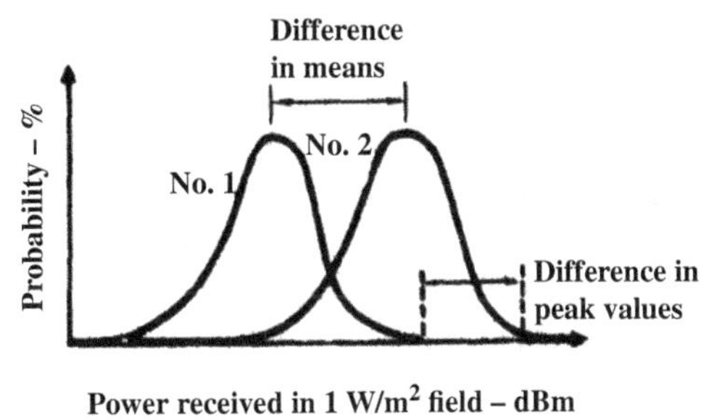

Figure 5.6 Probability density functions for shielded and unshielded wires

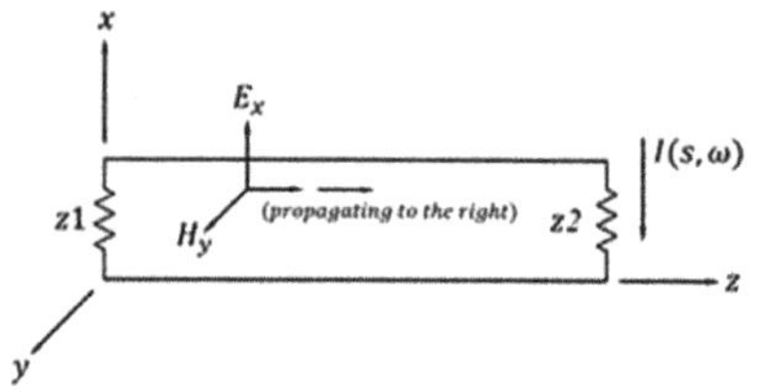

Figure 5.7 Construction for determination of FTW

In addition to antenna currents, loop currents that obey Faraday's law may also be induced by an incident electromagnetic field. These currents are differential mode currents that appear on the cable or wire and the nearby structure via capacitive or inductive coupling at the termination point(s). Figure 5.7 illustrates a construction used to discuss classic FTW coupling in accordance with Faraday's law. In this construction, it is assumed the electric field is parallel to the transmission line height (separation between the conductors) and is directed upward, with the corresponding magnetic field penetrating the area between the line conductors at a right angle to the plane of the transmission line and directed outward from the page. The direction of propagation is to the right, in the positive z-direction. This construction maximizes the possible coupled potential of the transmission line because the B field penetrates the loop area at a right angle.

In this figure, the length of the line is defined as $z = s$, the separation between the two conductors is defined as $x = b$, and the diameter of the conductors is defined as a. The current in the circuit (loop) is determined by the terminating impedances of the transmission line. This excellent reference [22] provides significant detail regarding transmission line responses to radiated field exposures. Those who are keenly interested in diving deeper into this topic area are encouraged to obtain a copy.

The current is given by (5.92) [22]

$$I(s,\omega) = \frac{\boldsymbol{b}E_x^i(0,\omega)Z_0 - \boldsymbol{b}E_x^i(s,\omega)[Z_0\cos(\beta s) + (jZ_1\sin(2\beta s))]}{D} \tag{5.92}$$

After some manipulation, (5.92) becomes

$$\frac{I(s,\omega)}{E_x^i(\omega)} = \frac{b(Z_0 - Z_1)}{2D}\{[1 - \cos(2\beta s)] + [j\sin(2\beta s)]\} \tag{5.93}$$

where

b = the separation of the two wires in the line
s = the length of the line
Z_0 = the characteristic impedance of the line
Z_1 = the source impedance of the line
Z_2 = the termination impedance of the line
$D = (Z_0Z_1 + Z_0Z_2)\cos(\beta s) + j(Z_0^2 + Z_1Z_2)\sin(\beta s)$

which is the load current spectrum normalized to the magnitude of the electric field spectrum. Equation (5.93) is plotted in Figures 5.8 and 5.9 for variations in the terminating impedances and in the separation distance between the conductors.

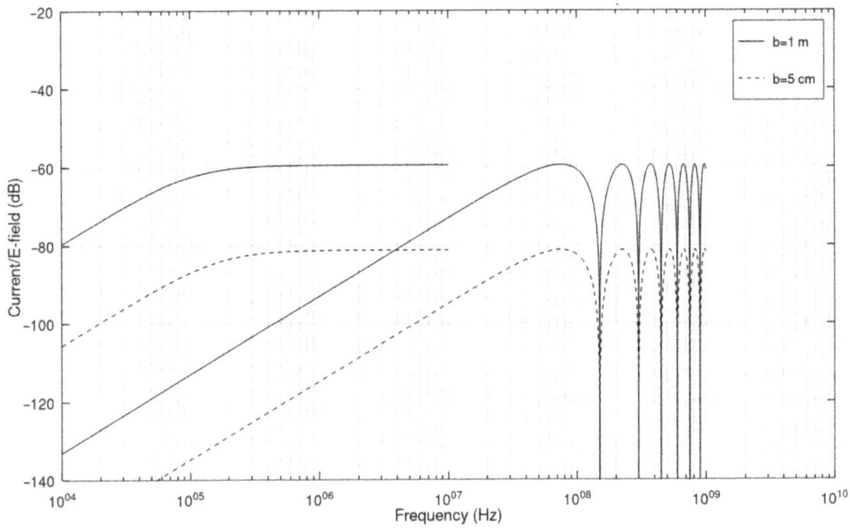

Figure 5.8 Current normalized to 1 m 20 AWG line with conductor separations of 1 m and 5 cm

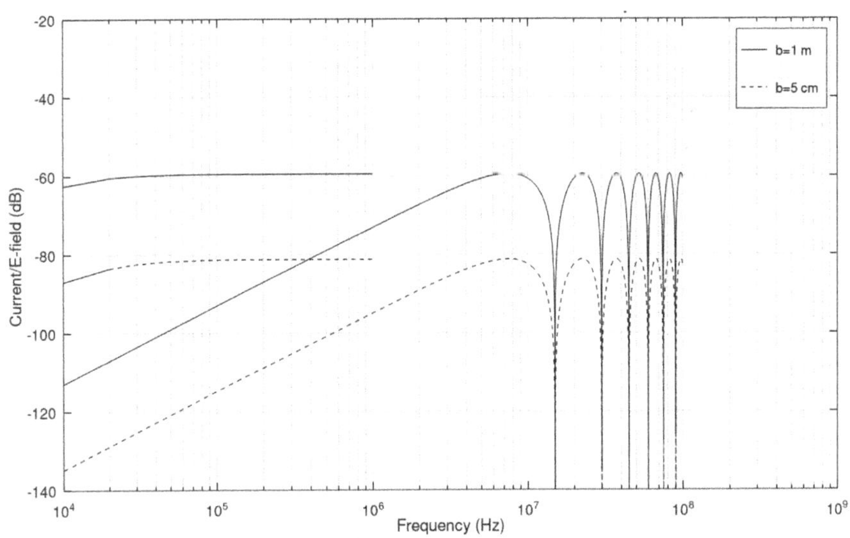

Figure 5.9 Current normalized to 10 m 20 AWG line with conductor separations of 1 m and 5 cm

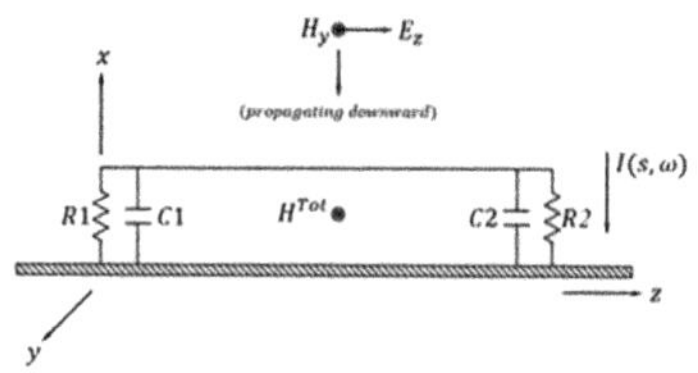

Figure 5.10 Configuration of a conductor above a common reference plane

As a final example, a conductor above a common reference will be examined briefly, again based on discussions in [22]. The configuration of interest is illustrated in Figure 5.10.

In the figure, the length of the line is again defined as $z = s$, the height of the conductor above the common reference is defined as $x = h$, and the diameter of the conductors is again defined as a. As was done previously, the current in the circuit (loop) normalized to the electric field is determined by the terminating impedances of the transmission line, and can be expressed as given by

$$\frac{I(s,\omega)}{E_z^i(\omega)} = \frac{4j\sin(\beta h)[Z_0\sin(\beta s) + jZ_1(1-\cos(\beta s))]}{\beta D} \tag{5.94}$$

where

h = the height of the transmission line above the common reference
s = the length of the line
Z_0 = the characteristic impedance of the line
Z_1 = the source impedance of the line
Z_2 = the termination impedance of the line
$D = (Z_0Z_1 + Z_0Z_2)\cos(\beta s) + j(Z_0^2 + Z_1Z_2)\sin(\beta s)$

Z_1 is defined using 5.95.

$$Z_1 = \frac{2R1}{1 + j\omega R1C1} \tag{5.95}$$

Z_2 is defined using 5.96.

$$Z_2 = \frac{2R2}{1 + j\omega R2C2} \tag{5.96}$$

The load current spectrum normalized to the magnitude of the electric field spectrum (Equation 5.94) is plotted in Figures 5.11 and 5.12 for variations in the length of the conductor and in the separation distance between the conductor and the common reference.

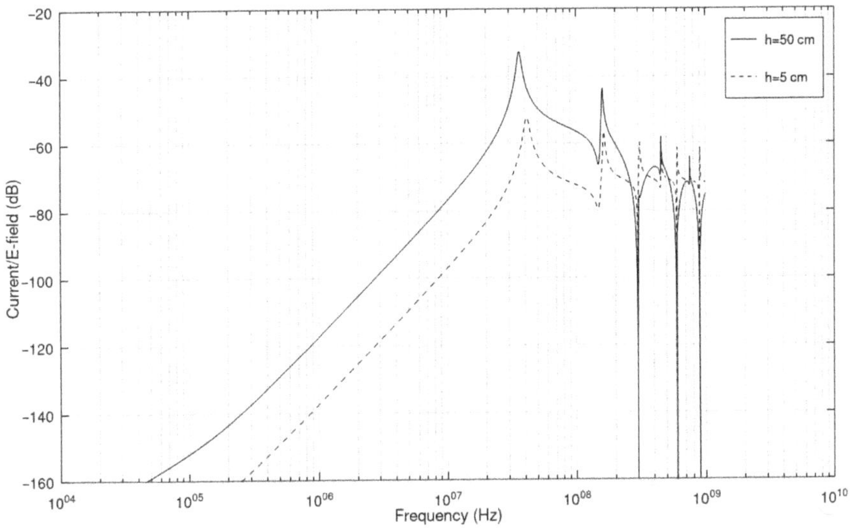

Figure 5.11 *Current normalized to E-Field for 1 m 20 AWG line above a reference plane at separations of 50 cm and 5 cm*

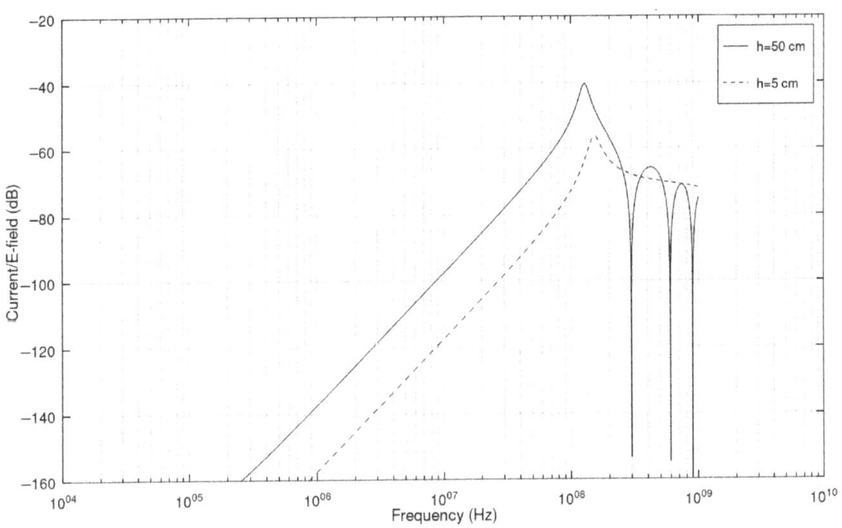

Figure 5.12 *Current normalized to E-Field for 0.1 m 20 AWG line above a reference plane at separations of 50 cm and 5 cm*

References

[1] W.L. Stutzman and G.A. Thiele, *Antenna Theory and Design*, John Wiley & Sons, Hoboken, NJ, 2013.

[2] R.S. Elliott, *Antenna Theory and Design*, John Wiley & Sons, Hoboken, NJ, USA, 2003.

[3] J.D. Krauss, and R.J. Marhefka, *Antennas for All Applications*, 3rd ed., McGraw-Hill, New York, NY, 2002.

[4] C.A. Balanis, *Antenna Theory Analysis and Design*, 4th ed., Hoboken, NJ, John Wiley & Sons, 2016.

[5] W.L. Weeks, *Antenna Engineering*, McGraw-Hill, New York, NY, 1968.

[6] J.L. Volakis, *Antenna Engineering Handbook*, 5th ed., McGraw-Hill, New York, NY, 2019.

[7] R.C. Johnson, Ed., *Antenna Engineering Handbook*, 3rd ed., McGraw-Hill, New York, NY, 1993.

[8] T.A. Milligan, *Modern Antenna Design*, 2nd ed., Wiley Interscience, Hoboken, NJ, 2005.

[9] S.A. Schelkunoff and H.T. Friis, *Antennas Theory and Practice*, John Wiley & Sons, New York, NY, 1952.

[10] R.W.P. King, *The Theory of Linear Antennas*, Harvard University Press, Cambridge, MA, 1956.

[11] B.D. Popovic, M.B. Dragovic, and A.R. Djordjevic, *Analysis and Synthesis of Wire Antennas*, Research Studies Press, a Division of John Wiley & Sons, New York, NY, 1982.

[12] J.D. Heys, *Practical Wire Antennas Effective HF Designs for the Radio Amateur*, Radio Society of Great Britain, 1994.

[13] G. Green, "An Essay on the Application of Mathematical Analysis to the Theories of Electricity and Magnetism," self-published by the author and printed by T. Wheelhouse at Nottingham, 1828, later reprinted in the Journal für die reine und angewandte Mathematik (also known as Crelle's Journal), vol. 44, pp. 356–374, 1852, available at: http://eudml.org/doc/147512, and more recently transcribed from its reprint in Crelle's Journal (1850–54), with several typographical corrections and a reference section added. https://arxiv.org/pdf/0807.0088.pdf, DOI: 10.48550/arXiv.0807.0088.

[14] J.D. Jackson, *Classical Electrodynamics*, 3rd ed., John Wiley & Sons, New York, NY, 1999.

[15] O. Jefimenko, *Electricity and Magnetism: An Introduction to the Theory of Electric and Magnetic Fields*, 2nd ed., Electret Scientific Company, Star City, WV, 1989.

[16] J. Larmor, "On the Theory of the Magnetic Influence on Spectra; and on the Radiation from Moving Ions," *London, Edinburgh and Dublin Philosophical Magazine and Journal of Science*, 5th Series, vol. 44, no. 63, pp. 503–512, 1897.

[17] G.F. Fitzgerald, "On the Energy Lost by Radiation from Alternating Electric Currents," *Report of the 53rd Meeting of the British Association for the Advancement of Science, Transactions of the Sections, Section A, Mathematical and Physical Science*, no. II, pp. 404–405, 1883.

[18] G.F. Fitzgerald, "On the Quantity of Energy Transferred to the Ether by a Variable Current," *The Scientific Transactions of the Royal Dublin Society*, vol. III, Series 2, no. 2, pp. 57–60, 1884.

[19] H.R. Hertz, "Über sehr schnelle elektrische Schwingungen," *Wiedemann's Annalen der Physik und Chemie 31*, pp. 421–448; 543–544, 1887.

[20] A. Vukicevic, F. Rachidi, M. Rubenstein, and V. Tkachenko, "On the Evaluation of Antenna-mode Currents Along Transmission Lines," *IEEE Transactions on Electromagnetic Compatibility*, vol. 48, no. 4, pp. 693–700, 2006.

[21] J.M. Roe, *et al.*, *Integrated Circuit Electromagnetic Susceptibility Handbook*, McDonnell Douglas Astronautics Company, St. Louis, MO, USA, Contract N60921-76-C-A030, Report MDC-E1929, Integrated Circuit Electromagnetic Susceptibility Investigation – Phase III, 1978.

[22] A.A. Smith, *Coupling of External Electromagnetic Fields to Transmission Lines*, John Wiley & Sons, New York, NY, 1977.

Chapter 6
Basic shielding theory

Shielding is a major tool in the arsenal of the electromagnetic compatibility (EMC) Engineer, aiding in the suppression of radiation of electromagnetic (EM) emissions generated inside electrical/electronic hardware and the protection of susceptible electrical/electronic hardware from interference caused by the penetration of external EM-radiated fields. A shield may be designed specifically to control either electric fields or magnetic fields, or both. Shielding may assume many different forms, including single barriers, multiple-layer stack-ups, perforated screens, wire mesh, and honeycomb vents. The variations of forms enable the engineer to provide a custom-tailored electric and magnetic shielding capability.

The term shielding as employed in this text refers to a physical barrier that inhibits the passage of EM radiation from one side of the barrier, or shield, to the other. The degree to which a shield performs this action is referred to as shielding effectiveness (SE).

The engineering considerations for shielding design include the basic shield material characteristics such as conductivity, permeability, and thickness; consideration of all sources of EM radiation addressed, both internal and external; and the field characteristics of the threats, whether they are in the near field or the far field. Additional considerations for shielding include the size and shape of the enclosure, the number and dimensions of any apertures, seams, penetrations, or connectors, the type and spacing of fasteners used to assemble the enclosure, as well as other environmental design constraints such as thermal/humidity/vibration requirements and desired/required surface finish/treatments. Of course, the minimization of mass, volume, cost, and schedule impacts is mandatory for any successful design.

Shielding as a topic has been around for as long as EM waves have been known. Early technical papers on the topic of shielding were written and published by Maxwell [1–3], Rayleigh [4], Larmor [5], and William Thomson (Lord Kelvin) [6]. Maxwell first explored the manner in which eddy currents are produced in conducting sheets by the incidence of magnetic fields. In accordance with Lenz' Law, induced eddy currents would produce opposing magnetic fields that tended to cancel the incident field. Since that time, multiple authors have addressed the use of eddy currents. This technique is still widely used for low-frequency magnetoquasistatic applications. In 1889, in performing experiments to assess the theory of skin effect as put forward by Maxwell and as interpreted by Heaviside and Poynting, Hertz discussed at some length the theory of skin depth and addressed the action of shielding. Hertz used a grating of parallel wires to surround a spark gap located in the length of a wire along which he attempted to propagate an alternating signal of approximately 100 MHz, radiated from a secondary conductor run in parallel, and found that with the grating in place, he was unable to produce any sparks on the "shielded" conductor. He later continued his experiments with a single wire run through a long conductive cylinder using the same grating of wires extended to a greater length [7]. Over time, many hundreds of papers were written on the topic, but there was little consistency amongst the authors, as the topic was largely considered academic in nature. As an example, in 1898 Lamb published an exact solution to the quasi-static problem of a parallel grating of thin strips [8,9], but the work was largely ignored until 1946 when the problem was revisited, and re-solved, by MacFarlane [10]. The Circuit Approach, useful for low-frequency applications, was first brought forward explicitly by Morecroft and Turner in 1925 as a parallel arrangement of three coils. The approach typically models the shielding configuration as a relatively simple resistor-capacitor (RC) or resistor-inductor (RL) circuit [11]. Schelkunoff's so-called Transmission Line Shielding Theory [12–14] and Kaden's work [15] are still recognized as primary sources of fundamental engineering application guidance. Multiple other authors have explored and developed complimentary theories, notably in Moeller [16] and Vasseur [17], with more recent work in the late 1950s, through the 1960s, and into the late 1980s by Miedzinski [18], Miller and Bridges, who revisited and improved on the Circuit Approach [19,20], Schulz [21,22], Moser [23], Monroe [24–27], and Vance [28–30], to name a few. More recent still are references by White and Mardiguian [31–33], Celozzi, Araneo, and Lovat [34], Paul [35], and Ott [36]. The interested reader is invited to seek out some or all of these references in the literature.

6.1 Basic shielding theory

In general, SE varies as a function of frequency, the materials from which the shield is fabricated, the geometry of the shield, the position where the fields are measured with respect to the shield, whether the field is electric or magnetic, how close to the source is the shield located, and the angle of the incident wavefront.

The approach pioneered by Schelkunoff in his 1934 [12] and 1938 [13] papers was based on the behaviour of EM waves propagating along transmission lines and the transmission and reflection of those waves upon encountering impedance discontinuities in the line. This approach has been adopted and effectively redeveloped many times. What follows will echo and expand on Schelkunoff's work, as well as include material from adopted and redevelopmental works.

Figure 6.1 compares an EM wave propagating through a transmission line to an EM wave encountering a shield of finite thickness.

The upper image in Figure 6.1 is a simple depiction of a transmission line whose length is l and through which the x-oriented wave indicated to the right of the transmission line is propagating, the line being terminated into Z_t. The lower image in Figure 6.1 is taken from Schelkunoff's 1938 paper [13] and was his transmission line representation of a shield in that paper.

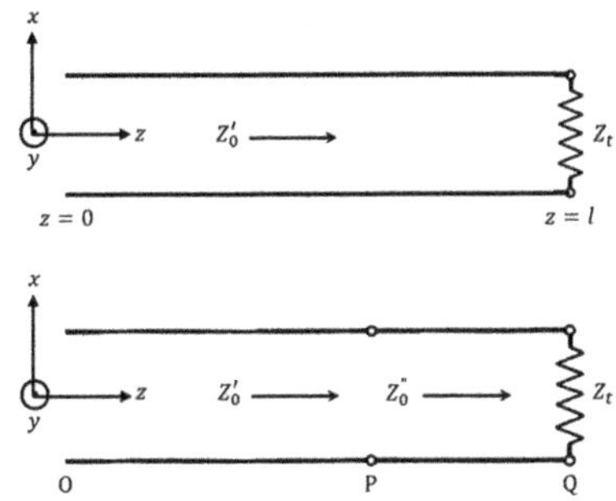

Figure 6.1 Comparison of transmission line to shield of finite thickness

Consider a semi-infinite uniform transmission line terminated in an impedance Z_t, such as is shown in the upper image of Figure 6.1. An EM wave is incident upon the transmission line at $z = 0$. The intrinsic, or wave, impedance of the transmission line is Z_0. The terminating impedance, Z_t, may or may not be equal to Z_0. If $Z_t = Z_0$, the two impedances are referred to as being matched, and no reflection will occur when the EM wave encounters the terminating impedance, Z_t. If $Z_t \neq Z_0$, some portion of the incident EM wave that encounters the terminating impedance, Z_t, will be reflected and will travel in the opposite direction to that of the travel of the originally incident EM wave, while some portion of the incident EM wave will appear across Z_t. The sum of the incident and reflected waves is equal to their total value at the terminating impedance, expressed as

$$E_i + E_r = E_t \text{ and } H_i + H_r = H_t \tag{6.1}$$

The following relationships exist between the field components and Z_0:

$$E_i = Z_0 H_i, E_r = -Z_0 H_r, \text{ and } E_t = Z_t H_t \tag{6.2}$$

We can also define the ratio between the wave impedance of the transmission line to the terminating impedance as

$$k = \frac{Z_0}{Z_t} \tag{6.3}$$

Using (6.2) and (6.3), we can then express (6.1) as

$$H_i - H_r = \frac{H_t}{k} \text{ and } H_i + H_r = H_t \tag{6.4}$$

After some algebraic manipulations, we can define the reflection and transmission coefficients for the currents and voltages as

$$q_H = \frac{H_r}{H_i} = \frac{k-1}{1+k} \text{ and } q_E = \frac{E_r}{E_i} = \frac{1-k}{1+k} \tag{6.5}$$

$$p_H = \frac{H_t}{H_i} = \frac{2k}{1+k} \text{ and } p_E = \frac{E_t}{E_i} = \frac{2}{1+k} \tag{6.6}$$

It is clear from (6.5) and (6.6) that the amounts of reflection and transmission are entirely dependent on the ratio of the wave impedance of the transmission line to the terminating impedance. Observe that neither the propagation constants nor the wave velocities on either side of the interface between the transmission line and the terminating impedance have any impact on the degree of reflection that occurs.

Drawing our attention to the lower image, we observe the difference between this image and the upper image is the addition of a small section of transmission line that has been inserted between the original transmission line and the terminating impedance, Z_t, with boundaries identified as P and Q. Both the original transmission line and the inserted section are uniform lines; that is, their wave impedances are the same "looking" in opposite directions. The inserted section of transmission line has a wave impedance different from that of the original transmission line, and for our purposes represents a shield of finite thickness and conductivity, the length of which, l, is equal to the distance between P and Q. By inserting this section, a second interface has been added between the source and the terminating impedance. The simplest case to consider is that of an electrically thick shield in which the attenuation is sufficiently high so that any EM waves reflected from Q have negligible effect on the originally incident EM wave at P.

The generator represents the source of the incident EM wave in the transmission line. For the purposes of this discussion, it will be assumed the EM wave generated by the source appears with negligible attenuation at the P interface, regardless of the type of transmission line. The ratio between the electric and magnetic fields generated and impressed on the transmission line will then be assumed to solely be constrained by the wave impedance, Z_0', of the OP section. The section from P to Q, representing a shield, has an intrinsic impedance, Z_0'', that differs from Z_0'. Z_t is the impedance of the volume of space immediately outside of the shield into which any transmitted waves will emerge. This impedance, Z_t, may or may not be equal to Z_0' or Z_0''. The construction as described will result in a minimum of two separate and independent reflections at each interface, or side of the shield, governed by the ratios of the two impedances on either side of each interface and combined with attenuation and phase change as a function of the length of the inserted section.

The transmission coefficients for the electric and magnetic fields across the shield are given by the products of the individual transmission coefficients at each interface multiplied by an exponential term to account for attenuation across the shield. The individual transmission coefficients are given by

$$p_{E,P} = \frac{2Z_0''}{Z_0' + Z_0''} \text{ and } p_{E,Q} = \frac{2Z_t}{Z_t + Z_0''} \tag{6.7}$$

$$p_{H,P} = \frac{2Z_0'}{Z_0' + Z_0''} \text{ and } p_{H,Q} = \frac{2Z_0''}{Z_t + Z_0''} \tag{6.8}$$

$$p_E = p_{E,P}p_{E,Q}e^{-\gamma_{PQ}l} = \left(\frac{2Z_0''}{Z_0' + Z_0''}\right)\left(\frac{2Z_t}{Z_t + Z_0''}\right)e^{-\gamma_{PQ}l}$$

$$= \frac{4Z_0''Z_t}{(Z_0' + Z_0'')(Z_t + Z_0'')} e^{-\gamma_{PQ}l} \tag{6.9}$$

$$p_H = p_{H,P}p_{H,Q}e^{-\gamma_{PQ}l} = \left(\frac{2Z_0'}{Z_0' + Z_0''}\right)\left(\frac{2Z_0''}{Z_t + Z_0''}\right)e^{-\gamma_{PQ}l}$$

$$= \frac{4Z_0'Z_0''}{(Z_0' + Z_0'')(Z_t + Z_0'')} e^{-\gamma_{PQ}l} \tag{6.10}$$

where

$p_{E,P}$ = the transmission coefficient for E at P
$p_{E,Q}$ = the transmission coefficient for E at Q
$p_{H,P}$ = the transmission coefficient for H at P
$p_{H,Q}$ = the transmission coefficient for H at Q

If $Z_t = Z_0'$, the transmission coefficients become

$$p_{E,P} = \frac{2Z_0''}{Z_0' + Z_0''} \text{ and } p_{E,Q} = \frac{2Z_0'}{Z_0' + Z_0''} \tag{6.11}$$

$$p_{H,P} = \frac{2Z_0'}{Z_0' + Z_0''} \text{ and } p_{H,Q} = \frac{2Z_0''}{Z_0' + Z_0''} \tag{6.12}$$

$$\begin{aligned} p_E &= p_{E,P}p_{E,Q}e^{-\gamma_{PQ}l} = \left(\frac{2Z_0''}{Z_0' + Z_0''}\right)\left(\frac{2Z_0'}{Z_0' + Z_0''}\right)e^{-\gamma_{PQ}l} \\ &= \frac{4Z_0''Z_0'}{(Z_0' + Z_0'')^2} e^{-\gamma_{PQ}l} \end{aligned} \tag{6.13}$$

$$\begin{aligned} p_H &= p_{H,P}p_{H,Q}e^{-\gamma_{PQ}l} = \left(\frac{2Z_0'}{Z_0' + Z_0''}\right)\left(\frac{2Z_0''}{Z_0' + Z_0''}\right)e^{-\gamma_{PQ}l} \\ &= \frac{4Z_0'Z_0''}{(Z_0' + Z_0'')^2} e^{-\gamma_{PQ}l} \end{aligned} \tag{6.14}$$

$$p_E = p_H = \frac{4k}{(k+1)^2} e^{-\gamma_{PQ}l} \tag{6.15}$$

where $k = \frac{Z_0'}{Z_0''}$, the ratio of the impedances across the interfaces.

The corresponding reflection coefficients for the electric and magnetic fields across the shield are given by the product of the reflection coefficient at P and the negative of the reflection coefficient at Q. Each reflection coefficient is multiplied by an exponential term to account for attenuation across the shield. The individual reflection coefficients are given by

$$q_{E,P} = \frac{Z_0'' - Z_0'}{Z_0' + Z_0''} \text{ and } q_{E,Q} = \frac{Z_t - Z_0''}{Z_t + Z_0''} \tag{6.16}$$

$$q_{H,P} = \frac{Z_0' - Z_0''}{Z_0' + Z_0''} \text{ and } q_{H,Q} = \frac{Z_0'' - Z_t}{Z_t + Z_0''} \tag{6.17}$$

$$\begin{aligned} q_E &= q_{E,P}(-q_{E,Q})e^{-\gamma_{PQ}l} \\ &= \left(\frac{Z_0'' - Z_0'}{Z_0' + Z_0''}\right)\left(\frac{Z_0'' - Z_t}{Z_t + Z_0''}\right)e^{-\gamma_{PQ}l} \end{aligned} \tag{6.18}$$

$$\begin{aligned} q_H &= q_{H,P}(-q_{H,Q})e^{-\gamma_{PQ}l} \\ &= \left(\frac{Z_0' - Z_0''}{Z_0' + Z_0''}\right)\left(\frac{Z_t - Z_0''}{Z_t + Z_0''}\right)e^{-\gamma_{PQ}l} \end{aligned} \tag{6.19}$$

where

$q_{E,P}$ = the reflection coefficient for E at P
$q_{E,Q}$ = the reflection coefficient for E at Q
$q_{H,P}$ = the reflection coefficient for H at P
$q_{H,Q}$ = the reflection coefficient for H at Q

If $Z_t = Z_0'$, the reflection coefficients become

$$q_{E,P} = \frac{Z_0'' - Z_0'}{Z_0' + Z_0''} \text{ and } q_{E,Q} = \frac{Z_0' - Z_0''}{Z_0' + Z_0''} \tag{6.20}$$

$$q_{H,P} = \frac{Z_0' - Z_0''}{Z_0' + Z_0''} \text{ and } q_{H,Q} = \frac{Z_0'' - Z_0'}{Z_0' + Z_0''} \tag{6.21}$$

$$\begin{aligned} q_E &= q_{E,P}(-q_{E,Q})e^{-\gamma_{PQ}l} \\ &= \left(\frac{Z_0'' - Z_0'}{Z_0' + Z_0''}\right)\left(\frac{Z_0'' - Z_0'}{Z_0' + Z_0''}\right)e^{-\gamma_{PQ}l} \end{aligned} \tag{6.22}$$

$$\begin{aligned} q_H &= q_{H,P}(-q_{H,Q})e^{-\gamma_{PQ}l} \\ &= \left(\frac{Z_0' - Z_0''}{Z_0' + Z_0''}\right)\left(\frac{Z_0' - Z_0''}{Z_t + Z_0''}\right)e^{-\gamma_{PQ}l} \end{aligned} \tag{6.23}$$

$$q_E = q_H = \frac{(k-1)^2}{(k+1)^2}e^{-\gamma_{PQ}l} \tag{6.24}$$

If the shield is electrically short and the original transmission line and the inserted section are uniform lines, we will have multiple reflected waves moving back and forth inside the shield. This is illustrated in Figure 6.2.

We can write an equation for each of the transmitted waves as follows (the magnetic field is shown):

$$\frac{H_t^2}{H_i^1} = p_{H,1}e^{-\gamma_2 l}p_{H,2} = p_H e^{-\gamma_2 l} \tag{6.25}$$

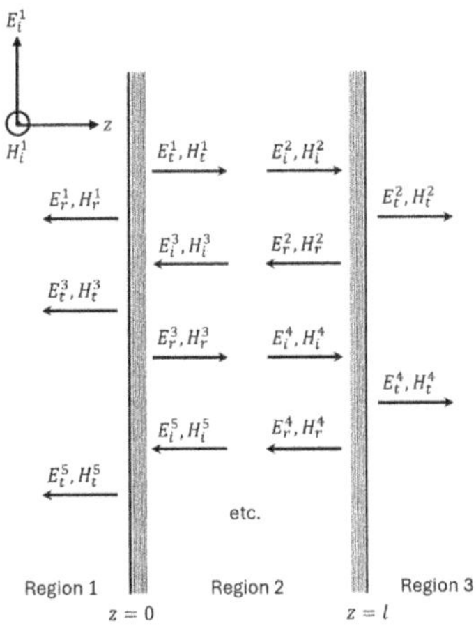

Figure 6.2 Multiple reflections inside a shield

$$\frac{H_t^4}{H_i^1} = p_{H,1}e^{-\gamma_2 l}q_{H,2}e^{-\gamma_2 l}q_{H,1}e^{-\gamma_2 l}p_{H,2} = p_H e^{-\gamma_2 l}q_H e^{-2\gamma_2 l} \tag{6.26}$$

$$\begin{aligned}\frac{H_t^6}{H_i^1} &= p_{H,1}e^{-\gamma_2 l}q_{H,2}e^{-\gamma_2 l}q_{H,1}e^{-\gamma_2 l}q_{H,2}e^{-\gamma_2 l}q_{H,1}e^{-\gamma_2 l}p_{H,2} \\ &= p_H e^{-\gamma_2 l}\left(q_H e^{-2\gamma_2 l}\right)^2\end{aligned} \tag{6.27}$$

where

$$p_H = p_{H,1}p_{H,2}$$

$$q_H = q_{H,1}q_{H,2}$$

Equations (6.25)–(6.27) and subsequent terms form an infinite geometric series

$$T_H = p_H e^{-\gamma_2 l} + p_H q_H e^{-3\gamma_2 l} + p_H q_H^2 e^{-5\gamma_2 l} + \cdots \tag{6.28}$$

If we let

$$a = p_H e^{-\gamma_2 l} .. \quad \text{and } r = q_H e^{-2\gamma_2 l} \tag{6.29}$$

we have

$$T_H = a\left(1 + r + r^2 + \cdots\right) \tag{6.30}$$

T_H is then given by

$$T_H = \sum\left[a\left(1 + r + r^2 + \cdots\right)\right] = \frac{a}{1-r} = \frac{p_H e^{-\gamma_2 l}}{1 - q_H e^{-2\gamma_2 l}} \tag{6.31}$$

where $p_H = \frac{4Z_0' Z_0''}{\left(Z_0' + Z_0''\right)\left(Z_t + Z_0''\right)}$ and $q_H = \left(\frac{Z_0' - Z_0''}{Z_0' + Z_0''}\right)\left(\frac{Z_t - Z_0''}{Z_t + Z_0''}\right)$.

If $Z_t = Z_0'$, the transmission and reflection coefficients become

$$p_H = \frac{4k}{(k+1)^2} \text{ and } q_H = \frac{(k-1)^2}{(k+1)^2} \tag{6.32}$$

The total magnetic field SE is then given by

$$\begin{aligned} SE_{H,TOT} &= A + R + B \\ &= 20\log_{10}\left|\frac{1}{T_H}\right| = 20\log_{10}\left|\frac{\left(1 - q_H e^{-2\gamma_2 l}\right)e^{\gamma_2 l}}{p_H}\right| \\ &= 20\log_{10}\left|e^{\gamma_2 l}\right| - 20\log_{10}|p_H| + 20\log_{10}\left|\left(1 - q_H e^{-2\gamma_2 l}\right)\right| \end{aligned} \tag{6.33}$$

where

$$A = 20\log_{10}\left|e^{\gamma_2 l}\right| \tag{6.34}$$

$$R = 20\log_{10}\left|\frac{1}{p_H}\right| \tag{6.35}$$

and

$$B = 20\log_{10}\left|\left(1 - q_H e^{-2\gamma_2 l}\right)\right| = 20\log_{10}\left|\left(1 - \frac{(k-1)^2}{(k+1)^2}e^{-2\gamma_2 l}\right)\right| \tag{6.36}$$

The total electric field SE is then given by

$$SE_{E,TOT} = \frac{Z_0'}{Z_t} SE_{H,TOT}$$

If $Z_t = Z_0'$, then

$$p_H = p_E = p \text{ and } q_H = q_E = q$$

$$T_H = T_E = T$$

$$\begin{aligned} SE_{E,TOT} &= SE_{H,TOT} = A + R + B \\ &= 20\log_{10}\left|e^{\gamma_2 l}\right| - 20\log_{10}|T| + 20\log_{10}\left|\left(1 - q e^{-2\gamma_2 l}\right)\right| \end{aligned} \tag{6.37}$$

where

$$A = 20\log_{10}\left|e^{\gamma_2 l}\right| \tag{6.38}$$

$$R = 20\log_{10}\left|\frac{1}{T}\right| \tag{6.39}$$

$$B = 20\log_{10}\left|\left(1 - q e^{-2\gamma_2 l}\right)\right| = 20\log_{10}\left|\left(1 - \frac{(k-1)^2}{(k+1)^2}e^{-2\gamma_2 l}\right)\right| \tag{6.40}$$

Let's examine each of these terms a bit more closely, beginning with the absorption term. Absorption loss is governed by the physical characteristics of the shield material, independent of the source, and the ratio of the thickness of the shield to the skin depth of the shield. Put another way, it is directly proportional to the thickness of the shield and to the square root of the permeability and the conductivity of the shield material. A highly conductive material provides a high loss as a function of frequency and is an effective shield to electric fields even if it is very thin.

Generally, absorption loss is low at low frequencies and increases in proportion to an increase in frequency. For a given thickness, materials that exhibit a permeability greater than that of free space, so-called magnetic materials, provide a greater amount of absorption than non-magnetic materials. By restating the absorption loss term in (6.34) and (6.38) in terms of skin depth, the contribution of frequency, conductivity, and permeability can be more clearly illustrated.

$$A = 20\log_{10}\left|e^{\gamma_2 l}\right|$$

In a metal, the propagation constant, γ, is generally given by

$$\gamma = \alpha + j\beta = \sqrt{j\omega\mu\sigma + j\omega\varepsilon} \cong \sqrt{j\omega\mu\sigma}$$

Thus,

$$\gamma = \alpha + j\beta \cong (1+j)\sqrt{\pi f\mu\sigma} \xrightarrow{yields} \alpha = \beta = \sqrt{\pi f\mu\sigma}$$

Now,

$$A = 20\log_{10}\left|e^{\gamma_2 l}\right| = 20\log_{10}\left|e^{\sqrt{\pi f\mu\sigma}(l)}\right|$$

$$\sqrt{\pi f\mu\sigma} = 1/\delta \text{ and } l = t$$

and so

$$A = 20\log_{10}\left|e^{t/\delta}\right| = 8.686(t/\delta) \tag{6.41}$$

Equation (6.41) indicates that a wave traversing a shield will suffer approximately 9 dB of loss per single pass through a shield thickness equal to 1 skin depth.

Reflection loss is governed by the ratio of the wave impedance across the interfaces of the shield:

$$R = -20\log_{10}|T| = 20\log_{10}\left|\frac{1}{T}\right| = 20\log_{10}\left|\frac{(k+1)^2}{4k}\right| \tag{6.42}$$

If $k \geq 100$,

$$R \cong 20\log_{10}\left|\frac{k}{4}\right| \tag{6.43}$$

For metals we might choose for shielding purposes, the foregoing case will in general be true. The higher the conductivity of the shield, the greater will be the magnitude of the reflected wave. Taken together with absorption characteristics, this strongly supports the concept that even thin shields that are highly conductive can provide very effective electric field shielding.

Practical effective shields for electric fields can be formed from many different conductive materials, including copper and aluminium. For example, household aluminium foil with typical thickness between 6 and 25 microns can be used for effective electric field shielding. The skin depth at 100 MHz in common household aluminium foil, an 8011-aluminium alloy with an electrical conductivity between 30% and 45% of IACS (1.74×10^7S/m to 2.61×10^7S/m) and a permeability equal to that of free space, is equal to

$$\delta = \sqrt{\frac{1}{\pi f \mu \sigma}} = \sqrt{\frac{1}{\pi (100 \times 10^6) \mu_0 \sigma}}$$

$$12.07\ \mu\text{m} \leq \delta \leq 9.85\ \mu\text{m}$$

At 1.0 GHz, these values are

$$3.82\ \mu\text{m} \leq \delta \leq 3.12\ \mu\text{m}$$

At 25 MHz, these values are

$$24.13\ \mu\text{m} \leq \delta \leq 19.7\ \mu\text{m}$$

Clearly as the skin depth increases, the attenuation as a function of penetration depth of the incident wave tends to decrease, in accordance with (6.41) and the curves illustrated in Figure 6.3. To further emphasize the additional relationships at

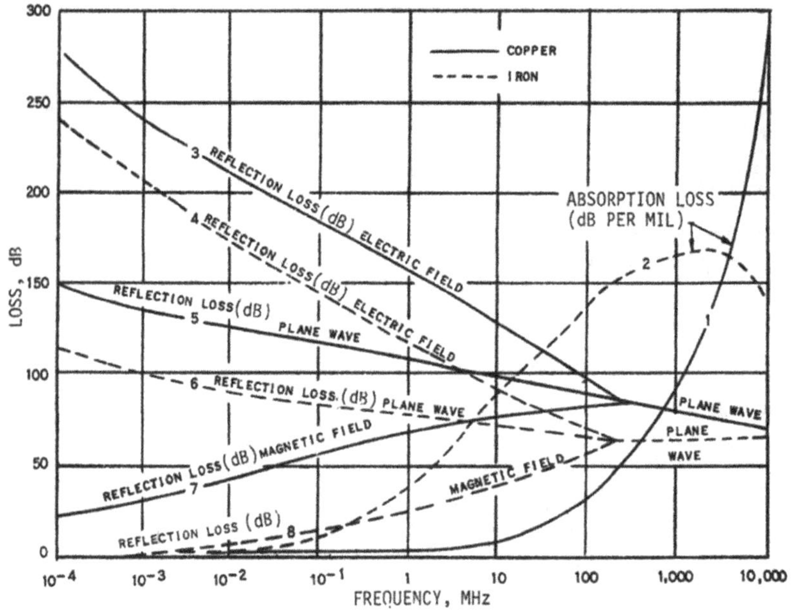

Figure 6.3 Comparison of reflection and absorption loss for Fe versus Cu [37]

play, a comparison of reflection loss for copper versus iron is provided in Figure 6.3. Absorption loss is included for comparison.

In this figure, the curves labelled as "Electric Field" and "Magnetic Field" represent near-field shielding effectivity associated with a source exhibiting a comparatively high wave impedance ($\gg 120\pi\ \Omega$), and a source exhibiting a comparatively low wave impedance ($\ll 120\pi\ \Omega$), respectively. In this figure, the two types of curves combine with that of a plane wave at a distance of separation in free space from the source equivalent to $\lambda/2\pi$, or $\cong 0.16\lambda$. Those labelled as "Plane Wave" represent far-field shielding effectivity, regardless of the source wave impedance. For plane wave incidence, the wave impedance is the same as that of free space, approximately $120\pi\ \Omega$. Refer to Section 5.7 of Chapter 5 for more details on wave impedance.

Note that in each case, near or far field, copper exhibits the higher reflection loss up to about 10 GHz. Curves 3–8 are for shields that are at least 1 skin depth thick at the frequency of interest. Curve 2 is comparable to the curve for 1-mil steel in Figure 6.3. Finally, note the magnitude of the reflection loss strongly depends on the wave impedance of the source.

Three examples of computations for reflection loss follow that are likely helpful for understanding, the first of which is for a plane wave in free space, and the second and third of which are for near-field exposures to a high-impedance electric field and a low-impedance magnetic field, respectively. Recall that for a near-field electric source, the wave impedance becomes larger than the free space value as the distance from the source becomes less than $\lambda/2\pi$. Schelkunoff expressed this wave impedance as

$$Z_E = 1/i\omega\epsilon\rho \tag{6.44}$$

where $\epsilon = \epsilon_0 = \frac{1}{36\pi} \times 10^{-9}$ F/m, and ρ is the distance from the source in metres.

Conversely, for a near-field magnetic source, the wave impedance becomes smaller than the free space value as the distance from the source becomes less than $\lambda/2\pi$. Schelkunoff expressed this wave impedance as

$$Z_H = i\omega\mu\rho \tag{6.45}$$

where $\mu = \mu_0 = 4\pi \times 10^{-7}$ H/m, and ρ is the distance from the source in metres.

For plane wave incidence, using (6.43) and $k = K_0/K_s$
where $K_0 = 120\pi$, and $K_s = \sqrt{\omega\mu/\sigma}$, we have

$$\begin{aligned} R(dB) &= 20\log_{10}\left|\frac{k}{4}\right| = 20\log_{10}\frac{K_0}{4K_s} = 20\log_{10}\frac{120\pi}{4\sqrt{\frac{\omega\mu}{\sigma}}} \\ &= 168 + 10\log_{10}\left(\frac{\sigma_r}{f\mu_r}\right) \end{aligned} \tag{6.46}$$

For near-field electric wave incidence, using (6.43) and $k = K_0/K_s$

where $K_0 = 1/\omega\epsilon\rho$, and $K_s = \sqrt{\omega\mu/\sigma}$, we have

$$R(dB) = 20\log_{10}\left|\frac{k}{4}\right| = 20\log_{10}\frac{K_0}{4K_s} = 20\log_{10}\frac{1}{8\pi f\epsilon\rho\sqrt{\frac{\omega\mu}{\sigma}}}$$
$$= 322 + 10\log_{10}\left(\frac{\sigma_r}{r^2 f^3 \mu_r}\right) \tag{6.47}$$

For near-field magnetic wave incidence, using (6.43) and $k = K_0/K_s$, where $K_0 = \omega\mu\rho$ and $K_s = \sqrt{\omega\mu/\sigma}$, we have

$$R(dB) = 20\log_{10}\left|\frac{k}{4}\right| = 20\log_{10}\frac{Z_0'}{4Z_0''} = 20\log_{10}\frac{2\pi f\mu r}{4\sqrt{\frac{\omega\mu}{\sigma}}}$$
$$= 14.6 + 10\log_{10}(r^2 f\mu_r\sigma_r) \tag{6.48}$$

Re-reflection loss results from repeated internal reflections, as discussed in (6.25)–(6.36), and the attenuation of each passage of the wave through the shield. With reference to Figure 6.2, each subsequent re-reflection will tend to reduce the effect of the reflection term R by increasing the total transmitted magnitude of the original incident wave. Conversely, each subsequent re-reflection will tend to increase the magnitude of the absorption of the original incident wave as well as each of the re-reflected wave components. The total effect is, however, an aggregate reduction in the effectiveness of the shield. This is a rather intuitive result when considering the re-reflection term is only of significance when considering shields with a thickness less than that of a skin depth at the frequency(ies) of interest. Referring to (6.40), we have

$$B = 20\log_{10}\left|\left(1 - qe^{-2\gamma_2 l}\right)\right| = 20\log_{10}\left|\left(1 - qe^{-2t/\delta}\right)\right|$$
$$= 20\log_{10}\left|\left(1 - \frac{(k-1)^2}{(k+1)^2}e^{-2t/\delta}\right)\right| \tag{6.49}$$

If $k \geq 100$,

$$B \cong 20\log_{10}\left|\left(1 - e^{-2t/\delta}\right)\right| \tag{6.50}$$

As was true for the reflection term, for metals we might choose for shielding purposes, the foregoing case will in general be true. In such a case, the magnitude of the re-reflection term is dominated by the ratio of the thickness of the shield and the skin depth in the shield as a function of frequency.

Note that if $(t/\delta) \geq 1$ in the A term, the absorption term will correspondingly be greater than 8.686 dB. In such a case for practical purposes, the B term may thus be neglected.

For example, with a shield whose thickness is equivalent to 1 skin depth, the exponential factor in (6.50) is approximately equal to 0.135; the factor inside the absolute value brackets is then approximately equal to 0.865, and the overall re-reflection term is approximately equal to -1.26 dB. As the shield becomes thicker, the magnitude of the B term approaches zero.

If the shield has a thickness equivalent to 2 skin depths, the overall re-reflection term is approximately equal to -0.160 dB. If the shield has a thickness equivalent to 10 skin depths, the overall re-reflection term is approximately equal to -17.9×10^{-9} dB. As the shield grows thicker, the re-reflection term grows smaller, and the corresponding absorption becomes greater.

Conversely, if the shield is much less than 1 skin depth, say 0.1 skin depths, the exponential factor in (6.45) is approximately equal to 0.818; the factor inside the absolute value brackets is then approximately equal to 0.181, and the overall re-reflection term is approximately equal to -14.83 dB, a much greater magnitude that cannot be ignored, with a corresponding reduction in SE. As the shield grows thinner, the re-reflection term grows larger, and the corresponding absorption becomes smaller.

From these examples, we can draw the following two rather general conclusions: (1) with large absorption loss (i.e. thickness greater than 1 skin depth), the re-reflection term will tend to be small and will likely be insignificant and (2) with small absorption loss (i.e. thickness less than 1 skin depth), the re-reflection term will tend to be large and will likely be significant.

6.2 Impacts to parameters involved in skin depth

6.2.1 Impacts to electrical conductivity

In general, the electrical resistivity of metals decreases with temperature (or, conversely, the electrical conductivity increases with a decrease in temperature). As the temperature of a metal is reduced, the temperature dependence of resistivity follows a power law function of temperature. Mathematically the temperature dependence of the intrinsic resistivity, ρ, of a metal can be approximated through the Bloch–Grüneisen formula:

$$\rho(T) = \rho(0) + A\left(\frac{T}{\Theta_R}\right)^n \int_0^{\frac{\Theta_R}{T}} \frac{x^n}{(e^x - 1)(1 - e^{-x})} dx \tag{6.51}$$

where

$\rho(0)$ = the residual resistivity due to defect scattering
A = a constant that depends on the velocity of electrons at the Fermi surface the Debye radius and the number density of electrons in the metal
Θ_R = the Debye temperature as obtained from resistivity measurements
n = an integer that depends upon the nature of interaction:
- $n = 5$ implies the resistance is caused by scattering of electrons by phonons (as it is for simple metals)
- $n = 3$ implies the resistance is caused by s-d electron scattering (as is the case for transition metals)
- $n = 2$ implies the resistance is caused by electron/electron interaction.

The Bloch–Grüneisen formula is an approximation. If more than one source of scattering is simultaneously present, Matthiessen's rule states that the total resistance can be approximated by adding up several different terms, each with the appropriate value of n. As the temperature of the metal decreases, its resistivity usually asymptotically approaches a constant value known as the residual resistivity, the value of which depends on the type of metal, its impurity concentration, and its thermal history.

Perhaps a more familiar expression, also an approximation, for resistivity versus temperature is given by

$$\rho(T) = \rho(0)[1 + \alpha_0(T - T_0)] \tag{6.52}$$

where ρ_0 is the resistivity at the reference temperature T_0.

$$\alpha_0 = \frac{1}{\rho_0}\left(\frac{d\rho}{dT}\right)_{T_0}$$

The expressions above are applicable over a range of temperatures for many metals, but not all. Some metals, such as tungsten, exhibit a power law relationship given by

$$\rho(T) = \rho(0)\left(\frac{T}{T_0}\right) \tag{6.53}$$

Figure 6.4 illustrates the variation of resistivity with temperature of a typical electrical conductor in accordance with (6.51) [38].

Note that the resistivity follows a relatively straight line down to roughly 100 K, and then becomes asymptotic as temperature continues to decrease.

Figure 6.5 illustrates the behaviour of certain metals at low temperatures. [39].

Note in Figure 6.5 that base metals are plotted with solid lines, metallic alloys with dot-dashed lines, and two superconducting alloys with dotted lines. The takeaway here is that for an application where the surrounding temperature may be below 100 K, it is necessary to determine the range of electrical resistivity that should be considered in any calculations that involve skin depth.

If the operating temperature is expected to remain between 100 K and 400 K, within which range the temperature coefficient is expected to be linear, it is possible to calculate the temperature coefficient for a different temperature using

$$\alpha(T_2) = \frac{\alpha(T_1)}{1 + [\alpha(T_1)](T_2 - T_1)} \tag{6.54}$$

6.2.2 Impacts to magnetic permeability

The governing physics of a material's permeability is a complex topic that is out of the scope of this text. The reader is encouraged to access the several references on the topic for details. Nonetheless, some general comments and engineering

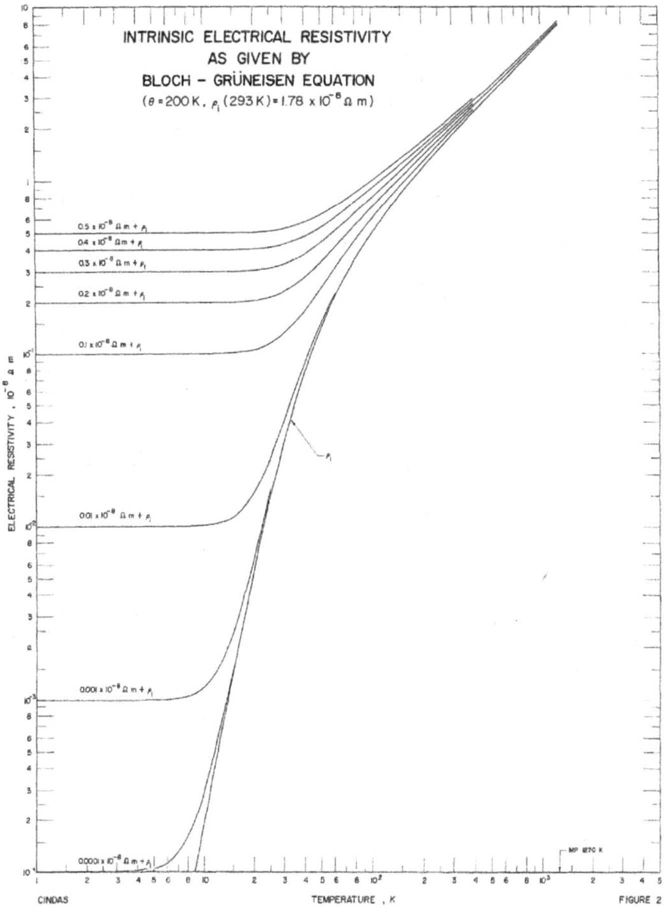

Figure 6.4 Intrinsic resistivity of a metal per Grüneisen

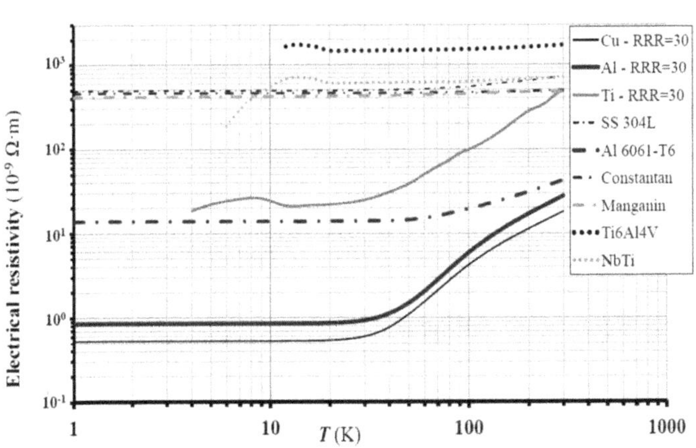

Figure 6.5 Electrical resistivity of various materials versus temperature

guidance are in order. The relative permeability of diamagnetic and paramagnetic materials is generally near unity and is relatively constant with temperature changes. In contrast, the relative permeability of ferromagnetic materials tends to be much higher than unity, often ranging up to 10,000 and occasionally higher. This higher relative permeability is dependent on many interrelated factors, including grain size, grain and domain cooperation and exchange interaction between atoms, annealing processes, crystalline structure, applied magnetic field strength, magnetization history that includes characteristics such as remanence and coercivity, operating temperature range, and, for alloys particularly, the percentage of different components admixed into the alloy. Multiple factors combine to produce the magnetic permeability of a material:

1. Material composition and microstructure: The intrinsic properties of the material, such as its atomic structure and the types of atoms present, significantly influence magnetic permeability. Ferromagnetic materials tend to exhibit a permeability greater than that of free space, while diamagnetic and paramagnetic materials tend to exhibit a permeability very close to that of free space. The physical structure of the material, including grain size, phase distribution, and the presence of impurities, plays a significant role in the determination of the magnetic properties.
2. Temperature: Magnetic permeability is a function of temperature, and in general, as the temperature increases, permeability decreases. Ferromagnetic materials exhibit a phenomenon called the Curie point, above which they lose their ferromagnetic properties and become paramagnetic. The Curie point temperature varies between materials and is of concern if the shield is to be used in high-temperature conditions.
3. Magnetic field strength: Magnetic permeability is a function of applied magnetic field strength. If the applied magnetic field strength is sufficiently high, it can saturate the material, and in such a case, the permeability will drop precipitously.
4. Frequency of the magnetic field: Magnetic permeability is a function of applied frequency. At high frequencies, the permeability can change as a result of effects related to the presence of eddy currents and hysteresis losses, particularly in conductive materials. The magnetic domains in ferromagnetic materials exhibiting high values of permeability are unable to respond to fast-changing magnetic fields at frequencies greater than a few GHz, and relative permeability for such materials typically decreases rapidly towards unity. Nickel, for example, with an initial relative permeability of between 100 and 600, will drop to somewhere between 1 and 2 at 10 GHz. In many cases, in the frequency range within which the relative permeability drops from several hundred to near unity, the frequency only increases ten-fold. Because permeability and frequency are both in the denominator of the skin-depth formula, in this frequency range the skin depth may increase over part of that frequency range.
5. Magnetization history: The previous magnetization of a material can strongly influence the materials' permeability because of residual magnetization. This

is a direct result of remanence, and materials that behave in this fashion will also have a characteristic coercive magnetic field strength to eliminate any such residual magnetization.

6. External conditions: External factors such as pressure and mechanical stress can also affect the magnetic permeability of materials. When they are subjected to cold working, a portion of the austenite in stainless steels such as 304 and 316, which normally do not exhibit an elevated permeability, can be converted to martensite [40]. When this happens, the steel can develop an elevated permeability and can thus be magnetized. The effect is less marked in alloys with high concentrations of austenite stabilizers such as nickel, nitrogen, and carbon. Magnetic attraction effects are most often apparent in wire or the dished end of pressure vessels that have been formed by cold-working deformation. It is possible to remove the magnetic effects by solution annealing and water quenching, but this can reduce the tensile properties of the steel. As mentioned, the addition of Ni, Mn, C, Cu, and/or N can act to stabilize the austenite in the steel and also act to inhibit the formation of martensite during the cold-working process. For example, stainless steel alloy 1.4401 (AISI 316),

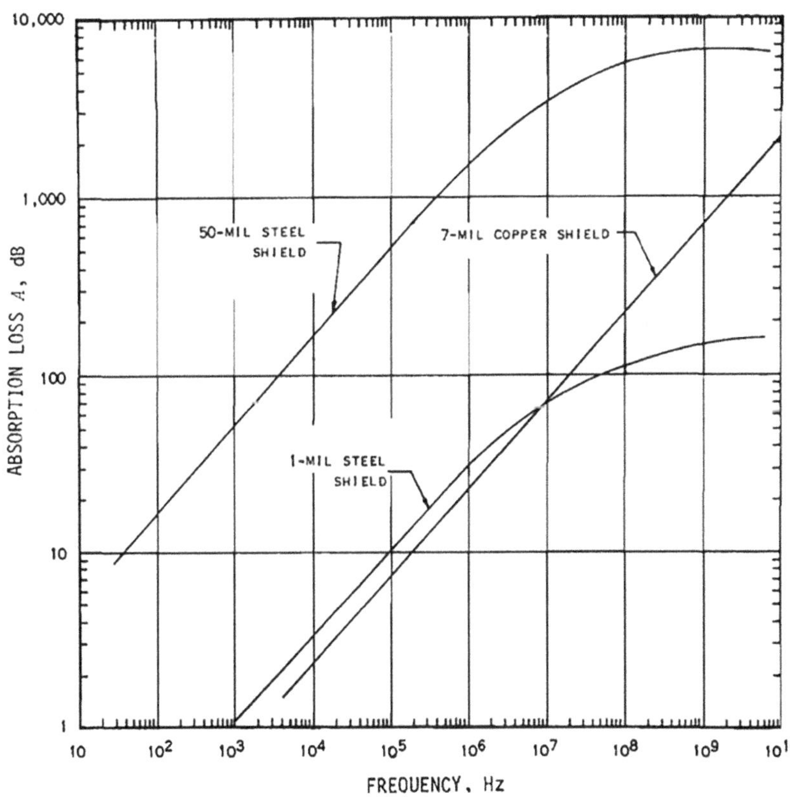

Figure 6.6 Absorption loss for steel and copper shields from 30 Hz to 10 GHz

having a carbon percentage of 0.06%, a chromium percentage of 17.70 %, a nickel percentage of 10.50%, exhibits a baseline relative permeability of 1.08. Even after an 80% reduction, the permeability remains low and normally does not exceed 1.30.

Increased permeability resulting in magnetic attraction effects is most often noticed in heavily cold-worked fabrications such as wire or the dished end of a pressure vessel. It is possible to reduce increased permeability as a function of

SUB-NOTE 1.1(1) Electrical Properties of Shielding Materials (Part II)†

MATERIAL (MAGNETIC)	FREQUENCY (Hz)	CONDUCTIVITY σ	PERMEABILITY μ	$\sigma \times \mu$	σ/μ
Conetic*	0 to 1 k	0.0304	25 115	764	1.2×10^{-6}
Hypernick	0 to 1 k	0.035	4 529	159	7.7×10^{-6}
Iron (Commercial)	0 to 150 k	0.044	54.1	2.4	8.1×10^{-4}
Iron (Purified)	0 to 150 k	0.17	1 000	170	1.7×10^{-4}
	1M	0.17	700	119	2.4×10^{-4}
	3 M	0.17	600	102	2.8×10^{-4}
	10 M	0.17	500	85	3.4×10^{-4}
	15 M	0.17	400	68	4.3×10^{-4}
	100 M	0.17	100	17	0.0017
	1 G	0.17	50	8.5	0.0034
	1.5 G	0.17	10	1.7	0.017
	10 G	0.17	1.0	0.17	0.17
Mumetal	0 to 1 k	0.03	19 833	595	1.5×10^{-6}
Ni-Fe 50	0 to 1 k	0.12	3 162	380	3.8×10^{-5}
Netic (Blue)	0 to 1 k	0.1116	570	63.6	2.0×10^{-4}
Netic (Special)	0 to 1 k	0.1263	440	55.6	2.9×10^{-4}
Netic S3-6	0 to 1 k	0.1263	1 000	126	1.3×10^{-4}
Nichrome	0 to 1 k	0.02	18.2	0.364	0.0011
Nickel	0 to 1 k	0.2	100	20	0.002
Permalloy 4/79	0 to 1 k	0.03	20 667	620	1.45×10^{-6}
Permalloy 45	0 to 1 k	0.04	2 450	98	1.6×10^{-5}
Permalloy 78	0 to 1 k	0.035	2 692	94	1.3×10^{-5}
Primag 40*	0 to 1 k	0.0116	1 700	19.7	6.8×10^{-6}
Primag 90*	0 to 1 k	0.33	780	257	4.2×10^{-4}
Si-Fe 4%	0 to 1 k	0.0247	425	10.5	5.8×10^{-5}
Si-Fe 4% (Oriented)	0 to 1 k	0.11	4 787	527	2.3×10^{-5}
Steel (Cold Rolled)	0 to 200 k	0.17	224	38	7.6×10^{-4}
	1 k	0.17	212	36	8.0×10^{-4}
	6 k	0.17	153	26	1.1×10^{-3}
	8 k	0.17	147	25	1.16×10^{-3}
	10 k	0.17	135	23	1.26×10^{-3}
	15 k	0.17	97	16.5	1.75×10^{-3}
	20 k	0.17	86	14.6	1.97×10^{-3}
	30 k	0 17	59	10	2.88×10^{-3}
	40 k	0.17	49	8.3	3.47×10^{-3}
Steel (Galvanized)	0 to 1 k	0.1766	227	40	7.7×10^{-4}
Steel (Hot rolled)	0 to 1 k	0.1603	160	25.6	0.001
Steel (Terne)	0 to 1 k	0.1517	157	23.8	9.7×10^{-4}
Supermalloy	0 to 1 k	0.029	100 000	2900	2.9×10^{-7}

*Values given hold for magnetic induction between 3 2 and 159.2 A/m, above 318 4 A/m the permeabilities at 1 kHz are as follows:

Primag 40 48 000
Primag 90 60 000

NOTE Inquire at the manufacturers of magnetic materials for permeabilities and other magnetic properties (saturation, etc) if used above 1 kHz! Permeabilities generally decrease with increasing frequencies. The minimum permeabilities of some materials at undefined frequencies are:

Conetic 3000
Netic (Blue) 240
Netic (Special) 125
Netic (S3-6) 220
Primag 40 1300. above 318 4 A/m 1500
Primag 90 560 above 318 4 A/m 3000

Figure 6.7 Table from AFSC design handbook 1-4, DN 5F2

cold-working by solution annealing and water quenching, but this will also very likely reduce the tensile properties and may give rise to distortion.

Most metals with a relative permeability greater than 1 also exhibit a relatively low conductivity. Nickel, iron, and Mu metal are all examples of such metals. Such materials can be very thin and still offer significant absorption capability at a low frequency, precisely what we want for magnetic shielding.

To emphasize the effects of changing frequency, the curves in Figure 6.6 illustrate the variation in absorption as a function of frequency for copper, a diamagnetic material, and for steel, a ferromagnetic material with typically low-to-moderate permeability. Note the absorption loss of 7-mil copper is less than that of 1-mil steel at frequencies below approximately 10 MHz and far below the absorption of 50-mil steel into the GHz range [41].

Additional data for the variation of permeability as a function of frequency can be found in many sources. One such is shown below in Figure 6.7, taken from [42], ppg. 1 and 2, Sub-Note 1.1 (1) parts I and II.

Two more examples of data available are provided in Figures 6.8 and 6.9. These are taken from [41] ppg. 4-112 and 4-113, Figures 4-125 and 4-126.

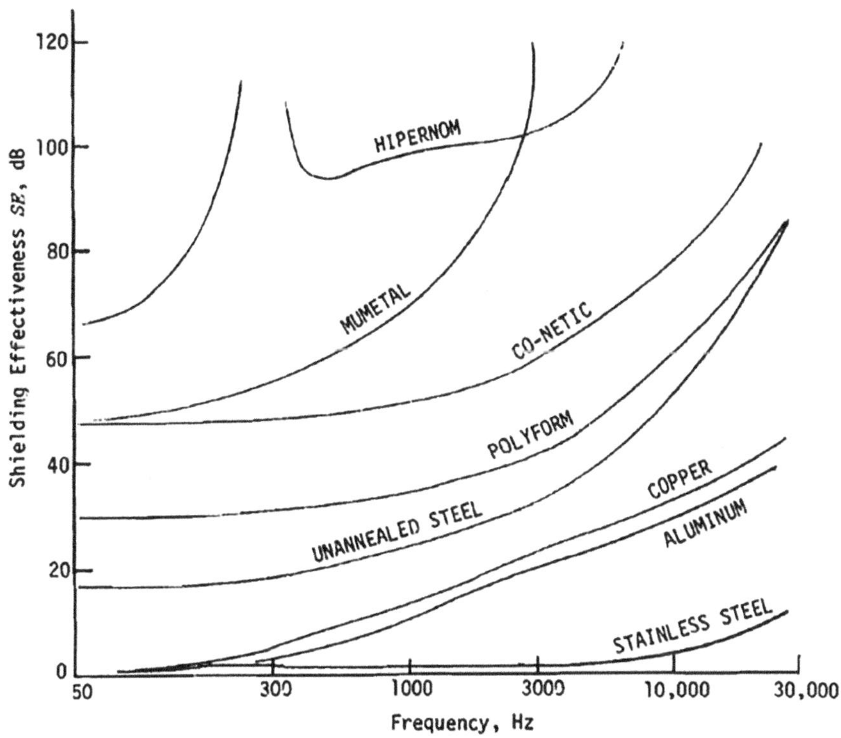

Figure 6.8 Shielding effectiveness of various magnetic materials versus frequency

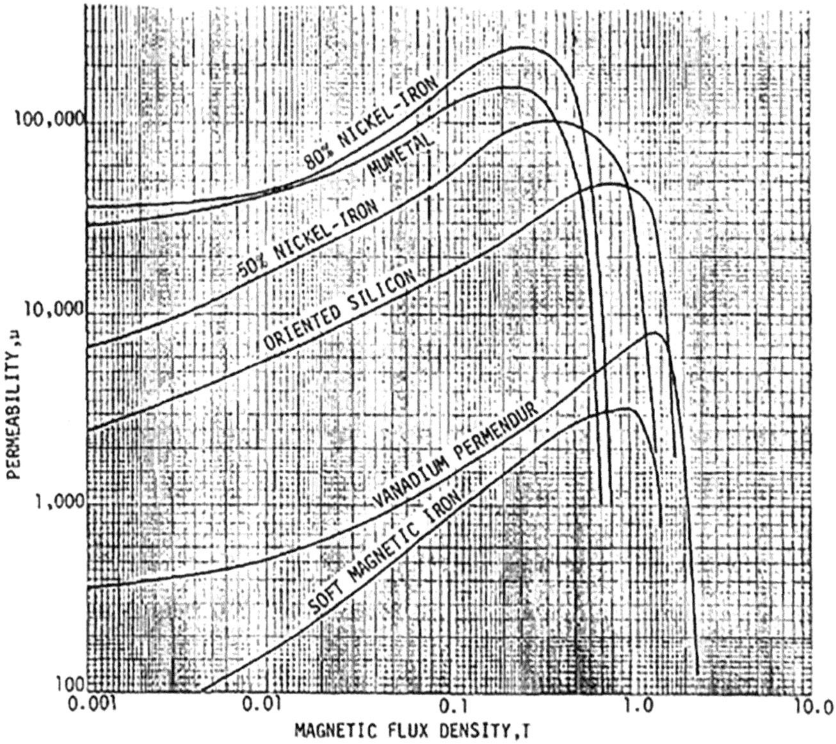

Figure 6.9 Permeability of various magnetic materials versus magnetic flux density

6.3 General rules for selection of shielding materials

Considering the foregoing information, some general rules for selection of shielding materials may be formulated:

1. Materials that exhibit high conductivity such as copper, aluminium, and magnesium should be used for high-frequency shields to obtain the highest reflection loss across frequency.
2. Materials that exhibit high permeability such as iron and Mu metal should be used for low-frequency shields to obtain the highest absorption loss across frequency.
3. To provide a given degree of shielding, reference to documented or calculated absorption loss curves for the selected material will permit quick estimates of the thickness required for the application.
4. Any structurally sound shielding material will usually be thick enough for shielding electric fields at any frequency.

In addition, here are some traits of magnetic shielding materials to keep in mind:

1. Permeability is non-linear.
2. Permeability is dependent on multiple factors and characteristics such as applied field strength, is generally inversely proportional to frequency, and may be complex.
3. Published permeability values are usually initial DC values.
4. Magnetic shielding materials can exhibit widely variant hysteresis curves and coercivity values.
5. Not all nickels are permeable; electroless nickel exhibits a wide range of electrical conductivity and may or may not have any measurable permeability.
6. Pure iron has a relative conductivity equal to some aluminium alloys, but when alloyed 50% with pure nickel, it becomes almost 5× less conductive.
7. The magnetic properties of some magnetic shielding materials can change as a result of machining or forming processes. Some austenitic stainless steels can become highly permeable as a result of cold-working.

6.4 Shielding effectiveness for magnetostatic or very low-frequency magnetoquasistatic applications

The SE of a shield for magnetostatic or very low-frequency magnetoquasistatic shielding is dependent primarily on the permeability of the shield material and the thickness of the shield material, and indirectly on the separation distance between the source and the shield.

Assuming the shield is much thinner than the skin depth, the impedance of the metal is much greater than that of the incident wave, and the absorption is very low, so we have $\Gamma_B l$, k, $2\Gamma_B l$ all $\ll 1$. This allows the following expressions, starting from the fundamental Schelkunoff equation:

$$SE_{mag} = 20\log_{10}\left(\left|\frac{(k+1)^2}{4k}\right| + \left|e^{\Gamma_B t}\right| + \left|1 - \frac{(k-1)^2}{(k+1)^2} e^{-2\Gamma_B t}\right|\right)$$

$$SE_{mag} = 20\log_{10}\left(\left|\frac{1}{4k}\right| |1| |1 - (1 - 4k)(1 - 2\Gamma_B l)|\right)$$

$$SE_{mag} = 20\log_{10}\left(\left|\frac{1}{4k} - \frac{1}{4k}(1 - 2\Gamma_B t - 4k + 8k\Gamma_B t)\right|\right)$$

$$SE_{mag} = 20\log_{10}(1 + \Gamma_B t/2k)$$

Substituting $\Gamma_B = \sqrt{j\omega\mu_0\mu_r\sigma}$; $Z_w = j\omega\mu_0 r$; $Z_m = \sqrt{j\omega\mu_0\mu_r/\sigma}$; $k = Z_w/Z_m$

$$SE_{mag} = 20\log(1 + \mu_r t/2r)$$

Flux shunting, or ducting, is another technique used for low-frequency magnetoquasistatic applications. The process is governed by both Gauss' and Ampere's

Laws, requiring the tangential component of the magnetic field and the normal component of the magnetic induction to be continuous across material discontinuities. The interested reader is encouraged to review the available literature for further information on this topic.

6.5 Multiple shields

Up to this point, we have been discussing single-layer shielding. Another approach employs multiple layers. A layered shield may be configured as a laminated section of multiple layers or as two or more layers each separated from the next by free space. Illustrations of multilaminar/layered shields are provided in Figure 6.10.

Careful bookkeeping of interface impedance ratios will generate equations similar to those previously derived for the R, A, and B terms. For the case of a simple two-layer laminated shield with no space between the layers, such as that illustrated on the right-hand side of Figure 6.10, we have

$$\begin{aligned} S(dB) &= 20\log_{10}\left(\left|\frac{1}{p}\right| + |e^{\Gamma t}| + \left|1 - qe^{-2\Gamma t}\right|\right) \\ &= 20\log_{10}\left(\left|\frac{1}{8}\left(1+\frac{\eta_1}{Z_w}\right)\left(1+\frac{\eta_2}{\eta_1}\right)\left(1+\frac{Z_w}{\eta_2}\right)\right| + |e^{\Gamma_1 t_1} + e^{\Gamma_2 t_2}| \right. \\ &\quad \left. + \left|(1 - q_1 e^{-2\Gamma_1 t_1})(1 - q_2 e^{-2\Gamma_2 t_2})\right|\right) \end{aligned} \tag{6.55}$$

where $Z_w = 120\pi$, and η_1 and η_2 are, respectively, the impedances of each of the two layers in the shield configuration.

Absorption and reflection losses for the two-layer case shown here are fundamentally sums of the individual losses at each interface. Note that η_1 and η_2 both vary as a function of the square root of frequency, so the interaction at the metal-to-metal interface is independent of frequency. The re-reflection term is not a simple sum of terms, since the first reflection coefficient involves the impedance looking into the second sheet.

Equation (6.55) can be broken down into the familiar three components, R, A, and B, as follows:

$$\begin{aligned} R(\mathrm{dB}) &= 20\log_{10}\left|\frac{1}{8}\left(1+\frac{\eta_1}{Z_w}\right)\left(1+\frac{\eta_2}{\eta_1}\right)\left(1+\frac{Z_w}{\eta_2}\right)\right| \\ &= -20\log_{10}\frac{\left|1+\frac{\eta_1}{Z_w}\right|}{2} + 20\log_{10}\frac{\left|1+\frac{\eta_2}{\eta_1}\right|}{2} + 20\log_{10}\frac{\left|1+\frac{Z_w}{\eta_2}\right|}{2} \end{aligned} \tag{6.56}$$

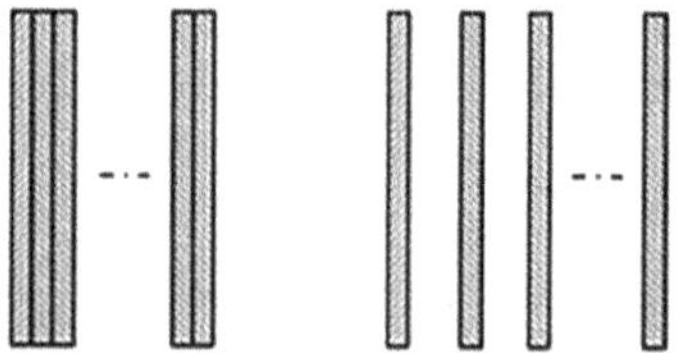

Figure 6.10 Laminated and layered shields

$$A(\text{dB}) = 20\log_{10}\left|e^{\Gamma_1 t_1} + e^{\Gamma_2 t_2}\right| \tag{6.57}$$

$$\begin{aligned} B(\text{dB}) &= 20\log_{10}\left|(1 - q_1 e^{-2\Gamma_1 t_1})(1 - q_2 e^{-2\Gamma_2 t_2})\right| \\ &= 20\log_{10}\left|(1 - q_1 e^{-2\Gamma_1 t_1})\right| + 20\log_{10}\left|(1 - q_2 e^{-2\Gamma_2 t_2})\right| \end{aligned} \tag{6.58}$$

For the case of a double shield with interstitial free space, such as that illustrated on the left-hand side of Figure 6.10, we have entirely similar expressions accounting for the air layer between the shield layers, as follows:

$$\begin{aligned} R(\text{dB}) = &-20\log_{10}\frac{\left|1 + \dfrac{\eta_1}{Z_w}\right|}{2} + 20\log_{10}\frac{\left|1 + \dfrac{Z_w}{\eta_1}\right|}{2} \\ &+20\log_{10}\frac{\left|1 + \dfrac{\eta_3}{Z_w}\right|}{2} + 20\log_{10}\frac{\left|1 + \dfrac{Z_w}{\eta_3}\right|}{2} \end{aligned} \tag{6.59}$$

$$A(\text{dB}) = 20\log_{10}\left|e^{\Gamma_1 t_1} + e^{\Gamma_3 t_3}\right| \tag{6.60}$$

$$\begin{aligned} B(\text{dB}) &= 20\log_{10}\left|(1 - q_1 e^{-2\Gamma_1 t_1})(1 - q_2 e^{-j2\beta_2 t_2})(1 - q_3 e^{-2\Gamma_3 t_3})\right| \\ &= 20\log_{10}\left|(1 - q_1 e^{-2\Gamma_1 t_1})\right| + 20\log_{10}\left|(1 - q_2 e^{-j2\beta_2 t_2})\right| \\ &\quad + 20\log_{10}\left|(1 - q_3 e^{-2\Gamma_3 t_3})\right| \end{aligned} \tag{6.61}$$

6.6 Difficulties with Schelkunoff's theory

Much work has been done to adopt and expand Schelkunoff's theory over the years. It is notable that so much work has been done by so many authors for low-frequency applications, motivated by the apparent failure of Schelkunoff's transmission line method (TLM) to properly predict behaviour in the magnetoquasistatic region. Possible reasons for this include the need to be very accurate with the definition of the wave impedances associated with sources and victims in the low-frequency regime, as well as being very accurate with the behaviour of any magnetic materials as they are impacted by operating conditions. Keeping these thoughts in mind, let's review some of this work.

In 1967, Moser [43] reviewed three separate solutions for low-frequency shielding of a plane shield of infinite extent interposed between two circular loop antennas axially perpendicular to the shield surface. He examined an application of Maxwell's eddy current thin shield analysis credited to Levy in 1936 [44], derived a detailed vector wave equation, but reported that no solutions had been obtained at the time of publication because of difficulties associated with programming skill to hold calculation results within upper and lower machine limits, and then explored the use of Schelkunoff's TLM for this application.

All of the work shown in Moser's 1967 paper was captured in a series of 19 detailed curve plots. Moser concluded

1. The eddy-current solution was inaccurate for shields having a permeability much greater than unity.
2. The vector wave equation solution, while presumed to be the most accurate, was the least useful, since it required numerical solution in all cases. Moser further stated the vector wave equation solution predicted SE was independent of source-to-shield spacing in the low-frequency regime.
3. The Schelkunoff approach proved to be the most useful of the three, but to obtain agreement between calculation and experimental data required arbitrary adjustment of the spacing between the source and the shield such that the source-to-shield distance was equal to 1/3 of the distance between the source and receiving loops.

Bannister followed Moser closely with a paper published in 1968 [45] that examined the same configuration as did Moser. Bannister applied two so-called quasi-near approximations to Moser's exact vector wave equation solution:

1. The propagation constant in air is very nearly equal to zero (true when the measurement distance is $< \lambda/20$).
2. He equated one of the variables in Moser's expression to the propagation constant in the shield (true when the measurement distance is $>> \delta$ in the shield and the shield is thicker than 2δ).

and was able to generate the simple result (for materials with near unity permeability):

$$S(dB) = 8.686\tau_r t + 20\log_{10}\left(\frac{K}{8.485}\right) \tag{6.62}$$

where $K = \tau_r z/\mu_r$, $\tau_r = 1/\delta$, $z =$ distance between loop centres, $t =$ shield thickness, and $\mu_r =$ relative permeability of the shield material.

Remarkably, when Bannister applied the same quasi-near approximations and inserted Moser's 1/3 spacing criteria into Schelkunoff's TLM expression for absorption and reflection terms, he generated an expression identical to (6.62). Note that Bannister excluded the re-reflection term given the assumption the shield is thicker than 2δ.

For materials with permeability greater than unity, Bannister shows an equally simple expression, entirely similar to (6.62)

$$S(\mathrm{dB}) = 8.686\tau_r t + 20\log_{10}\left(\frac{1}{2.828K}\right) \tag{6.63}$$

where all variables are the same as used in (6.62).

Here Bannister was again able to show exact agreement with Schelkunoff's TLM expression for absorption and reflection terms, but this required the source loop to be spaced from the shield by 1/2 the loop centre-to-centre distance. The common thread here is that Schelkunoff's theory is applicable at low frequencies, but with spacing and material constitutive parameter constraints.

Bannister went on to identify that when the quotient of the product of the centre-to-centre spacing of the loop antennas and the skin depth, and the relative permeability of the shield, that the expression for the reflection term becomes more complicated. Even so, Bannister was able to show a simple expression based on Schelkunoff's TLM, yielding:

$$R(\mathrm{dB}) = 20\log_{10}\left(\frac{0.345}{K} + 0.118K + 0.408\right) \tag{6.64}$$

where $K = \tau_r z/\mu_r = 0.768\sqrt{f\sigma_r/\mu_r}r_1$ (r is in inches, and is still equal to 1/2 the loop centre-to-centre distance).

Bannister then converted his expression into

$$R(\mathrm{dB}) = 20\log_{10}\left(\frac{0.462}{\sqrt{\frac{f\sigma_r}{\mu_r}}r_1} + 0.091\sqrt{\frac{f\sigma_r}{\mu_r}}r_1 + 0.408\right) \tag{6.65}$$

He compared this result to that of Schulz *et al.* from 1963 [46] and updated in 1988 [22]:

$$R(\mathrm{dB}) = 20\log_{10}\left(\frac{0.462}{\sqrt{\frac{f\sigma_r}{\mu_r}}r_1} + 0.136\sqrt{\frac{f\sigma_r}{\mu_r}}r_1 + 0.354\right) \tag{6.66}$$

Here, (6.65) is a nearly perfect match.

In spite of the success of Moser and Bannister in their analysis, Bridges and Miller published a letter in the Correspondence section of the IEEE Transactions on EMC in 1968 [47]. In their letter, they bemoaned the apparent and seemingly rampant misapplication of Schelkunoff's TLM, citing multiple instances of large errors, approaches that ignored distant sources, lack of attention to the importance of proper electrical bonding for enclosures, and the generation of extremely thick shields where more rigorous solutions had shown careful selection of materials could produce an equal or better performance with a comparatively thin shield. They closed their letter by stating the preponderance of the data favoured outright rejection of the Schelkunoff approach, particularly for low frequencies, distant sources, and practical enclosures.

Schulz, who conducted extensive research into the application of Schelkunoff's theory, responded to Bridges and Miller's correspondence with agreement in three major problem areas:

1. The use of an incorrect wave impedance for the incident wave.
2. The assumption that in a practical enclosure, the wave emerging into the enclosure volume exhibits the same wave impedance as the incident wave outside the enclosure.
3. The belief that energy transmitted from a source to a receiver through a shield in a practical enclosure passes only through the shield material itself.

Schulz concluded the lack of agreement between Schelkunoff's TLM and measured performance resulted from the application of ideal conditions in the theory that did not exist in the "real" world.

6.6.1 Schulz Point 1

Also in 1968, W.S. Adams [48] acknowledged the importance of the wave impedance of the incident field and the difficulties associated with the determination of this factor. He developed a generalized set of equations derived from Schelkunoff's relationships and then generated a series of both real and imaginary wave impedance components for Hertzian electric and magnetic dipoles, and for finite electric dipoles of various lengths and at three aspect angles, 90°, 60°, and 30°. He also included a BASIC program to calculate the magnitudes of the absorption term, both reflection terms, and the re-reflection term. Adams made an interesting statement in this paper, to wit, that loop antennas were seldom encountered in practice, but that low-impedance, current-carrying conductors could be modelled using the provided curves for the Hertzian magnetic dipole.

Whitehouse published a paper in 1969 [49] addressing the wave impedance to be used for magnetic dipole sources for shielding calculations. Whitehouse derived two "new" expressions for the wave impedance and then went on to show excellent agreement down to about 70 Hz using Schelkunoff's TLM with results from Moser's 1967 rigorous expression from a full vector wave approach. Whitehouse agreed with Moser in that the SE for the applicable low-frequency case was independent of the spacing between the source and the shield.

A paper in 1995 by Nagao *et al.* [50] addressed errors encountered in the application of Schelkunoff's TLM for near-field use. They compared a rigorous vector wave solution, Schelkunoff's TLM, and a relationship using shield material surface resistance (they referred to this as the "resistive film approximation"). The authors determined two separate correction factors, based on three dimensionless parameters derived from Schelkunoff's TLM relationships, and proceeded to show that appropriate application of these correction factors to both Schelkunoff's TLM and the resistive film approximation could generate results in excellent agreement with the results of the rigorous solution.

In 1999, Hansen and Moser published an update to Moser's 1967 paper [51], extending the approach employed by Moser in his 1967 paper to add an electrically small receiving antenna in the near-field. In this paper, Hansen and Moser were able to show a monotonic variation of coupling with loop spacing, unlike the original results in Moser's earlier work.

In 2000, Hansen published a short paper [52] essentially replicating the findings of Whitehouse.

In 2008, Celozzi *et al.* published a volume titled *Electromagnetic Shielding*. In this text, they summarized wave impedance issues very well, as follows:

> High impedance sources in general are producing high frequency electric fields. Even if the source is assumed to be a Hertzian electric dipole, the wave impedance of the incident field is generally much higher than the

intrinsic impedance of the conductive shield itself. This results in a large reflection term. The absorption term is proportional to frequency, so as the frequency increases, so does the absorption term (i.e., the shield is electrically thick). The combination of these two terms is typically of such high magnitude that any errors introduced by the assumption of a Hertzian electric dipole wave impedance are negligible.

Not so in the case of the low impedance source. In general, these sources are producing low frequency magnetic fields. At low frequencies, a conductive shield is electrically thin, so the absorption term in this case is very small. Although the wave impedance of the source is still larger than that of the shield, the difference is much less than for the high impedance source. The re-reflection term contributes a negative impact to the overall shielding effectiveness, and its absolute value can be nearly as large as the reflection term. Because of this situation, even a small error in the determination of the wave impedance can generate a very large error.

Celozzi *et al.* provided in their text expressions usable for low-frequency wave impedances for current loops perpendicular and parallel to a shield and for wires parallel to a shield.

6.6.2 Schulz Point 2

Waves having passed through a shield and entering into an enclosure are extremely likely to encounter near-field effects from currents and voltages existing in internal circuitry and from cavity effects associated with the dimensions of the enclosure. The sum total of these effects can, and will, dramatically impact the wave impedance of the volume inside the enclosure, and thus the wave impedance encountered by the wave emerging into that volume.

Given what we now know about wave impedance, it should be clear that the presence of electric and magnetic fields associated with the normal operation of electronic circuitry can directly impact the wave impedance of the volume inside an electronic enclosure. It goes without saying that such effects are extremely difficult to model or predict.

Cavity theory will not be addressed in this text, but suffice it to say that enclosure cavities may very well exhibit resonant behaviour that can dramatically impact shielding effectivity. A large number of papers have been published on this topic in the literature, and many of them are specific to the shielding problem.

6.6.3 Schulz Point 3

Practical electronic enclosures are often not solidly seamed shut, although in some cases they may seem nearly that well designed. Useful enclosures have covers or access plates and often have connectors and other penetrations, all of which can leak EM energy to some degree. Up to this point, we have been talking about simple plane wave incidence on solid shields. In real life, we rarely encounter solid shields. Practical equipment enclosures have air passages for cooling, access panels, bulkhead-mounted connectors, transparencies, illuminated displays, and so forth. At the simplest level, any solid shield that has even a single aperture in it has an SE dominated by that of the aperture. Multiple apertures only make it worse. The amount of emission "leakage" through an aperture depends on three main items:

1. The maximum linear dimension (not area) of the aperture,
2. The wave impedance of the incident field, and
3. The frequency of the source.

Conversely, the SE of an aperture is dependent on its reflection term, which in turn depends on the impedance of the aperture. The impedance of an aperture may be determined from Babinet's Principle:

$$Z_{ap}Z_{comp} = \frac{Z_0^2}{4} \tag{6.67}$$

where Z_0 is the impedance of the medium within which the shield and the aperture of interest resides, Z_{ap} the impedance of the aperture, and Z_{comp} the impedance of the complementary antenna (usually taken as a dipole structure). As an example, assume the shield is irradiated by a wave approaching from free space such that $Z_0 = 120\pi$, and the aperture's complementary antenna is a half-wavelength dipole with an impedance of $Z_{comp} = 73 + j42.5$. The aperture impedance is then given by

$$\begin{aligned} Z_{ap} &= \frac{Z_0^2}{4Z_{comp}} = \frac{(120\pi)^2}{4(73 + j42.5)} \\ &\cong \frac{35530.58}{73 + j42.5} \cong 363.51 - j211.63 \end{aligned} \tag{6.68}$$

Knowing the aperture impedance will generate the far-field reflection component straightaway, with

$$R(\text{dB}) = 20\log_{10}\left|\frac{1}{p}\right| = 20\log_{10}\left|\frac{(k+1)^2}{4k}\right| \tag{6.69}$$

In this case, since the aperture is equivalent to a half-wavelength dipole, the resulting reflection term is

$$R(\text{dB}) = 20\log_{10}|0.935 - j0.028| \cong -0.58\text{dB} \cong 0\text{dB} \tag{6.70}$$

If the maximum linear dimension $< \lambda/2$, the SE of a single aperture may be estimated by

$$SE_{ap} = 100 - 20\log(lf) + 20\log(1 + \ln(l/h)) + 30(t/l) \tag{6.71}$$

where l is the length of the aperture in mm, f the frequency of the incident wave in MHz, h the height of the aperture in mm, and t the thickness of the aperture in mm.

Appendix A

A.1 Brief refresher of applicable transmission line theory

The fundamental shielding theory discussed in this chapter is based heavily on transmission line theory as it pertains to rectangular and circular waveguides. Fundamental and advanced transmission line theory is well documented, and the interested reader is encouraged to review the many excellent references available. In similar fashion, the reader who has skipped previous material to arrive here is encouraged to return to Chapter 3 should some of what follows seem a bit unfamiliar.

Our discussions for this section will begin with Maxwell's Equations in time harmonic form to represent the transverse electric and magnetic field components in rectangular waveguide, to wit:

$$\frac{\partial \mathcal{E}_z^{\pm}}{\partial x} \pm \gamma \mathcal{E}_y^{\pm} = -j\omega\mu \mathcal{H}_x^{\pm} \tag{A.1}$$

$$\pm\gamma \mathcal{E}_y^{\pm} - \frac{\partial \mathcal{E}_z^{\pm}}{\partial x} = -j\omega\mu \mathcal{H}_y^{\pm} \tag{A.2}$$

$$\frac{\partial \mathcal{E}_y^{\pm}}{\partial x} - \frac{\partial \mathcal{E}_x^{\pm}}{\partial y} = -j\omega\mu \mathcal{H}_z^{\pm} \tag{A.3}$$

$$\frac{\partial \mathcal{H}_z^{\pm}}{\partial y} \pm \gamma \mathcal{H}_y^{\pm} = j\omega\epsilon \mathcal{E}_x^{\pm} \tag{A.4}$$

$$\pm\gamma \mathcal{H}_x^{\pm} - \frac{\partial \mathcal{H}_z^{\pm}}{\partial x} = j\omega\epsilon \mathcal{E}_y^{\pm} \tag{A.5}$$

$$\frac{\partial \mathcal{H}_y^{\pm}}{\partial x} - \frac{\partial \mathcal{H}_x^{\pm}}{\partial y} = j\omega\epsilon \mathcal{E}_z^{\pm} \tag{A.6}$$

These six equations can be reduced to a set of four by successively eliminating $\mathcal{H}_y^{\pm}$ from (A.2) and (A.4) and solving for $\mathcal{E}_x^{\pm}$, eliminating $\mathcal{H}_x^{\pm}$ from (A.1) and (A.5) and solving for $\mathcal{E}_y^{\pm}$, eliminating $\mathcal{E}_x^{\pm}$ from (A.2) and (A.4) and solving for $\mathcal{H}_x^{\pm}$, and finally eliminating $\mathcal{E}_y^{\pm}$ from (A.1) and (A.5) and solving for $\mathcal{H}_y^{\pm}$, to yield

$$\mathcal{E}_x^{\pm} = -\frac{1}{k_c^2}\left[\pm\gamma \frac{\partial \mathcal{E}_z^{\pm}}{\partial x} + j\omega\mu \frac{\partial \mathcal{H}_z^{\pm}}{\partial y}\right] \tag{A.7}$$

$$\mathcal{E}_y^{\pm} = \frac{1}{k_c^2}\left[\mp\gamma \frac{\partial \mathcal{E}_z^{\pm}}{\partial y} + j\omega\mu \frac{\partial \mathcal{H}_z^{\pm}}{\partial x}\right] \tag{A.8}$$

$$\mathcal{H}_x^{\pm} = \frac{1}{k_c^2}\left[j\omega\epsilon \frac{\partial \mathcal{E}_z^{\pm}}{\partial y} \mp \gamma \frac{\partial \mathcal{H}_z^{\pm}}{\partial x}\right] \tag{A.9}$$

$$\mathcal{H}_y^{\pm} = -\frac{1}{k_c^2}\left[j\omega\epsilon \frac{\partial \mathcal{E}_z^{\pm}}{\partial x} \pm \gamma \frac{\partial \mathcal{H}_z^{\pm}}{\partial y}\right] \tag{A.10}$$

Note in (A.7)–(A.10), the term $k_c^2 \equiv \gamma^2 + \omega^2\mu\epsilon$.

We are most interested in transverse electric (TE) modes, in which $\mathcal{E}_z^{\pm} = 0$. With this condition, (A.7)–(A.10) become

$$\mathcal{E}_x^{\pm} = -\frac{j\omega\mu}{k_c^2}\frac{\partial \mathcal{H}_z^{\pm}}{\partial y} \tag{A.11}$$

$$\mathcal{E}_y^{\pm} = \frac{j\omega\mu}{k_c^2}\frac{\partial \mathcal{H}_z^{\pm}}{\partial x} \tag{A.12}$$

$$\mathcal{H}_x^{\pm} = \mp\frac{\gamma}{k_c^2}\frac{\partial \mathcal{H}_z^{\pm}}{\partial x} \tag{A.13}$$

$$\mathcal{H}_y^{\pm} = \mp\frac{\gamma}{k_c^2}\frac{\partial \mathcal{H}_z^{\pm}}{\partial y} \tag{A.14}$$

from which we can express an intrinsic wave impedance for the waveguide as

$$\eta_{TE} \equiv \frac{j\omega\mu}{\gamma} = \frac{\mathcal{E}_x^+}{\mathcal{H}_y^+} = -\frac{\mathcal{E}_x^-}{\mathcal{H}_y^-} = -\frac{\mathcal{E}_y^+}{\mathcal{H}_x^+} = \frac{\mathcal{E}_y^-}{\mathcal{H}_x^-} \tag{A.15}$$

For TE mode solutions, we need to make some assumptions to facilitate our progress:

1. The waveguide is several wavelengths long and has dimensions of a for width and b for height.
2. The dielectric material inside the waveguide is lossless and has parameters μ and ϵ, and has $\rho_v = \boldsymbol{J} = 0$.
3. The walls of the waveguide are fabricated of perfect conducting material.
4. All field quantities vary as $e^{j\omega t \pm \gamma z}$.
5. $\mathcal{E}_z^{\pm} = 0$.

We proceed by using an homogeneous wave equation for **H** (refer to Section 3.2.8 in Chapter 3) from which we can form a scalar wave (A.16) to obtain a solution for $\mathcal{H}_z^{\pm}$.

$$\frac{\partial^2 \mathcal{H}_z^{\pm}}{\partial x^2} + \frac{\partial^2 \mathcal{H}_z^{\pm}}{\partial y^2} + k_c^2 \mathcal{H}_z^{\pm} = 0 \tag{A.16}$$

Equation (A.16) can be solved using a separation of variables technique with application of suitable boundary conditions to yield

$$\mathcal{H}_z^{\pm} = H_{z,mn}^{\pm} \cos\frac{m\pi}{a}x \cos\frac{n\pi}{b}y \quad m, n = 0, 1, 2, \ldots \tag{A.17}$$

where $H_{z,mn}^{\pm}$ represents the complex amplitude of any positive z or negative z travelling $\mathcal{H}_z$ component associated with the specific values of the mode numbers m and n.

Substituting (A.17) into (A.11)–(A.14) yields

$$\mathcal{E}_x^{\pm} = -\frac{j\omega\mu}{k_c^2}\frac{\partial \mathcal{H}_z^{\pm}}{\partial y} \equiv E_{x,mn}^{\pm}\cos\frac{m\pi}{a}x\sin\frac{n\pi}{b}y \tag{A.18}$$

$$\mathcal{E}_y^{\pm} = \frac{j\omega\mu}{k_c^2}\frac{\partial \mathcal{H}_z^{\pm}}{\partial x} \equiv E_{y,mn}^{\pm}\sin\frac{m\pi}{a}x\cos\frac{n\pi}{b}y \tag{A.19}$$

$$\mathcal{H}_x^{\pm} = \mp\frac{\gamma}{k_c^2}\frac{\partial \mathcal{H}_z^{\pm}}{\partial x} \equiv H_{x,mn}^{\pm}\sin\frac{m\pi}{a}x\cos\frac{n\pi}{b}y \tag{A.20}$$

$$\mathcal{H}_y^{\pm} = \mp\frac{\gamma}{k_c^2}\frac{\partial \mathcal{H}_z^{\pm}}{\partial y} \equiv H_{y,mn}^{\pm}\cos\frac{m\pi}{a}x\sin\frac{n\pi}{b}y \tag{A.21}$$

Note in (A.8)–(A.11), the term $k_c^2 \equiv \gamma^2 + \omega^2\mu\epsilon = \left(\frac{m\pi}{a}\right)^2 + \left(\frac{n\pi}{b}\right)^2$, where a and b represent the width and height of the rectangular waveguide, respectively.

The propagation constant is then given by

$$\gamma_{mn} = \sqrt{\left[\left(\frac{m\pi}{a}\right)^2 + \left(\frac{n\pi}{b}\right)^2\right] - [\omega^2\mu\epsilon]} \tag{A.22}$$

This propagation constant is either pure real, or pure imaginary, depending on the value of the terms in the brackets in (A.22). The frequency at which these two terms is equal is known as the cut-off frequency, and is given by

$$\omega_{c,mn}\sqrt{\mu\epsilon} = \sqrt{\left[\left(\frac{m\pi}{a}\right)^2 + \left(\frac{n\pi}{b}\right)^2\right]} \tag{A.23}$$

$$2\pi f_{c,mn}\sqrt{\mu\epsilon} = \sqrt{\left[\left(\frac{m\pi}{a}\right)^2 + \left(\frac{n\pi}{b}\right)^2\right]} \tag{A.24}$$

$$f_{c,mn} = \frac{1}{2\pi\sqrt{\mu\epsilon}}\sqrt{\left[\left(\frac{m\pi}{a}\right)^2 + \left(\frac{n\pi}{b}\right)^2\right]} \tag{A.25}$$

Now, by definition,

$$\gamma = \alpha + j\beta \tag{A.26}$$

Using this, we can rewrite (A.22) to identify the real part of γ_{mn} for a condition wherein $f < f_c$ (this condition is known as either below cutoff or beyond cutoff).

$$\begin{aligned}\gamma_{mn} &= \alpha = \sqrt{\left[\omega_{c,mn}^2\mu\epsilon\right] - [\omega^2\mu\epsilon]} \\ &= \omega\sqrt{\mu\epsilon}\sqrt{\left[\frac{\omega_{c,mn}}{\omega}\right]^2 - 1} = \omega\sqrt{\mu\epsilon}\sqrt{\left[\frac{f_{c,mn}}{f}\right]^2 - 1}\end{aligned} \tag{A.27}$$

Now we can substitute (A.27) into (A.15) to yield

$$\eta_{TE} \equiv \frac{j\omega\mu}{\gamma} = \frac{j\omega\mu}{\alpha} = \frac{j\omega\mu}{\omega\sqrt{\mu\epsilon}\sqrt{\left[\frac{f_{c,mn}}{f}\right]^2 - 1}} \tag{A.28}$$

We know the lowest TE mode for rectangular waveguide is the TE_{10} mode, so that m and n in (A.17) through (A.25) are equal to 1 and 0, respectively. This leads to (A.25) equating to

$$f_{c,mn} = \frac{1}{2\pi\sqrt{\mu\epsilon}}\sqrt{\left(\frac{\pi}{a}\right)^2} = \frac{1}{2a\sqrt{\mu\epsilon}} \tag{A.29}$$

Substituting this into (A.28) yields

$$\eta_{TE} = \frac{j\omega\mu}{\omega\sqrt{\mu\epsilon}\sqrt{\left[\frac{1}{2af\sqrt{\mu\epsilon}}\right]^2 - 1}} \tag{A.30}$$

$$= \frac{j\omega\mu}{\left[\left(\omega\sqrt{\mu\epsilon}\right)/\left(2af\sqrt{\mu\epsilon}\right)\right]\sqrt{1 - \left(2af\sqrt{\mu\epsilon}\right)^2}} \tag{A.31}$$

Recalling that $f < f_{c,mn}$, as f becomes smaller with respect to $f_{c,mn}$

$$\frac{1}{\sqrt{1 - \left(2af\sqrt{\mu\epsilon}\right)^2}} = \frac{1}{\sqrt{1 - \left(\frac{f}{f_{c,mn}}\right)^2}} \approx 1$$

so that (A.31) becomes

$$\eta_{TE} \approx \frac{j\omega\mu a}{\pi} \tag{A.32}$$

Here, (A.32) is the wave impedance of the aperture of a rectangular waveguide operating beyond cutoff.

An entirely similar process may be used to derive the wave impedance for TE modes in circular waveguide, given here without derivation.

$$\eta_{TE,Circ} \approx \frac{j\omega\mu d}{3.682} \tag{A.33}$$

The attenuation of the waveguide can be estimated by taking the ratio of the input wave to the output wave. The time domain expression for the electric field in a rectangular waveguide operating beyond cutoff can be given by

$$E(t) = Ee^{-\beta_0 z}\sin(kx)\cos(\omega t)\widehat{a}_y \tag{A.34}$$

Operation beyond cutoff produces what are sometimes referred to as evanescent waves. An evanescent wave is defined as a non-propagating wave that decreases exponentially over some distance. Note that an evanescent wave also does not produce a net energy flow. The electric and magnetic fields that exist in the reactive near-field region of an antenna are evanescent fields, as they collapse back into the source every other cycle.

The attenuation ratio of interest can be determined by comparing the field in the waveguide at the same x-position (width) at the input and the output to obtain

$$attenuation = \frac{E(t, x_0)}{E(t, x_l)} = \frac{Ee^{-\beta_0(0)} \sin(kx) \cos(\omega t) \widehat{a}_y}{Ee^{-\beta_0(l)} \sin(kx) \cos(\omega t) \widehat{a}_y} \tag{A.35}$$

which reduces to

$$attenuation = \frac{1}{e^{-\beta_0(l)}} \tag{A.36}$$

Equation (A.35) may be written in terms of decibels as follows:

$$\begin{aligned} attenuation_{dB} &= 20\log_{10}\left(\frac{1}{e^{-\beta_0(l)}}\right) = -20\log_{10}\left(e^{-\beta_0(l)}\right) \\ &\approx 8.686\beta_0(l) \end{aligned} \tag{A.37}$$

Now for a waveguide operating in TE_{10} mode, so that m and n in (A.7) through (A.25) are equal to 1 and 0, respectively, (A.37) may be written as

$$\begin{aligned} attenuation_{dB} &= 8.686(l)\sqrt{\left(\frac{m\pi}{a}\right)^2 - \left(\frac{\omega}{c}\right)^2} = 8.686(l)\sqrt{\left(\frac{\omega_c}{c}\right)^2 - \left(\frac{\omega}{c}\right)^2} \\ &= \frac{8.686(l)\omega}{c}\sqrt{\left(\frac{\omega_c}{\omega}\right)^2 - 1} \end{aligned} \tag{A.38}$$

where $\omega_c = \dfrac{m\pi}{a}c$.

As the ratio in (A.38) increases (ω becomes smaller), (A.39) approaches

$$attenuation_{dB,rect} = \frac{8.686(l)\omega_c}{c} = \frac{8.686(l)m\pi}{a} \approx 27.29\frac{l}{a} \tag{A.39}$$

where a = the width (or longest side) of the waveguide.

An entirely similar process may be used to derive the wave attenuation for TE modes in circular waveguide, given here without derivation.

$$attenuation_{dB,circ} \approx 31.93\frac{l}{d} \tag{A.40}$$

where d = the diameter of the waveguide.

References

[1] J.C. Maxwell, *A Treatise on Electricity and Magnetism, Vol. 1*, Clarendon Press, London, UK, 1881, sections 203–205.

[2] J.C. Maxwell, "On the Induction of Electric Currents in an Infinite Plane Sheet of Uniform Conductivity," *Proceedings of the Royal Society of London*, vol. 20, no. 132, pp. 160–168, 1872.

[3] J.C. Maxwell, *A Treatise on Electricity and Magnetism, Vol. 1*, Clarendon Press, London, UK, 1881, sections 654–667.

[4] Rayleigh, "On the Passage of Waves Through Apertures in Plane Screens and Allied Problems," *London, Edinburgh, and Dublin Philosophical Magazine and Journal of Science*, series 5, vol. 43, no. 263, pp. 259–272, 1897.

[5] Larmor, "Electromagnetic Induction in Conducting Sheets and Solid Bodies," *The London, Edinburgh, and Dublin Philosophical Magazine and Journal of Science*, series 5, vol. 17, no. 103, pp. 1–23, 1884.

[6] Sir W. Thomson, "On Electrostatic Screening by Gratings, Nets, or Perforated Sheets of Conducting Metals," *Proceedings of the Royal Society of London*, vol. 49, no. 300, pp. 405–418, 1891.

[7] H. Hertz, "On the Propagation of Electric Waves by Means of Wires," *Wiedemann's Annalen* vol. 37, p. 39, 1889. English translation by DE jones, in "Electric Waves being Researches on the Propagation of Electric Action with Finite Velocity Through Space," London and New York: Macmillan, 1893

[8] H. Lamb, "On the Reflection and Transmission of Electric Waves by a Metallic Grating," *Proceedings of the London Mathematical Society*, vol. s1-29, no. 1, pp. 523–546, 1897. DOI:10.1112/plms/s1-29.1.523.

[9] H. Lamb, "Passages of E.M. Waves through Wire Grids," *Proceedings of the London Mathematical Society*, vol. 29, pp. 523–543, 1898.

[10] G.G. MacFarlane, "Surface Impedance of an Infinite Wire Grid at Oblique Angles of Incidence," (parallel polarization), *Journal of the Institution of Electrical Engineers*, vol. 93, no. 10 (III A), pp. 1523–1527, 1946.

[11] J.H. Morecroft and A. Turner, "The Shielding of Electric and Magnetic Fields," *Proceedings of the Institute of Radio Engineers*, vol. 13, no. 4, pp. 477–505, 1925.

[12] S.A. Schelkunoff, "Electromagnetic Theory of Coaxial Transmission Lines," *Bell Systems Technical Journal*, vol. 13, no. 4, pp. 532–579, 1934.

[13] S.A. Schelkunoff, "The Impedance Concept and its Application to Problems of Reflection, Refraction, Shielding and Power Absorption," *Bell Systems Technical Journal*, vol. 17, iss. 1, pp. 17–48, Jan 1938.

[14] S.A. Schelkunoff, *Electromagnetic Waves*, 2nd ed., D. Van Nostrand, Boston, MA, USA, 1943, particularly sections 8.18 and 8.19, pp. 303–315.

[15] H. Kaden, *Wirbelstrome und Schirmung in der Nachrichtentechnik*, Springer, Berlin,1959.

[16] F. Moeller, "Die elektromagnetische Abschirmung," *Archiv für Elektrotechnik*, p. 328, 1948.

[17] J.P. Vasseur, "Diffraction des Ondes Electromagnetiques par des Ouvertures dans les Ecrans Plans Conducteurs," *Annales de Physique*, vol. 12, no.7, pp. 506–563, 1952.

[18] J. Miedzinski, "Electromagnetic Screening, Theory and Practice", *The British Electrical and Allied Industries Research Association*, Thorncroft Manor, Dorking Road, Leatherhead, Surrey, UK: Tech Report M/TI35, 1959.

[19] D.A. Miller J.E. Bridges, "Review of Circuit Approach to Calculate Shielding Effectiveness," *IEEE Transactions on Electromagnetic Compatibility*, vol. EMC-10, no. 1, pp. 52–62, 1968.

[20] J.E. Bridges, "An Update on the Circuit Approach to Calculate Shielding Effectiveness," *IEEE Transactions on Electromagnetic Compatibility*, vol. 30, no. 3, pp. 211–221, 1988.

[21] R.B. Schulz, G.C. Huang, W.L. Wiliams, "RF Shielding Design," *IEEE Transactions on Electromagnetic Compatibility*, vol. EMC-10, no. 1, pp. 168–175, 1968.

[22] R.B. Schulz, V.C. Plantz, and D.R. Brush, "Shielding Theory and Practice," *IEEE Transactions on Electromagnetic Compatibility*, vol. 30, no. 3, pp. 187–201, 1988. DOI:10.1109/15.3297.

[23] J.R. Moser, "Low-Frequency Shielding of a Circular Loop Electromagnetic Field Source," *IEEE Transactions on Electromagnetic Compatibility*, vol. 9, no. 1, pp. 6–18, 1967.

[24] R.L. Monroe, "EMP Shielding Effectiveness and Mil-STD-285," Harry Diamond Laboratories, Washington DC, USA, Tech. Rep. HDL-TR-1636, Jul 1973.

[25] R.L. Monroe, "A Theory of Electromagnetic Shielding with Applications to MIL-STD-285, IEEE-299 and EMP Simulation," *Harry Diamond Laboratories*, Washington DC, USA, Tech. Rep. HDL-CR-85-052-1, Feb 1985.

[26] R.L. Monroe, "*Research in Electromagnetic Shielding Theory: Part 1. Shielding by Rectangular Enclosures*," Harry Diamond Laboratories, Washington DC, USA, Tech. Rep. HDL-CR-88-052-1, Mar 1988.

[27] R.L. Monroe, "*Research in Electromagnetic Shielding Theory: Part 2. EMP Solution Using Small Loops*," Harry Diamond Laboratories, Washington DC, USA, Tech. Rep. HDL-CR-88-052-2, Mar 1988.

[28] E.F. Vance, "Shielding Effectiveness of Braided Wire Shields," Interaction Note 172, 1974, [Online]. https://ece-research.unm.edu/summa/notes/In/0172.pdf (accessed May 6, 2024.).

[29] E.F. Vance, "Shielding Effectiveness of Braided Wire Shields," *IEEE Transactions on Electromagnetic Compatibility*, vol. EMC-17, no. 2, pp. 71–77, 1975.

[30] E.F. Vance, *Coupling to Shielded Cables*, John Wiley & Sons, New York, NY, 1978.

[31] D.R.J. White, *Electromagnetic Shielding Materials and Performance*, 2nd ed., Don White Consultants, Inc., Gainesville, VA, 1980.

[32] D.R.J. White, *A Handbook on Shielding Design Methodology and Procedures*, Interference Control Technologies, Inc., Gainesville, VA, 1986.

[33] D.R.J. White and M. Mardiguian, *Electromagnetic Shielding, Volume 3 of A Handbook Series on Electromagnetic Interference and Compatibility*, Interference Control Technologies, Inc., Gainesville, VA, 1988.

[34] S. Celozzi, R. Araneo, and G. Lovat, *Electromagnetic Shielding*, Wiley Interscience, Hoboken, NJ, 2008.

[35] C.R. Paul, R.C. Scully, and M. Steffka, *Introduction to Electromagnetic Compatibility*, 3rd ed., John Wiley & Sons, Hoboken, NJ, 2023, ch. 10 and sections 9.5 and 9.6.

[36] H.W. Ott, *Electromagnetic Compatibility Engineering*, John Wiley & Sons, Hoboken, NJ, 2009, chs. 2 and 6.

[37] C.S. Vasaka, "Theory, Design, and Engineering Evaluation of Radio-Frequency Shielded Rooms," *Aero. Electronic and Electrical Lab., Bureau of Aeronautics, US Naval Air Development Center*, Johnsville, PA, Report No NADC-EL-54129, Aug. 13, 1956, page 16.

[38] R.A. Matula, "Electrical Resistivity of Copper, Gold, Palladium and Silver", *Journal of Physical and Chemical Reference Data*, Vol. 8, no. 4, 1979, pp. 1147–1298. DOI:10.1063/1.555614.

[39] P. Duthil, "Material Properties at Low Temperature", *Proceedings of the CAS-CERN Accelerator School: Superconductivity for Accelerators*, Erice, Italy, Apr 24–May 4, 2013.

[40] C.B. Post and W.S. Eberly, "Stability of Austenite in Stainless Steels," *Transactions of the American Society for Metals*, vol. 39, pp. 868–890, 1947.

[41] *Engineering Design Handbook, Electromagnetic Compatibility*, 1977, pp. 4–79, Figure 4-88.

[42] *Air Force Systems Command Design Handbook*, 1-4, 1984, Design Note 5F2, page 2, Sub-Note 1.1 (1) part II.

[43] J.R. Moser, "Low Frequency Shielding of a Circular Loop Electromagnetic Field Source," *IEEE Transactions on Electromagnetic Compatibility*, vol. EMC-9, no. 1, 1967.

[44] S. Levy, "Electromagnetic Shielding Effect of an Infinite Plane Conducting Sheet Placed Between Circular Coaxial Coils," *Proceedings of the Institute of Radio Engineers*, vol. 24, no 6, 1936.

[45] P.R. Bannister, "New Theoretical Expressions for Predicting Shielding Effectiveness for the Plane Wave Case," *IEEE Transactions on Electromagnetic Compatibility*, vol. EMC-10, no. 1, 1968.

[46] R.B. Schulz, V.C. Plantz and D.R. Brush, "Shielding theory and practice", *Proceedings of 9th Tri-Service Conference on Electromagnetic Compatibility*, pp. 597–636, 1963.

[47] J.E. Bridges and D.A. Miller, "Comparison of Shielding Calculations," *IEEE Transactions on Electromagnetic Compatibility*, vol. EMC-10, no. 1, pp. 175–177, 1968. DOI:10.1109/TEMC.1968.302926.

[48] W.S. Adams and A.H. Mills, "Electromagnetic Shielding in the Near Field" IEEE International Symposium on Electromagnetic Compatibility, Jul 23–25, 1968, Seattle, WA.

[49] A.C.D. Whitehouse, "Screening: New Wave Impedance for the Transmission Line Analogy," *Proceedings of the Institution of Electrical Engineers*, vol. 116, no. 7, 1969, pp. 1159–1164. DOI:10.1049/piee.1969.0216.

[50] H. Nagao, A. Nishikata, and Y. Shimizu, "Errors and Correction Factors for Schelkunoff's Shielding Effectiveness Formula," *Electronics and Communications in Japan* (part 1: Communications), 1995. DOI: 1002/ecja.4410781004.

[51] R.C. Hansen and J.R. Moser, "Loop-Shield-Loop Shielding Effectiveness," *IEEE Transactions on Electromagnetic Compatibility*, vol. 41, no. 2, 1999.

[52] R.C. Hansen, "Shielding Formulas for Near Fields," *Microwave Journal*, vol. 11, 2000, pp. 154. https://www.microwavejournal.com/articles/3087-shielding-formulas-for-near-fields

Chapter 7
Practical shielding theory for wire structures

The previous six chapters have provided a discussion of the electromagnetic behaviour of conductors and dielectrics most often used in cable design; some basic electromagnetic theory and a bit of basic mathematics to aid in understanding; some additional electromagnetic theory with emphasis on the interactions between wire structures; a bit of basic antenna theory with emphasis on dipoles and their behaviour; and the description and discussion of shielding theory developed by Schelkunoff, together with some discussions of the shortfalls of this theory that have been addressed over time, all of which have led us to this final chapter. Certainly, it is possible to regard each chapter as a stand-alone measure of information, and it is hoped the content of these previous chapters will prove to be of some practical use to the reader as such. Better still, perhaps, it is possible to integrate these stand-alone measures of information in a fashion that makes it clear proper cable construction and installation are critical aspects of system electromagnetic compatibility (EMC) design. A number of general cable shielding references are included for the reader [1–36].

Common practice within many organizations is to take the direction that new harness design will simply duplicate previous design. In a manufacturing environment that is churning out multiple copies of the same product all of which are intended for use in relatively the same electromagnetic environment, this can be a reasonable approach, presuming that previous design was proven through certification and customer experience.

Unfortunately, the electromagnetic environment is in constant flux, as evidenced by the continuous release and reissue of widely accepted standards for the control of undesirable electromagnetic environmental effects. The evolution of Mil-STD-461 is a case in point. In 1968, the radiated susceptibility limit in Mil-STD-461A [37] was 1 V/m. In 1980, when revision B [38] was released, the singular limit had become a table (see Table 7.1).

By 2015, revision G [39], the table had grown further (see Table 7.2).

Table 7.1 Mil-STD-461B RS03 table

Frequency range	Electric field strength (V/m)
14 kHz to 2 MHz	10 (Army procurements 1 V/m)
2 MHz to 30 MHz	10
30 MHz to 10 GHz	5
10 GHz to 40 GHz	20

Table 7.2 Mil-STD-461G RS103 table

Frequency range	Platform	Limit levels (volts/metre)							
		Aircraft (external or safety critical)	**Aircraft internal**	**All ships (above deck and exposed below deck) and submarines (external)***	**Ships (metallic) (below decks)**	**Ships (non-metallic) (below deck) ****	**Submarine (internal)**	**Ground**	**Space**
2 MHz to 30 MHz	A	200	200	200	10	50	5	50	20
	N	200	200	200	10	50	5	10	20
	AF	200	20	–	–	–	–	10	20
30 MHz to 1 GHz	A	200	200	200	10	10	10	50	20
	N	200	200	200	10	10	10	10	20
	AF	200	20	–	–	–	–	10	20
1 GHz to 18 GHz	A	200	200	200	10	10	10	50	20
	N	200	200	200	10	10	10	50	20
	AF	200	60	–	–	–	–	50	20
18 GHz to 40 GHZ	A	200	200	200	10	10	10	50	20
	N	200	60	200	10	10	10	50	20
	AF	200	60	–	–	–	–	50	20

Key: A, Army; N, Navy; AF, Air Force.

*For equipment located external to the pressure hull of a submarine but within the superstructure, use ships (metallic) (below deck).

**For equipment located in the hanger deck of aircraft carriers.

Within the plethora of international EMI standards, this same trend is visible over time. IEC 61000-4-3 was first released in 1995, with three predefined levels, beginning with 1 V/m, then 3 V/m, and then 10 V/m. The most recent revision of this document, released in 2020, contains four predefined levels: 1, 3, 10, and 30 V/m. While one could say these levels have been essentially static across time, the addition of the 4th level reflects that even within the commercial sector, the recognized electromagnetic threat has increased. Further insight can be gleaned from a review of IEC TR 61000-2-5, first released in 1995 and most recently in 2017. The document has always provided guidance for the determination of various immunity levels to be applied under a wide variety of conditions, conducted and radiated. Some sense of the scope of change in the content of this document can be gained from the realization that in 1995, the document only had roughly 100 pages (50 pages in French, 50 pages in English), while the 2017 revision has 150 pages in English alone.

Cable shielding is of course not restricted to protection from the external electromagnetic environment. Cable shielding can be very effective in preventing undesirable radiated emissions related to time-varying currents that necessarily need to flow in some circuit applications. An equally voluminous number of standards compared to references in the previous paragraph address radiated emissions, and shielding plays a central role in the control and mitigation of said radiated emissions.

Aside from satisfying one or more electromagnetic environmental effects standards, at a more basic level cable shielding may prove to be necessary for simple functionality of a design. Here's a case in point [40]:

> The Wake Shield Facility (WSF) was a NASA experimental science platform that was placed in low Earth orbit by the Space Shuttle on 3 separate missions. The purpose of the WSF was to explore use of the ultrahigh vacuum extant inside the science platform to study epitaxial film growth. The first WSF mission was to be deployed on an extensible cable and towed behind STS-60, launched February 3, 1994. Unfortunately, EMI in the attitude control system prevented full cable deployment and the WSF was restricted to remaining secured to the end of the Canadarm robotic boom. The EMI was determined to be caused by inductive coupling (crosstalk) between the unshielded attitude control sensor cable and the power bus of the spacecraft. The control systems cable was redesigned and shielding added prior to the next mission.

The fact is that even with a manufacturing process that has been established for some time and has been successful over that time, it is prudent to periodically examine what changes have occurred in both internal and external electromagnetic environments and exercise due diligence in addressing any changes that might impact existing cable design or that might impact any new cable designs being explored. This concept may seem trivial to a manufacturer of some small tabletop device intended to be operated in a private home setting, but perhaps not so for the manufacturer of a large-body commercial air transport aircraft intended to carry hundreds of passengers on international flight itineraries. This becomes even more important for those operations that are designing and building one-off products, such as a satellite manufacturer constructing vehicles with specific missions in interplanetary space.

Regardless of the potential for reuse, from an EMC perspective, the process of proper cable shielding design cannot be wholly based on some set of cookbook ideas or memorized recipes derived from or developed for some previous experience. No single simple approach can be applied universally to large cross sections of products or across huge swaths of the frequency spectrum. While it is true, as stated previously, that in some cases cable designs may not need to evolve or may only need minor changes from one version to the next. In many other cases cable designs may be unique from one product to the next. Because of the uniqueness of each design, what is found to work satisfactorily in one application may not work at all satisfactorily in another application and in fact may actually worsen the noise problem or problems being addressed. Designing effective cable shielding is in fact a rather complicated process that should be approached with the same care and consideration as other electrical engineering tasks.

What sort of cable shielding is required for what type of circuit is dependent on multiple factors, including exposure to internal and external EM environments, circuit characteristics such as length, operating voltages and currents, and terminating impedances. Some practical engineering considerations for cable and wire shielding design include

1. Material electromagnetic constitutive characteristics such as conductivity, permeability, and permittivity
2. Material physical characteristics such as thermal conductivity, outgassing, and operational temperature range
3. A potential host of environmental conditions, dependent on application
4. Mass and volume impact
5. Control of mechanical stress and vibration
6. The type of shield under consideration (Braided? Solid? Spiral Wrap? Cigarette Wrap?)
7. Harness flex, size, shape and routing
8. Apertures and penetrations
9. Connector and backshell design/selection
10. Desired/required surface finish/treatments

A critical aspect of cable shield design involves careful selection and segregation of different types of circuit characteristics, accounting for differences in voltages, currents, terminating impedances, and signal types. Knowledge of these parameters can be used to establish at the outset what sort of shielding design might be the “best place to start”. An example of this sort of approach is Table 7.3 [41].

As can be seen, classifications are defined in terms of operating frequency, rise and fall times for pulses, voltage or sensitivity, source and load impedances, circuit classification, wire type, and the shield “ground”.

Table 7.3 Wire EMC classification

Frequency *f*: Rise Fall times (ms) tr, tf	**Voltage or sensitivity**	**Source impedance (ohm)**	**Load impedance (ohm)**	**Circuit class**	**Wire type**	**Shield ground**
Analog	≤ 100 mV	All	<600 k	ML	TWS	MPG
(AC, DC)	≤ 100 mV	All	≥ 600 k	ML	TWDS	MPG
$f \leq 50$ kHz tr,	<6 V	All	All	ML	TWS	MPG
tf > 10 μs	6–40 V	All	All	HO	TW	None
	>40 V	All	All	EO	TW	None
50 kHz$< f$	<100 mV	All	All	RF	TWDS	MPG
$\leq$ 4MHz:tr,	≥ 100 mV	All	All	RF	TWS	MPG
tf ≤ 10 μs						
$f >$ 4MHz	All	All	All	RF	TWS	MPG

Acronyms and abbreviations: ML, HO, EO, Arbitrary nomenclature to define. circuit classification; MPG, Multiple point ground; RF, Radio frequency; TW, Twisted; TWDS, Twisted double shielded; TWS, Twisted shielded.

Cable shielding configurations come in many flavours:

1. No shields (YES, this is a shielding configuration)
2. Shields with all ends not connected to a common reference (same as #1!)
3. Shields with only a single end connected to a common reference
4. Shields with multiple ends connected to a common reference
5. Outer shields
6. Inner shields
7. Double shields
8. Circumferential terminations
9. Pigtail terminations
10. … and numerous other variations …

Once the decision has been made that a particular cable or cable harness requires some amount of shielding, additional and no less critical decisions need to be made focused on the shield's design. To be effective as an electromagnetic shield, cable and cable harness shields must be referenced to a common potential. The success of shielding cables and signal wires against electromagnetic environmental effects is critically dependent on selecting a shield referencing technique that complements the needs of the specific application. For example, where and how a shield is connected to a common reference can dramatically alter the shielding performance, regardless of any intrinsic "goodness" of the shield itself. Common experience suggests that implementation of this connection can be quite problematic and often rather puzzling. The key to "getting it right" is to identify the path or paths of current flow, recognizing where and how the current originates on the shield, where that current is going, and most importantly, how it returns to its source (including both intentional and unintentional paths!). Pay close attention to the current flow in sensitive circuits. If where current is flowing and a suitable means of controlling and directing that current flow can be determined, any number of undesirable coupling situations may be mitigated, either from internally or externally generated interference.

A frequent impediment to achieving this goal is one of the most stubbornly held beliefs among certain system engineers, particularly among instrumentation engineers, that being that cable shields should be connected to a common reference at only one end to prevent the creation of so-called "ground" loops. This topic seems to arise most frequently with respect to typical instrumentation circuits, common throughout applications in aerospace, automotive, and industrial facility settings.

In applications of comparatively low frequency wherein the shield cable is enclosed in an otherwise well-shielded environment, or in an environment known to experience exposure only to low-to-moderate levels of low-frequency external electromagnetic noise, a single-ended shield configuration can be an effective approach. In applications known to experience exposure to moderate to high frequency at most levels, or in applications within which strong external high-bandwidth electromagnetic interference is anticipated or known to exist, terminating a shield at only a single end can lead to all sorts of interference issues that can be exceedingly difficult to resolve. Note also that while it is certainly true connecting a shield's ends to a common reference can result in a so-called "ground loop", allowing a shield current to flow provides excellent shielding against low-frequency magnetic fields that would otherwise be completely unaffected by the presence of a single-ended shield. The use of a single-ended shield was thoroughly explored by both Faraday and Maxwell.

7.1 Faraday research and Maxwell's comments on shielding

Cylindrical conductors carrying electrical charges will have a series of equipotential surfaces surrounding the conductors, as illustrated in Figure 7.1.

If a cylindrical shield is placed such that its surface coincides with one of these equipotential surfaces and the shield is not connected to any other potential, the surrounding electric field will be completely undisturbed, and the shield will have no impact on the existing capacitive coupling between the two conductors [9].

If the shield is then connected to a common reference, electric charge will flow on the shield until the charges on the inside and the outside of the shield are balanced. This will result in the net electric field to which the shielded conductor is exposed being made equal to zero, reducing and possibly eliminating capacitive coupling between itself and any other conductor nearby. This type of shielding is most effective for electrostatic or low-frequency electric field control. A primary example of this concept was captured in 1843 in Faraday's ice pail demonstration [42] illustrated in Figure 7.2.

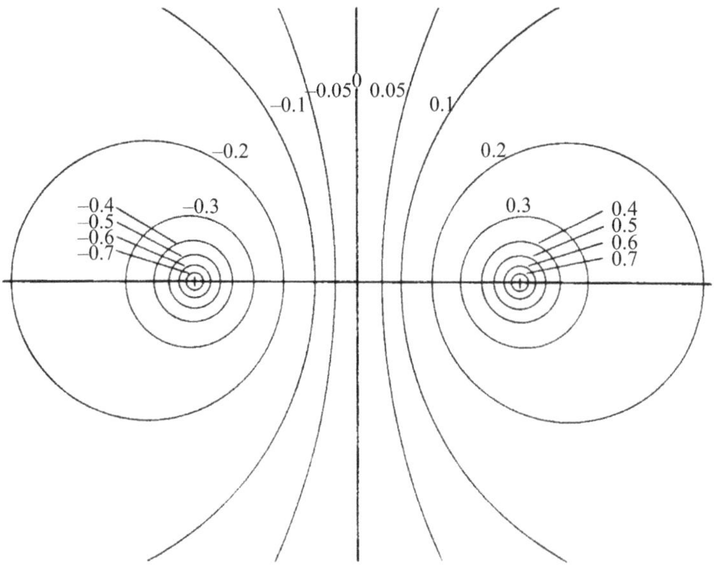

Figure 7.1 Two-dimensional view of equipotential surfaces surrounding two parallel cylindrical conductors

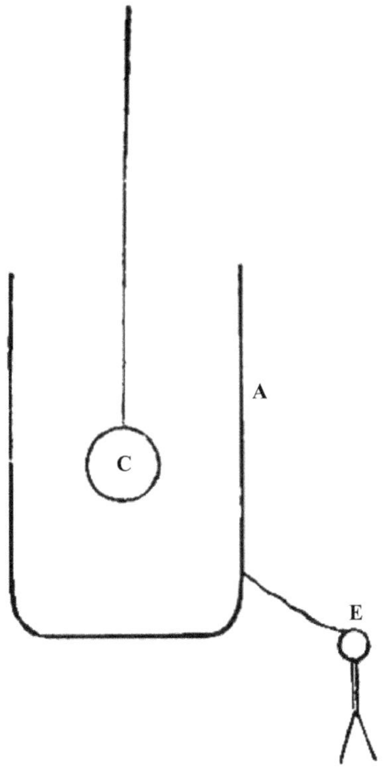

Figure 7.2 Illustration of Faraday's ice pail experiment

In Faraday's words [42]:

> Let A in the diagram represent an insulated pewter ice pail ten and a half inches high and seven inches diameter, connected by a wire with a delicate gold-leaf electrometer E, and let C be a round brass ball insulated by a dry thread of white silk, three or four feet in length, so as to remove the influence of the hand holding it from the ice-pail below. Let A be perfectly discharged, then let C be charged at a distance by a machine or Leyden jar and introduced into A as in the figure. If C be positive, E also will diverge positively; if C be taken away, E will collapse perfectly, the apparatus being in good order. As C enters the vessel A the divergence of E will increase until C is about three inches below the edge of the vessel and will remain quite steady and unchanged for any lower distance. This shows that at that distance the inductive action of C is entirely exerted upon the interior of A, and not in any degree directly upon external objects. If C be made to touch the bottom of A, all its charge is communicated to A; there is no longer any inductive action between C and A, and C, upon being withdrawn and examined, is found perfectly discharged.

James Clerk Maxwell repeated Faraday's experimentation and discussed his observations in [43,44]. Maxwell states in *A Treatise on Electricity and Magnetism* [44]:

> As the electrified ball approaches the vessel the indications of the electroscope continually increase, but when the ball is fairly within the vessel, that is when its depth below the opening of the vessel becomes considerable compared with the diameter of the opening, the indications of the electroscope no longer increase, but remain unchanged in whatever way the ball is moved about within the vessel.

Maxwell also states in the same chapter of *A Treatise on Electricity and Magnetism* [44]: "When an electrified body is placed within a vessel free of charge, the external electrification is equal to that of the body."

The takeaway here is a pail or open-ended cylinder could be nearly as effective an electrostatic shield as a sphere, given the depth of the pail or open-ended cylinder is "considerably" greater than the diameter of the opening.

Faraday also examined configurations of multiple concentric metallic vessels arranged as shown in Figure 7.3. Note that each cylinder is electrically isolated from any other cylinder in the stack-up.

Faraday concluded the number of vessels or the type of dielectric material, either partially or wholly, filling such vessels neither diminished nor increased the observed electrical charge appearing on the external surface of the outermost vessel. By extension, then, regardless of the number of external open-ended shields that might be placed around a conductor, the amount of electromagnetic shielding of that conductor is the same as if only a single open-ended shield were used.

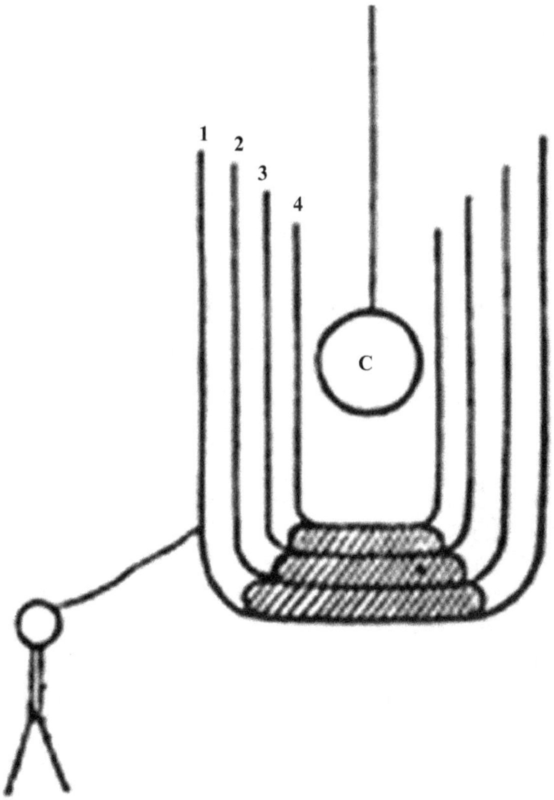

Figure 7.3 Illustration of Faraday's multiple metallic vessel experiment

7.1.1 What sort of problems might be expected if this sort of shielding configuration, with one end of the shield open, is adopted for a cable?

The charge on the shield surface will be dynamic when exposed to a varying external electric field such that charge may be moving on the shield surface much of the time to maintain a charge balance between the inner and outer surfaces.

This can be somewhat problematic, particularly if the shield is exposed to elevated frequencies of operation. As charge flows over the shield, the potential between the open end of the shield and the common reference varies with respect to the balanced condition, and this variable potential will ultimately appear on both sources and loads as common mode noise. This shield current can and will radiate. In fact, if the length of the shield terminated at only one end is $> \lambda/10$, it can become a very efficient radiating element as frequency increases.

Another undesirable possibility is created by the nearly inevitable extension of the shielded conductor beyond the shield confines (otherwise it might be exceedingly difficult to connect the wire to any terminating circuitry!). If this occurs, additional capacitances will arise between the exposed length of conductor and the surrounding circuitry and can modify and even effectively cancel any desired effects of the shield. To mitigate such complications, it is most desirable to minimize the length of any portion of the conductor that protrudes from the end of the shield to minimize undesirable capacitive coupling.

7.2 Where should the reference be connected?

Research performed by J.J. Goedbloed in 1992 [7,8] illustrated the impact of crosstalk between circuits employing no shielding, circuits employing a shield terminated at one end, and circuits employing a shield terminated at both ends. Data were collected using a fixture designed and constructed by Goedbloed in which 4 wires were embedded in a Perspex block above a common, conductive reference plane.

Goedbloed described a fixture comprising four 1 m lengths of 3-mm diameter wire. All four conductors were embedded in a 20-mm-thick block of Perspex acrylic ($\varepsilon_r \cong 2.65$) such that each conductor was supported above a common reference plane by 1.85 cm. A second Perspex block was laid over the top to assure all conductors were immersed in as nearly as possible the same dielectric environment. Each length of conductor formed an electromagnetic loop comprising a conductor suspended above a common reference plane. A diagram of the fixture, based on Goedbloed's description, is illustrated in Figure 7.4.

Loop 1 served as the source and is the uppermost conductor visible in the diagram. Loop 2 is 1 cm below Loop 1. Loop 3 is 3 cm below Loop 1, and Loop 4 is 10 cm below Loop 1. Loop 1 was left open at the far end to produce a capacitive coupling between itself and the other loops. Each of the remaining loops was either terminated to the underlying common reference plane by the characteristic impedance of the loop or was left open. Data were collected from the terminated loop. Only Loops 1, 2, and 3 are of interest in this discussion. Loop 1 was driven by

Figure 7.4 Fixture designed and built by J.J. Goedbloed for crosstalk investigation

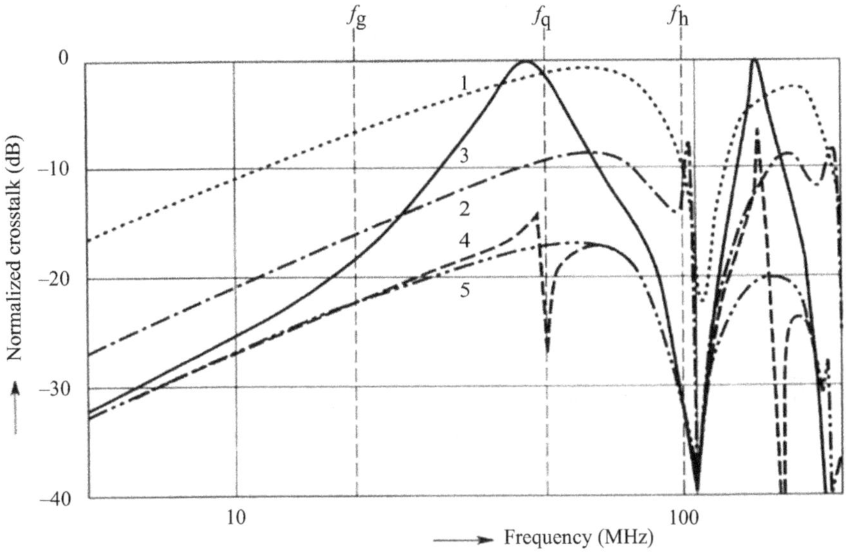

Figure 7.5 Data collected by Goedbloed using the fixture illustrated in Figure 7.4

a square wave generator injected on the left-hand side of the fixture (near end) as illustrated. The corresponding response to the voltage developed across Loop 1 was measured at the right-hand side (far end) of Loops 2 and 3, one at a time. Response data were collected for a total of 5 configurations using this fixture with the intent of highlighting the impact of the interposition of a screen between a source and a victim in a crosstalk pair.

Using the fixture described, Goedbloed collected data for each of the five configurations, illustrated here in Figure 7.5.

The results are briefly described as follows:

Plot 1: Loops 1 and 2. In this configuration Loop 2 is terminated at both ends by its Z_c. No other wire was interposed between Loops 1 and 2, such that Loop 2 represented an unshielded circuit exposed to the electric field emitted by Loop 1. The observed behaviour followed the expected response of an unshielded circuit, with resonant behaviour appearing as frequency neared and then exceeded 150 MHz.

Plot 2: Loops 1, 2, and 3. This plot captured the behaviour of Loop 3, terminated at both ends by its Z_c. In this configuration, the ends of Loop 2 were left unterminated. With Loop 2 interposed between Loops 1 and 3, Loop 2 represented an unterminated shield for Loop 3. The observed behaviour followed the expected response of an unshielded circuit, with resonant behaviour appearing as frequency neared and then exceeded 150 MHz. Crosstalk reduction improves by 10 dB well past 30 MHz simply because of the increased distance between Loops 1 and 3; the Loop 2 "shield" is doing nothing. Resonant behaviour appears as frequency nears and exceeds 150 MHz.

Plot 3: Loops 1, 2, and 3. In this configuration Loop 2 is shorted to the reference plane at the far end and is open at the near (source) end. Loop 2 is acting as a capacitive shield for Loop 3. Loop 3 is terminated at both ends by its Z_c. Compared to Plot 2 improved crosstalk reduction is visible below 30 MHz attributable to the shield, but this degrades quickly by 10's of dB with increasing frequency up to and beyond 30MHz towards a peak just below 75MHz. Resonant behaviour characterized by very large excursions appears as frequency nears and exceeds 150MHz. One might say the cure rapidly becomes worse than the disease as frequency increases.
Plot 4: Loops 1, 2, and 3. In this configuration Loop 2 is shorted to the reference plane at the near end and is open at the far (load) end. Loop 2 is again acting as a capacitive shield for Loop 3. Loop 3 is terminated at both ends by its Z_c. Compared to Plot 3, a significant improvement in crosstalk reduction is visible up to 150MHz but a sharp resonance appears in Plot 4 (and at the peak in Plot 3) near 75MHz resulting from inductively coupled crosstalk generated by current flowing in Loop 2. Loop 2 is behaving as a resonant LC circuit even though it is open at one end. As in all previous plots, resonant behaviour appears as frequency nears and exceeds 150MHz.

An approximate calculation for the sharp resonance near 46 MHz in Plot 4:

$$L = \frac{\mu_0 l}{2\pi} \ln\left\{\frac{h}{r} + \sqrt{\left[\left(\frac{h}{r}\right)^2 - 1\right]}\right\} \tag{7.1}$$

$$C \cong \frac{2\pi\epsilon_0\epsilon_r l}{\ln\left(\frac{l}{r}\right) + \ln\left(\frac{l}{h}\right) - 1} \tag{7.2}$$

For a length of 1 m in Perspex, we have

$$L \cong 640.8\text{nH}, \text{ and } C \cong 16.7 \text{ pF} \tag{7.3}$$

which yields

$$f_r = 1/\left(2\pi\sqrt{LC}\right) \cong 48.6\text{MHz} \tag{7.4}$$

Note in the plot, the frequencies are artificially adjusted by the square root of the relative permittivity: $f_g \cong 18.4$ MHz; $f_q \cong 46.1$ MHz; $f_h \cong 92.1$ MHz.
Plot 5: Loops 1, 2, and 3: In this configuration, Loop 2 is shorted to the reference plane at both ends and is acting as an inductive cable shield for Loop 3. Loop 3 is terminated at both ends by its Z_c. The most notable aspect of Plot 5 is the resonance at 75MHz has been eliminated, indicating Loop 2 is no longer resonant. As in all previous plots, resonant behaviour appears as frequency nears and exceeds 150MHz.

Specific observations of data collected for the described five configurations:

1. Baseline measurements of Loop 2 terminated at both ends in its characteristic impedance exhibited a typical high-pass capacitive coupling response with stimulus applied to Loop 1.

2. Measurements of Loop 3 terminated at both ends in its characteristic impedance with Loop 2 unterminated exhibited an approximate 10 dB reduction compared to the baseline measurements. This 10 dB reduction was attributed to the separation distance between the source (Loop 1) and the victim (Loop 3) now 3 cm rather than 1 cm ($20\log_{10}3$). The presence of Loop 2 acting as an unterminated shield showed no effect on the Loop 3 measurements.
3. Measurements of Loop 3 terminated at both ends in its characteristic impedance with Loop 2 shorted to the common reference at the load (far) end exhibited enhanced capacitive coupling to Loop 3, combined with inductive coupling associated with resonant current flow on Loop 2. The response curve of Loop 3 reflected that Loop 2 was behaving as a resonant LC circuit, with a strong resonance occurring at a quarter-wavelength of the stimulus equivalent to the length of Loop 2. This response was unique compared to all other collected data.
4. Measurements of Loop 3 terminated at both ends in its characteristic impedance with Loop 2 shorted to the common reference at the source (near) end exhibited a far more well-behaved coupling response compared to the previous configuration, approximately 6 dB lower across frequency than the second series of measurements. The response curve of Loop 3 again reflected that Loop 2 was behaving as a resonant LC circuit, with a minor resonance occurring at a quarter-wavelength of the stimulus equivalent to the length of Loop 2.
5. Measurements of Loop 3 terminated at both ends in its characteristic impedance with Loop 2 shorted to the common reference at both ends as well exhibited an almost identical response to the previous, with the minor resonance at a quarter-wavelength fully suppressed.

General observations of data collected for the described five configurations:

1. All five configurations exhibited continuous resonances above a half-wavelength of the stimulus.
2. Connecting one end of a capacitive shield to the common reference is only effective over a frequency range below the quarter-wave resonance of the shield.
3. Connecting one end of a capacitive shield to the common reference at the source end yields superior shielding performance to the alternative.
4. Connecting both ends of a capacitive shield to the common reference can be an effective control of undesirable resonances in the shield circuit.
5. The combination of separation distance and effective shielding is a powerful technique.

7.3 Are we the first to know about this?

The EMC community has in fact had long interest in this topic of shield "grounding", as might be expected. In 1961, Boeing published Interference Prevention Techniques Handbook [10], in which they examined various shield configurations used for the protection of circuits from exposure to external low-frequency

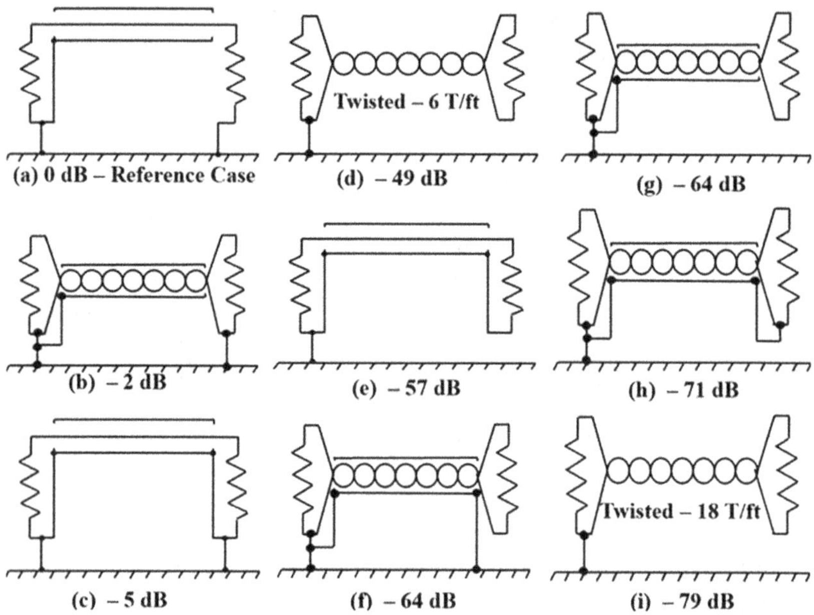

Figure 7.6 Various shield configurations examined by BOEING

time-varying fields (refer to Figure 7.6). Although the archive copy of this document was destroyed some years ago and to date no other copy has been located, fortunately some portions of the content were captured in four separate documents [45–48]. According to these 4 documents, the data were collected with conductors located 1 inch above a common reference plane, at 100 kHz, with the measurement parameters having to do with the type of cable used, whether or not the load was grounded, and how the cable shield was grounded. Specific comments included:

> The performance of tests indicates that, for low-frequency, low-level signals and low-impedance circuits where the distance between input connector and circuit input is small (on the order of 2 inches or less), the use of a twisted-pair may prove adequate. For long runs, the use of twisted-pair shielded wire becomes mandatory; this is true with balanced as well as unbalanced output devices. In the balanced case, there should be no grounding except for the shield; in the unbalanced case, it is often necessary to ground one of the conductors as well as the shield. The shield and conductor should be grounded at the same point. Single-point grounding of the shield should be used for short runs and multi-point grounding for long runs. Signals of high level will, in general, not be bothered by susceptibility of circuits. Rather, it may be a source of interference to lower-level signal lines. For this reason, and dependent on other characteristics of the signal, either a twisted-pair or a shielded lead should be used. Multiple shields may be required if the signal is of sufficient

level. Grounding should be applied at both ends to prevent electric field radiation from the cable.

The following configurations were examined in this study: although the primary commentary for each configuration is taken from the previous four references and is focused on circuit response to low-frequency fields, some additional commentary has been added to expand interpretation of the circuit configuration to include expected behaviour when exposed to electric fields:

7.3.1 Config (a)

This configuration is an unbalanced circuit, served as the baseline, and was assigned the value of 0 dB. Despite the presence of the shield, connected to the common reference at the source end, the combination of the closed "ground loop" formed of the conductor, the two terminating resistances, and the common reference plane proved to be highly susceptible to external low-frequency time-varying magnetic fields. The shield on the sample cable was made of woven copper wire having the same permeability as that of air and thus provided no shielding whatsoever to external low-frequency time-varying magnetic fields.

Time-varying current induced in this closed "ground loop" will cause a time-varying potential difference between the ends of the circuit, thereby creating electromagnetic interference with expected circuit behaviour. This configuration will also not reduce the effects of a low-frequency time-varying magnetic field created by intentional current flow in the circuit.

Conversely, this configuration provides excellent shielding of both external and internal low-frequency time-varying electric fields. A major potential drawback of this configuration is that as frequency increases, it will approach and eventually exceed the quarter-wave resonance of the shield, not only completely negating any electric field shielding but also behaving as an efficient radiator of an electric field. Goedbloed's measurements, as discussed in the previous section, strongly highlight these characteristic behaviours.

An equally concerning major potential drawback is that a fringing capacitance will exist between any extension of the internal conductor(s) and the end of the surrounding shield. This fringing capacitance is difficult to calculate, but it will increase as the centre conductor(s) extend further out of the shield confines. This fringing capacitance will provide a current path for high-frequency signals travelling on the outer shield surface directly onto the centre conductor(s), further degrading shield effectivity.

7.3.2 Config (b)

This configuration is also an unbalanced circuit and, except for the use of a twisted pair, is the same topologically as Config (a) with the shield connected to the common reference at the source end. This circuit is not expected to perform well when exposed to external low-frequency time-varying magnetic fields because of the ground loop formed by the circuit itself.

For balanced circuit applications exposed to time-varying magnetic fields, twisting the wire pair results in the induction of opposing potentials and currents in each successive loop. The net electromagnetic interference caused by these voltages and/or currents at either end of the cable is the algebraic sum of these opposing voltages and currents along the length of the cable. This configuration is not a balanced circuit, and it highlights a typical mistake made when an unbalanced pair is twisted together. The return wire is connected to the common reference at either end, effectively in parallel with the common reference plane. At low frequencies, the common reference will almost certainly have a much lower impedance as a function of frequency than will the twisted-pair return lead. Because of this, a large portion of return current may flow on the common reference. Instead of the small oppositely wound current loops formed by twisting the wire pair, the corresponding area of the dominant current-carrying loop is between the twisted pair and the common reference plane, largely negating any magnetic field shielding benefit gained by twisting the pair of wires. Because of the large current loop, interference voltages will appear across the load and source.

As was true for Config (a), this configuration provides excellent shielding of both external and internal low-frequency time-varying electric fields, but it also shares in toto the drawbacks of quarter-wave resonance and fringing capacitance discussed for Config (a). Note the slight attenuation reported for this configuration likely accounts for the attenuation of any accompanying low-frequency time-varying electric fields.

7.3.3 Config (c)

This configuration is an unbalanced circuit that has both circuit and shield "ground loops" and is identical to that of a coaxial cable. The reported immunity of this configuration to the effects of external low-frequency time-varying magnetic fields was only marginally better than that of Config (a), providing only 5 dB of attenuation.

Consider the distribution of low-frequency time-varying currents that will flow in this configuration. Regardless of the presence or absence of an external electromagnetic field, the return current will split between the shield and the common reference plane. The common reference plane most likely presents the lowest impedance as a function of frequency, such that the majority of return current would be most likely to flow in this common reference plane. This current loop involving the common reference plane is typically larger in cross section than that formed between the shield and the signal conductor. A penetrating magnetic field will induce a voltage differential between different points around both loops in accordance with Faraday's Law. The sum total of these induced voltages will appear as a potential difference between and across the terminating impedances. In addition, in accordance with Lenz's Law, the same penetrating magnetic field will induce currents in the shield and signal circuits that will tend to create an aggregate opposing magnetic field. The combined induced and intentional currents will further alter the voltages observed across the terminal impedances.

How can we know what path the current will take in this configuration?

An equivalent circuit for this configuration is illustrated in Figure 7.7. L_{cond} is the inductance of the coax centre conductor; L_s and R_s are the inductance and resistance

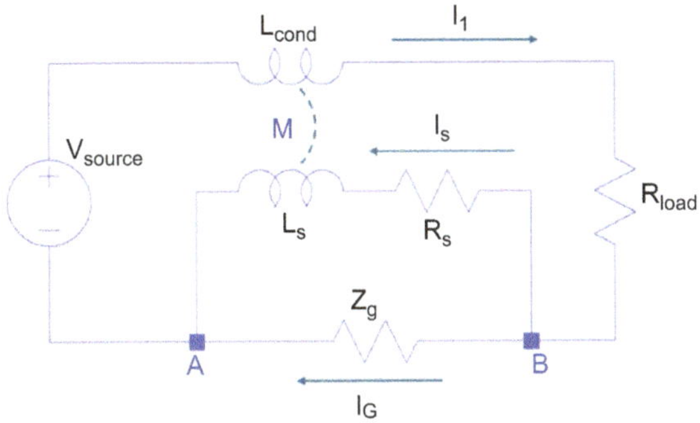

Figure 7.7 Equivalent circuit for Config (c)

of the shield, respectively. Zg is the impedance of the common reference plane, and M is the mutual inductance between the coax centre conductor and the shield.

The forward current in the coax centre conductor is Icoax (I_1), the return current in the shield is $\mathrm{I_{shield}}$ (I_s), and the return current in the ground plane is Iground (I_g).

The current in the common reference is given as

$$I_g = I_1 - I_s \tag{7.5}$$

We can now write an equation for the sum of the voltages around the loop $(A - L_s - R_s - B)$:

$$0 = -I_s(j\omega L_s + R_s) + I_1(j\omega M) + I_g Z_g \tag{7.6}$$

After substitution of (7.5) into (7.6), using the equivalency $L_s = M$, and some manipulation, we find

$$\frac{I_s}{I_1} = \left(\frac{j\omega}{j\omega + \frac{R_s + Z_g}{L_s}} + \frac{Z_g/L_s}{j\omega + \frac{R_s + Z_g}{L_s}} \right) \tag{7.7}$$

A plot of (7.7) is illustrated in Figure 7.8.

This plot shows that above the frequency defined as

$$\omega = \frac{R_s + Z_g}{L_s} \tag{7.8}$$

most if not all the return current will flow on the shield instead of the common reference plane. As frequency decreases below that defined in (7.8), an increasing amount of the return current will flow on the common reference plane.

Assuming that $Z_g \geq 0$, at very low frequencies, the shield current will be given by

$$I_s = I_1 \left(\frac{Z_g}{R_s + Z_g} \right) \tag{7.9}$$

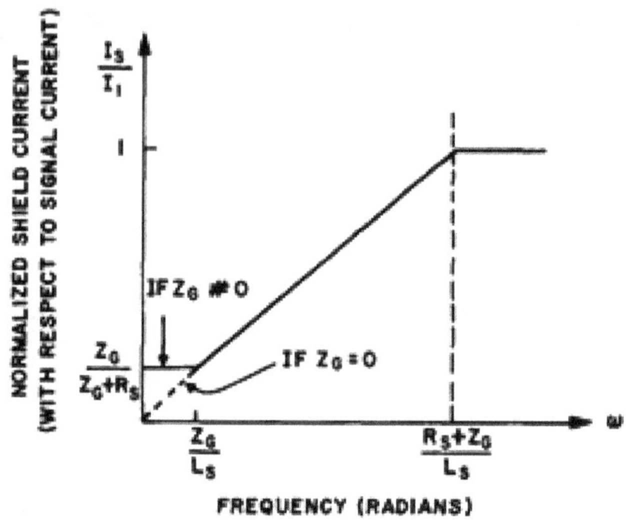

Figure 7.8 Plot of Equation (7.7) [49]

Setting $Z_g = 0$ yields

$$\omega_c = \frac{R_s}{L_s} \tag{7.10}$$

Equation (7.10) is known as the shield cutoff frequency.
The noise voltage that appears on the centre conductor is given by

$$V_N = j\omega M I_s \tag{7.11}$$

where

$$I_s = \frac{V_s}{L_s}\left(\frac{1}{j\omega + R_s/L_s}\right) \tag{7.12}$$

Substituting (7.12) into (7.11), and recalling that $M = L_s$ yields

$$V_N = j\omega M I_s = \frac{j\omega M V_s}{L_s}\left(\frac{1}{j\omega + R_s/L_s}\right) = V_s\left(\frac{j\omega}{j\omega + R_s/L_s}\right) \tag{7.13}$$

A plot of (7.13) illustrates the same shield behaviour as illustrated in Figure 7.8. Typical cutoff frequencies for cables using braided shields range from 0.6 to 2.2 kHz. Knowledge of the cutoff frequency for a given cable design will answer the question of which path the return current will take.

Regardless of which path the current chooses to follow, the shield in this configuration is part of the signal path. This means if currents whose frequencies are below the shield cutoff are induced by external sources and flow on the shield, the probability of interference with the intentional signal is increased, assuming the intentional signal is equally low in frequency.

As the skin depth of the shield decreases in its inverse relationship to frequency on either the inner side, the external side, or both sides of the shield, the electromagnetic isolation between external and internal fields increases proportionately. As a result, this configuration should work well in applications wherein the cable is exposed to predominantly electric fields, moderate-to-high-frequency time-varying electromagnetic fields and is likely the *best choice for an unbalanced circuit used in such applications*.

Aside from the electromagnetic interference to circuit operation created by the aforementioned activity, the larger current loop will radiate a magnetic field more efficiently than the smaller loop formed by the shield and the signal conductor, but the magnitude of the total radiated opposing magnetic field will be the algebraic sum of the two. A radiated opposing magnetic field created by shield current as described is likely the cause of the observed 5 dB reduction.

7.3.4 Config (d)

This was a typical balanced twisted pair, with 6 turns per foot ($\sim$ 18 turns per metre), terminated in a balanced load and isolated from the common reference at the load end. This configuration had no shield at all and was found to offer good immunity to the effects of external low-frequency time-varying magnetic fields in a differential mode sense by virtue of the reduction in circuit loop area created by the twist in the pair and in a common mode sense by virtue of there being no circuit or shield "ground loop".

By virtue of the balanced nature of the circuit and the twisting of the wire pair, this configuration will tend to reduce any radiated time-varying magnetic fields created by intentional current flow in the circuit.

At 100 kHz, it is likely that some low-frequency time-varying electric fields would have exposed the circuit, and since the circuit has no shielding, this exposure would have caused some degree of interference. This configuration will fail as frequency increases because both conductors are fully exposed to any external time-varying electric or electromagnetic fields, such as, for example, those associated with radio communications, radar, lightning, or electrostatic discharge (ESD).

7.3.5 Config (e)

This configuration comprises either a coaxial cable or single wire terminated in an isolated load with a shield that is connected to the common reference at the source end and only to the return side of the circuit at the load end. In either case, coax or single wire, the loop area is governed by the offset between the inner wire and the geometric centre of the shield, which can be quite small in high-quality coaxial cables. Immunity from the effects of external low-frequency time-varying magnetic fields for this configuration compared to Config (c) is much better because of the very small loop area and because all the return current flows in the shield.

Because of skin depth changes as discussed previously, immunity to external time-varying electric and electromagnetic fields is expected to approach that of Config (c) as frequency increases.

This configuration employs a technique referred to as "self-shielding". Magnetic field self-shielding occurs when the signal return current is allowed to flow through the shield such that the magnetic flux of the signal is completely contained within the magnetic flux of the signal return current flowing on the shield. The field containment is caused by the current distribution and not by the metal of the signal wires. Because of the "self-shielding" behaviour of this configuration, it is equally functional as a control for radiated magnetic field emissions from the circuit.

Electric field self-shielding occurs when electric charges produced by the signal voltage completely contain the electric flux within the signal cable. The shielding is produced by the charge distribution and not by the metal conductors. By virtue of this self-shielding effect, this configuration is also expected to effectively mitigate any electric fields that might be radiated by the circuit.

Compared to other shielding techniques, self-shielding can be the lowest cost and widest bandwidth shielding technique if it is designed into the system at the beginning.

7.3.6 Config (f)

This was a typical balanced twisted pair, with 6 turns per foot ($\sim$ 18 turns per metre), terminated in a balanced load and isolated from the common reference at the load end. This configuration combines Config (d) with a shield connected to the common reference at both ends. Although the "ground loop" formed by the shield would be considered an undesirable configuration for applications where differences of potential exist that will drive currents associated with power frequencies (50 Hz/ 60 Hz) or other external low-frequency time-varying magnetic fields, the balanced nature of the circuit and the fact the load is isolated provides excellent immunity to these low-frequency threats.

This configuration may, however, tend to radiate a magnetic field generated by low-frequency time-varying currents flowing on the shield regardless of source and is not recommended in applications where this radiated field might cause interference with sensitive hardware located nearby.

This configuration mitigates the quarter-wave resonance of Configs (a) and (b) below the shield resonance and is expected to present an increasingly equipotential surface in proportion to increasing frequency, thus offering an improved electric and/or electromagnetic field shielding capability. Skin depth decreases in proportion to increasing frequency serves to enhance the shielding action. For this reason, this is likely to be the *best choice for applications in areas of exposure to moderate- to high-frequency external electric or electromagnetic fields*.

7.3.7 Config (g)

This configuration is basically Config (f) with the shield connected only to the common reference at the source end, so no "ground loop" is formed, eliminating any shield current/voltage drop concerns. Unlike Config (b), this configuration is balanced and was found to offer good immunity to the effects of external low-frequency time-varying magnetic fields in a differential sense by virtue of the reduction in circuit loop area created by the twist in the pair, and in a common mode sense by virtue of there being no circuit or shield "ground loop".

By virtue of the balanced nature of the circuit and the twisting of the wire pair, this configuration will tend to reduce any radiated time-varying magnetic fields created by intentional current flow in the circuit.

The shield is expected to exhibit the same electric field behaviour as that discussed for Configs (a) and (b), and so this configuration is not recommended for applications wherein the cabling is exposed to predominantly electric fields, and moderate-to-high-frequency time-varying electromagnetic fields. Note that noise coupled via the fringing capacitance will constitute a common-mode threat, but because of the twisted pair and the balanced configuration, the circuit is less likely to be adversely affected as a result.

7.3.8 Config (h)

This configuration is a balanced circuit comprising a twisted pair terminated in an isolated load with a shield that is connected to the common reference at the source end and to the return side of the circuit at the load end. The topology of this circuit is essentially the same as that of Config (e) and shares the "self-shielding" characteristic.

The behaviour of this circuit is expected to be superior to that of Config (e) because of the use of the twisted pair and was found to be the *best choice for applications with exposure to external low-frequency time-varying magnetic fields*. This circuit is also expected to perform well in applications exposed to moderate-to-high-frequency external electric or electromagnetic fields.

7.3.9 Config (i)

This configuration is Config (d) with an increased number of turns per length, illustrating that twisting the conductors of a balanced circuit can be a very effective approach if shielding is not an attractive option. Be aware that this is not a universal panacea, as wires can only be twisted up to a point at which the conductors may be exposed to excessive stress and may fail in a manner not readily visible on the outside of the insulation, as discussed in Section 4.4.7 of Chapter 4.

Ott performed a closely similar series of measurements and produced results akin to those Boeing generated, with some differences. Ott collected his data using a 50-kHz frequency. He also explored a few additional configurations. The reader is urged to access Ott's material for comparison and additional understanding.

7.4 What about commonly known instrumentation circuits and the dreaded ground loop?

Figures 7.9–7.12 illustrate typical instrumentation circuit shielding configurations [16]. Each figure is accompanied by relevant commentary. These configurations resemble but in general differ from those illustrated in Figure 7.6.

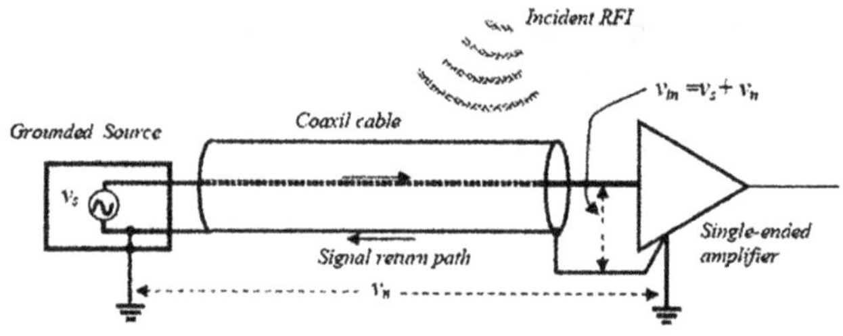

Figure 7.9 Configuration A of typical instrumentation circuit design

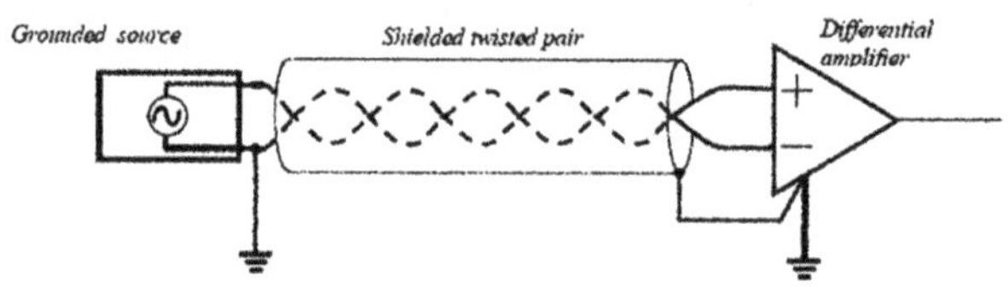

Figure 7.10 Configuration B of typical instrumentation circuit design

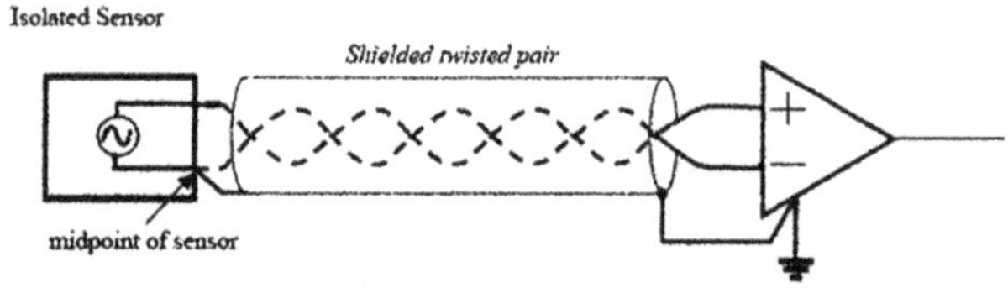

Figure 7.11 Configuration C of typical instrumentation circuit design

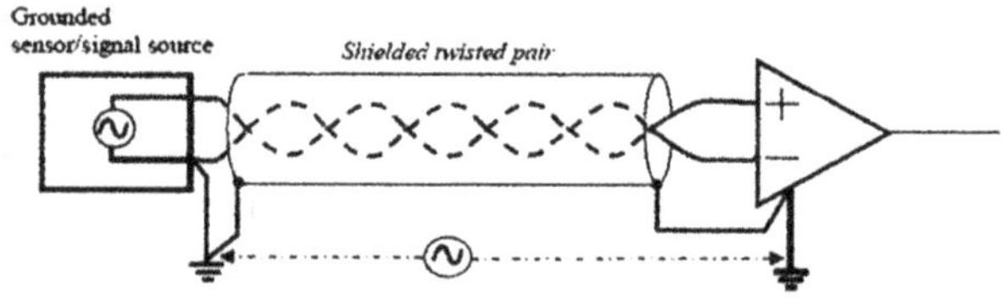

Figure 7.12 Configuration D of typical instrumentation circuit design

Config (A); illustrated in Figure 7.9, depicts a common implementation of a single-ended low-bandwidth, low-frequency sensor circuit, such as a pressure or temperature transducer or a stress/strain sensor, wherein the shield has been connected to the same common reference point as the sensor. Both the shield and the single-ended amplifier return circuit form "ground" loops, introducing possible elevated voltages across these "ground loops" created by otherwise normally expected circuit currents. Additionally, exposure to external low-frequency radio frequency interference (RFI) can induce currents and voltages in the circuit and on the shield. Regardless of the source, these undesirable currents and voltages will effectively negate any shielding action. Like Config (c) from Figure 7.6, this configuration will offer little to no low-frequency magnetic field coupling but should perform well as frequency increases, with the skin effect in the shield enhancing the shielding action. For this reason, this is likely to be the *best choice for applications in areas of exposure to moderate-to-high-frequency external RFI.*

Config (B); illustrated in Figure 7.10, depicts the same sensor as in Config (a), but this time a differential amplifier and a twisted shielded pair are used instead of a coax. Here we have a source that is connected to the common reference, providing a differential signal via a twisted shielded pair to a differential amplifier that is also connected to the common reference. Note that neither signal input is connected to the common reference, except through the cable to the source sensor. The circuit has a shield that is connected to the common reference at the amplifier (load) end.

Common-mode voltage will develop between the sensor connection to the common reference and the amplifier connection to the common reference. The distributed capacitance of the open-ended shield will act to shunt the amplifier input impedance to the common reference and create a path for noise currents to flow through the sensor, thereby degrading the signal quality. These noise currents are driven by the common-mode voltage. A preferred method of interconnection would remove the shield connection at the amplifier connection to the common reference and move it to the sensor connection to the common reference. By so doing, the effects of distributed capacitance in the shield are mitigated, eliminating the path for common-mode currents to flow. If either the amplifier or the sensor could be isolated, this would further aid in mitigating the effects of any common-mode currents. Unfortunately, regardless of which end the shield is connected to the common reference, as has been previously discussed this configuration will not work well for shield lengths equivalent to frequencies whose wavelength is equal to or greater than $\lambda/4$.

In configurations wherein the operational frequencies of the circuits to be shielded are comparatively low and the external RFI environment comprises equally low frequencies, a configuration like that illustrated in Figure 7.10 may prove a suitable approach. For example, bioelectric signals originating from patient monitoring in a typical hospital environment tend to be quite low in operating frequency, often less than 1 kHz. In such cases, the $\lambda/4$ length of the shield is likely to be less than the operational wavelengths of the circuits to be shielded (at 1 kHz, the corresponding $\lambda/4$ length equals 75 km!), so that the shield itself is not likely to break into resonance. Under such conditions, it is a relatively simple matter to filter any moderate-to-high-frequency RFI at the amplifier inputs that couples onto these signal lines without compromising the quality of the desired signal.

Config (C); in Figure 7.11, illustrates a sensor isolated from the common reference, but the shield is tied to the "mid-point" of the sensor, ostensibly to provide a DC reference for the amplifier. Because the shield is connected to the mid-point of the sensor, any undesirable external RFI that may couple onto the shield will likely appear on the sensor circuit DC reference, a form of ground-conducted interference. Any imbalance created by the presence of this noise in the sensor circuit can lead to unintentional current flow on the shield, as well as an unintended differential mode current on the signal lines. A preferred method of interconnection would be to incorporate a third conductor between the amplifier stage and the sensor to convey the sensor mid-point DC reference, rather than using the shield for this connection. As shown previously, the shield can begin to resonate at frequencies greater than $\lambda/4$.

Config (D); in Figure 7.12, has a "ground loop" through the shield and is very similar to Config (f) from Figure 7.6. The twisted pair should effectively reduce circuit loop area, acting to mitigate any magnetic field interference.

This configuration prevents the flow of any signal current through the shield. As the frequency of any external RFI increases the resulting coupled external shield current will tend to flow on the shield and not couple to the circuit because of skin effect in the shield. It also reduces any standing waves that may appear on the open shield, thereby preventing the shield from breaking into resonance. Performance measurements by instrumentation vendors on similar configurations, even for low-frequency applications, have shown this topology provides as much as a 40 dB higher noise immunity compared to the configurations illustrated in Figures 7.10 and 7.11. This is likely to be a "best" choice (limited by the twisted-pair bandwidth) for applications in areas of moderate to high exposure to external RFI.

"Accepted good shield grounding" practice dictates shields should always employ multiple connections to the common reference to control the effects of exposure to moderate-to-high-frequency RFI. Unfortunately, doing so inevitably creates multiple "ground" loops that may act to increase susceptibility to low-frequency RFI. Conversely, employing single connections to the common reference breaks the low-frequency loop, but this approach becomes increasingly less effective as operational and external RFI frequencies increase. As was previously discussed in this chapter, this dichotomy often presents what is perceived to be a dilemma by the designer. However, this dichotomy can be used to advantage by the designer who examines the application and makes a judicious choice in the cabling topology.

7.5 One more look at the dreaded ground loop

It is instructive to look closely at how currents flow in circuits having a single-ended shield. In Figure 7.13 are illustrated four circuit configurations (aa through dd) comprising a single conductor circuit with no shield that morphs into a coaxial structure with a shield connected to the common reference in three different ways [50]. Currents that flow in these configurations are indicated in the figure. The inclusion of the illustrated currents is intended to emphasize that where the current flows in a circuit without and with a shield is fundamental to how well (or poorly) the shield will act.

The first configuration is simply an unbalanced circuit consisting of a source (V_s) connected to a load (Z_l) with the source impedance (Z_s) in series. The second configuration is identical to the first, with the exception that a shield, connected to the common reference plane at the source, has been added. This configuration

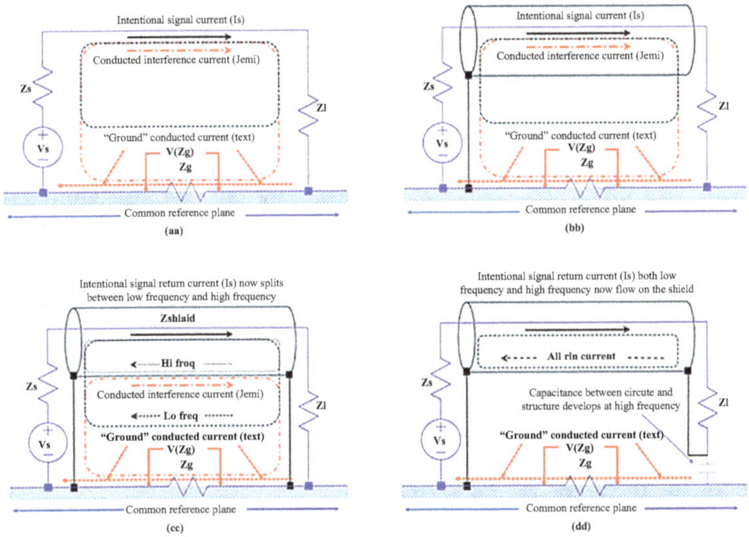

Figure 7.13 Successive shield configuration changes

should be compared to Config (a) in Figure 7.6. The third configuration is again identical, with the exception that the shield is now connected to the common reference plane at both ends of the circuit. This configuration should be compared to Config (c) in Figure 7.6. In the fourth configuration, the circuit has morphed from an unbalanced circuit to a balanced circuit. This configuration should be compared to Config (e) in Figure 7.6. Note this configuration could easily be converted to that illustrated in Config (h) in Figure 7.6 by use of a balanced twisted pair in place of the coaxial configuration.

In Configuration (aa), all of the signal return current is forced to flow on the common reference plane. Any other currents, identified as I_{emi}, that might also be flowing on the common reference plane can and will introduce interference into the desired signal by flowing through the intended signal path. The combined current flowing on the common reference plane will generate an interference voltage across Z_l given by

$$V_l = V_{(Zg)}\left(\frac{Z_l}{Z_l + Z_s}\right) = I_{ext}Z_g\left(\frac{Z_l}{Z_l + Z_s}\right) \tag{7.14}$$

This voltage represents the interference resulting from "ground"-conducted current, identified as I_{ext}. As frequency increases, the inductive reactance that exists in the large "ground loop" formed by the single conductor and the current return through the common reference plane will begin to dominate the interference voltage.

Configuration (bb) illustrates an ineffectual attempt to provide shielding for the circuit and will behave in the manner discussed for Config (a) in Figure 7.6. Basically, because this configuration lacks any circuit current in the shield, it will provide no magnetic field shielding. It will provide good to excellent electric field shielding up to the cutoff frequency of the Config (a) of Figure 7.6.

Configuration (cc) is another ineffectual attempt to provide shielding for the circuit and will behave in the manner discussed for Config (c) in Figure 7.6. Signal return current will split between the shield and the common reference plane, depending on the frequency of the current. This means that at high frequencies, signal return current in the shield will provide a self-shielding action, thus providing useful magnetic field shielding. Signal return current flowing at lower frequencies will more likely split and share the path through the common reference with any "ground"-conducted current as it did in Config (b) of Figure 7.6, with results similar to those highlighted by (7.14) appearing as voltages across Z_s and Z_l. Note that in this configuration at high frequencies the "ground"-conducted current will itself split, with a portion flowing on the outside of the shield. The portion of this current flowing on the outside of the shield will couple via the transfer impedance of the shield onto the internal circuit as a common-mode threat.

$$V_{coupled} \cong I_{emi} Z_T = \left(\frac{I_{ext} Z_g}{Z_{shield}} \right) Z_T \tag{7.15}$$

where

Z_{shield} = the intrinsic impedance of the shield (at low frequencies, this = R_{shield})
Z_T = the surface transfer impedance of the shield

Configuration (dd) is a far superior approach compared to the first three and is essentially the same as Config (e) in Figure 7.6.

7.5.1 What is cable surface transfer impedance?

The concept of Surface Transfer Impedance was first suggested by Schelkunoff [51] in his discussion of the surface impedances of a hollow cylindrical conductor. He introduced the concept with the following words:

> In the case of a hollow conductor whose inner and outer radii are respectively equal to *a* and *b*, the return coaxial path for the current may be provided either outside the given conductor or inside it or partly inside and partly outside. We designate by Z_{aa} the surface impedance with internal return, and by Z_{bb}, that with external return. These impedances are equal only at zero frequency; but if the conductor is thin, they are nearly equal at all frequencies. If the return is partly internal and partly external, we have in effect two transmission lines with a distributed mutual impedance Z_{ab} due to the mingling of the two currents in the hollow conductor common to both lines. However, since this quantity Z_{ab} is not the total mutual impedance between the two lines unless the hollow conductor is the only part of the electromagnetic field common to them, it is better to call Z_{ab} the transfer impedance from one surface of the conductor to the other.

Schelkunoff goes on at some length describing in mathematical detail a series of exact and approximate relationships to calculate the surface transfer resistance and impedance of a hollow cylindrical conductor. The surface transfer impedance yields the open-circuit voltage developed between the internal conductor(s) and the shield. Cable shield surface transfer impedance always contains a diffusion component dependent on the conductivity of the shield material. If the shield has gaps or apertures, such as a braid or tape-wrapped shield, shield currents will follow and penetrate gaps or apertures, adding inductive coupling. Literally hundreds of authors have published papers addressing the topic of cable shield transfer impedance in the years since 1934, in many of which "new" models of transfer impedance for braided shields have been put forth for community review. While these models tend to be discussed as being specific to braid shields, they can be modified appropriately for different styles of shields, such as foil, tape wrap, and others. All these papers include some form of a diffusion expression, while most of them also include one or more inductive components.

Surface transfer impedance is not the only measurement used when discussing EMI shielding products. Multiple measurements, including shielding effectiveness (SE), insertion loss, common impedance coupling, surface transfer inductance, surface transfer admittance, and charge transfer elastance, can provide helpful information in context. But, if you were to choose only one metric to rely on, surface transfer impedance is the most important and reliable measurement for characterizing the electromagnetic shielding properties of cables and connectors across a wide range of applications and frequencies. It provides a direct, intrinsic, and practical assessment of a shield's effectiveness, making it an indispensable tool for engineers in ensuring the reliability of electronic systems.

Most authors have concluded that cable surface transfer impedance is an inherent physical property of a shield that may be illustrated by the following construction illustrated in Figure 7.14 and defined in the mathematical expression (7.16):

$$Z_T \triangleq \frac{1}{I_0} \frac{\partial V}{\partial z}\bigg|_{I=0} \tag{7.16}$$

where

- $\partial V/\partial z =$ the RMS voltage per unit length of the uniform transmission line created by the inner surface of th e shielding layer and the cable core
- $I_0 =$ the induced current flowing on the outer surface of the shielded layer
- $I =$ the current flowing on the cable core

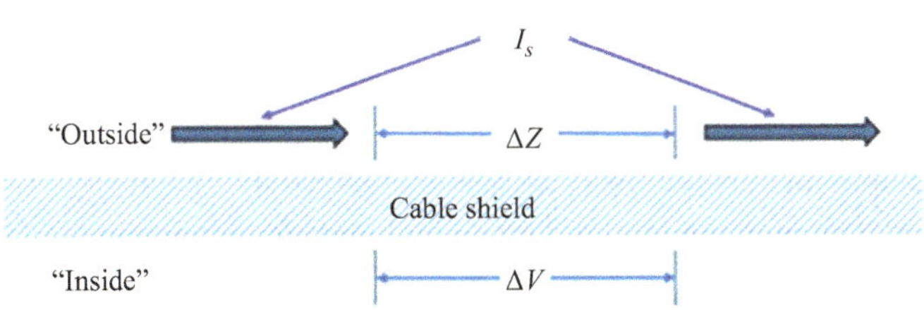

Figure 7.14 Graphical definition of transfer impedance (after Schelkunoff)

In a recent paper [52] that in its form is typical of the state of the art, a model was presented that includes a diffusion component and three inductive terms, accounting for (1) coupling of external magnetic fields through apertures (sometimes referred to as "hole" inductance); (2) coupling of the magnetic flux linkage between the inner and outer layers of the braid (sometimes referred to as "braid" inductance); and (3) the inductive effects of eddy currents induced by the presence of a tangential electric field on the surface of the shield (this term has also sometimes been referred to, or has incorporated, skin inductance, a term/concept introduced by Kley [53]). This model is given here:

$$Z_T = Z_d + j\omega(M_h \pm M_b) + M_e \tag{7.17}$$

where

Z_d = the diffusion component
M_h = the "hole" inductance component
M_b = the "braid" inductance component
M_e = the "eddy current" inductance component

The basic diffusion component developed by Schelkunoff, as reported by Vance, is given by

$$Z_d = \frac{1}{2\pi a\sigma t}\frac{(1+j)(t/\delta)}{\sinh[(1+j)(t/\delta)]} \tag{7.18}$$

where

a = the shield radius
σ = the electrical conductivity of the shield
t = the wall thickness of the shield
δ = the skin depth in the shield

Equation (7.18) represents the transfer impedance of a solid shield. This term is modified for braided shields to account for the specific physical characteristics of braided cable shielding. Vance examined braided shielding in multiple papers and identified these specific characteristics as illustrated in Figure 7.15 [54].

Using these images and the characteristics depicted therein, Vance was able to develop a specific diffusion component for a braided shield, given by

$$Z_d = \frac{4}{\pi d^2 NC\sigma \cos\alpha}\frac{(1+j)(d/\delta)}{\sinh[(1+j)(d/\delta)]} \tag{7.19}$$

where

a = the shield radius
σ = the electrical conductivity of the shield
d = the diameter of the wires in the shield
C = the number of carriers (belts of shield wires) in the shield
N = the number of wires in each carrier in the shield
σ = the electrical conductivity of the wires in the shield
α = the pitch, or weave, angle of the carriers in the shield
δ = the skin depth in the shield

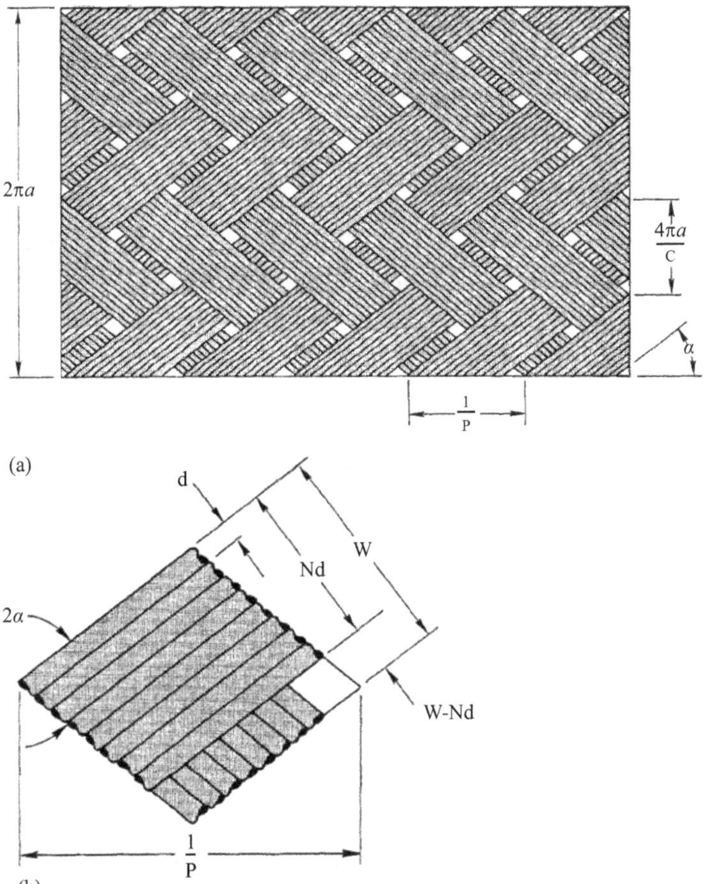

Figure 7.15 Braid shield characteristics, as identified by Vance. (a) Braid pattern developed on a plane and (b) One diamond of braid.

This is the diffusion component in (7.18) modified to account for the braid characteristics.

The basic expression for the "hole" inductance, M_h, through a single circular aperture with radius r in a zero-thickness wall has been interpreted and adopted by multiple authors. In a form that accounts for braid characteristics, the "hole" inductance in the recent paper is substantially the same as that given by [55] for "leakage inductance", and is given by

$$M_h' = \frac{2\mu_0 C}{\pi \cos\alpha}\left(\frac{b}{\pi D_m}\right)^2 e^{\frac{\pi d}{b}-2\alpha} \tag{7.20}$$

where

b = the braid density, represented as the width between two carriers
D_m = the mean diameter of the shield, $= D_0 + 2d + \widehat{h}$

$$\widehat{h} = \frac{2d}{1 + b/d}$$

Equation (7.20) fails to account for the thickness of the apertures, nor does it account for the curvature of the braid [56]. A factor of $0.875e^{-\kappa T}$ was determined to reasonably account for the effects of these characteristics [53,56], yielding

$$M_h = C_\kappa M_h' \tag{7.21}$$

where $C_\kappa = 0.875e^{-\kappa T}$, and T = shield wall thickness

The factor κ is approximated by the dominant TE-mode of a waveguide with elliptical or rhombic apertures. For an elliptical aperture, $\kappa \sim 1.84/r_{eq}$, where r_{eq} is the radius of a circular hole with surface equal to $\pi\omega l/4$. For a rhombic aperture, $\kappa \sim \pi/L$, where L is equal to the edge length of an equivalent rhombus with right angle $L = \sqrt{lw/2}$.

The expressions for "braid" inductance in [52] can be traced back to [55]. The expression Tyni provides is

$$M_b = \frac{\pm\mu_0 \widehat{h}}{4\pi D_m}\left(1 - \tan^2\alpha\right) \tag{7.22}$$

with (7.22) being negative for $\alpha \leq 45°$, positive otherwise.

As a side note, Kaden derived a formula for the configuration in which a longitudinal slit runs the length of the cable. This would apply for a foil braid, as an example. Note (7.18) would be used for the diffusion component.

$$M_1 = \frac{\mu_0 H_0}{D_m I}\left[\left(\frac{2b}{\pi}\right)^2 e^{\left(-\pi d/b^{-2}\right)}\right] \tag{7.23}$$

where

H_0 = the external magnetic field
I = the induced current flowing on the shield
b = the width of the slit
d = the thickness of the shield

Finally, research efforts into the "eddy current" term are not well distributed. In [53], the author states "Since no formula for *LSL* exists in literature, a qualitative approach ..." must be used to calculate the inductance associated with eddy currents in the shield. Demoulin provided an empirical formula [57,58] for surface cable transfer impedance based on measurements performed on cables having shields similar to those used for RG213, given by

$$Z_T = Z_D + j\omega L_h + k\sqrt{\omega}e^{j\pi/4} \pm j\omega L_b \tag{7.24}$$

In this equation, the second term represents the effects of eddy currents that may be present in the shield. Demoulin provides the equation and indicates the parameter k is a real coefficient whose value is dependent on braid parameters and has positive value if the pitch angle with respect to the cable axis is smaller than 45°. In 2013, Mushtaq *et al.* [57] published a paper in which (7.24) is simplified for pitch angles less than 45°:

$$Z_T = Z_D + k\sqrt{\omega}e^{j\pi/4} - j\omega L_b \tag{7.25}$$

where

$$k = -\left(\frac{1.16}{NCd}\right)\tan^{-1}\left(\frac{N}{3}\right)\sin\left(\frac{\pi}{2} - 2\alpha\right)\sqrt{\frac{\mu}{\sigma}}$$

Xiaoling *et al.* [52] include the second term of (7.25) with k defined as shown as representing the eddy current contribution to the transfer impedance.

7.5.2 One way to convert cable surface transfer impedance to shielding effectiveness

Surface transfer impedance is an excellent method for specifying cable shield performance but is poorly understood in terms of shielding effectivity. The difficulty with (7.16) is that ∂V is inaccessible for measurement. Several authors have approached this. The following is taken from the works from references [59,60], in which the outside of the cable shield and any nearby conductors are treated as transmission line #1, and the inside of the shield and any conductors contained therein as transmission line #2. They then applied a Green's function approach to develop a relationship between the terminal voltage of line #2 and the current flowing in line #1. This approach can be modified by (1) introducing the power flowing on the lines, (2) assuming the cable length is a small fraction of a wavelength, and (3) assuming any capacitive coupling is negligible. After much algebra, the result is given by

$$Z_T = \frac{\sqrt{Z_{01}Z_{02}}}{L(1+K)}\sqrt{\frac{P_L}{P_{OUT}}}\left(\frac{4(1-\rho_{12}\rho_{22})(1-\rho_{11}\rho_{21})}{(1-\rho_{22})(1-\rho_{11})(1-\rho_{12})(1-\rho_{21})}\right) \tag{7.26a}$$

where

Z_{01} = the characteristic impedance of transmission line #1
Z_{02} = the characteristic impedance of transmission line #2
L = the length of the cable

$K = Z_g/Z_{01}$; Z_g = the source impedance of the generator supplying P_{OUT}
$P_L = V_L^2/Z_{02}$ = the power on transmission line #2 at the load
$P_{OUT} = I_S^2 Z_{01}$ = the power on transmission line #1
ρ_{11} and ρ_{12} are respectively the near and far end reflection coefficients for transmission line #1

ρ_{21} and ρ_{22} are respectively the near and far end reflection coefficients for transmission line #2

All these are accessible and measurable quantities. (7.26a) can be expressed in a somewhat abbreviated manner as

$$Z_T = \sqrt{Z_{01}Z_{02}}\sqrt{\frac{1}{SE}\frac{F}{L}} \tag{7.26b}$$

where $SE = P_{OUT}/P_L$, the shielding effectiveness.

$$F = \left(\frac{4(1-\rho_{12}\rho_{22})(1-\rho_{11}\rho_{21})}{(1+K)(1-\rho_{22})(1-\rho_{11})(1-\rho_{12})(1-\rho_{21})}\right)$$

Solving (7.26b) for SE yields

$$SE = \frac{Z_{01}Z_{02}}{Z_T^2}\left(\frac{F}{L}\right)^2 \tag{7.27a}$$

which can be rewritten in logarithmic terms as

$$SE_{dB} = 10\log_{10}Z_{01} + 10\log_{10}Z_{02} + 20\log_{10}F - 20\log_{10}L - 20\log_{10}Z_T \tag{7.27b}$$

Equation (7.27b) characterizes the SE behaviour of the shielded cable immersed in its environment, and the terms characterize all the quantities of interest. The first term captures the relationship between the cable shield and conductors external to the cable, (e.g., wires, cables, structural members, etc.), and is equal to the characteristic impedance of the external circuit. A reasonable suggestion, according to Martin [59], for Z_{01} is 50 Ω. The second term captures the dependence of the SE on the characteristic impedance of the internal circuit (i.e., the physical construction of the cable). A reasonable suggestion for Z_{02} for a shielded twisted pair is 120 Ω. The third term captures the influence of cable conductor termination on the SE, and accounts for any mismatch between the source impedance and the characteristic impedance of the cable. Martin [59] suggests a reasonable approximation for the third term is equal to 4 dB. The fourth term captures a 20 dB/decade decrease of SE as a function of cable length. Note this term is a reasonable approximation for cables whose length is $< \lambda/4$. The final term indicates that as surface transfer impedance decreases, the SE increases. This last term describes the shield transfer behaviour and is the only term affected by the shield design itself. It is the independent nature of this term that highlights the importance of specifying the transfer impedance rather than the SE, which includes several other important system-level characteristics in addition to the transfer impedance.

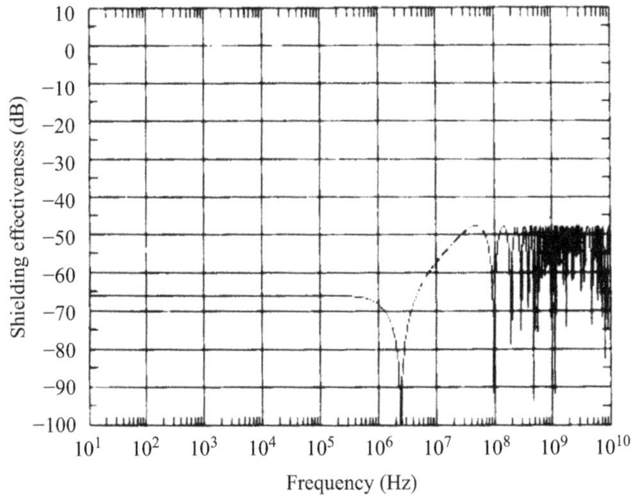

Figure 7.16 SE (ratio of currents) for a 1.2-m shielded twisted pair

We can use the approximations given in this paragraph to provide an estimate of SE for a typical 120 Ω twisted shielded pair based its length and on the measured cable surface transfer impedance. Inserting the foregoing approximations into (7.27b) yields

$$SE_{dB} = 50\text{dB} - 20\log_{10}L - 20\log_{10}Z_T \tag{7.27c}$$

For a 1-m long cable with $Z_T \cong 100\ \text{m}\Omega/\text{m}$, (7.27c) yields $SE_{dB} \cong 70$ dB. Compare this calculation result to the plotted SE (current ratio) performance, based on transfer impedance, shown in Figure 7.16 [61].

7.5.2.1 Misconception of cable shielding effectiveness behaviour

A commonly held misconception is that shielding effectivity decreases by 6 dB for every doubling in length, L, suggested by

$$SE \sim 20\log_{10}\left(\frac{L_0}{L}\right) \tag{7.28}$$

To explore the efficacy of (7.28), measured SE data collected by Martin and Emert [62] were superimposed on plots of equations from Merewether and Ezell [60] to illustrate the behaviour of SE as a function of length. For each plot, the stimulation frequency remained constant. Two of these plots are shown here in Figure 7.17. Note the setup used intentionally followed the basic premise of a triaxial setup described by Oakley [63] and later published the same year as standard practice 96-1A by the International Electrotechnical Commission (IEC).

The plot shown in Figure 7.17a was made by successively halving a length of triaxial cable. The outer braid shield of the triaxial cable was used to apply the

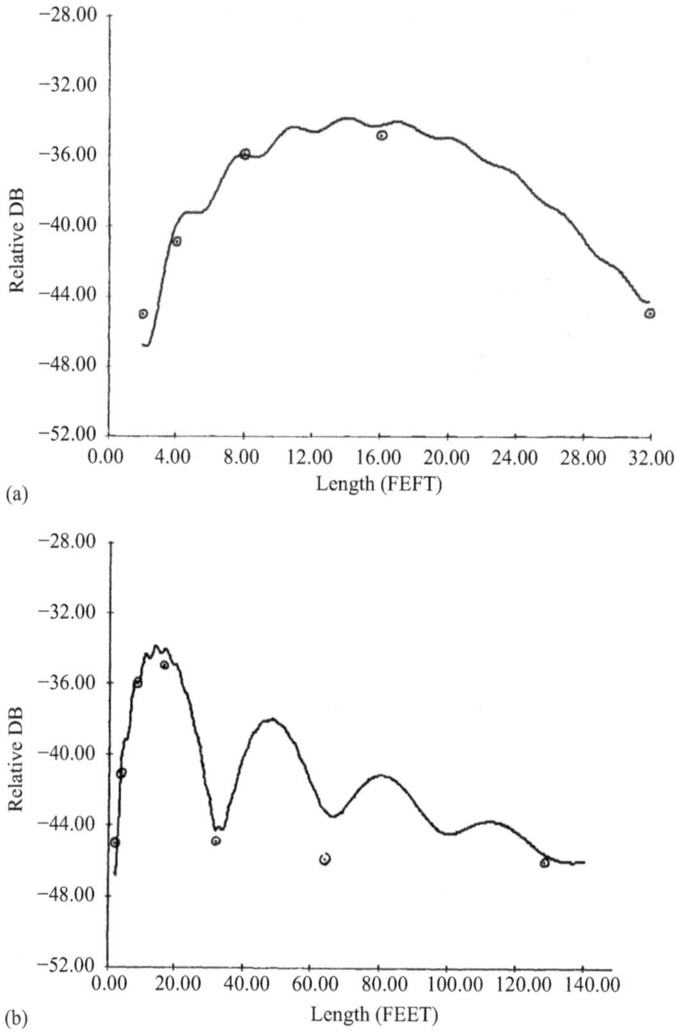

Figure 7.17 Relative SE data as a function of cable length

external stimulus, and the inner braid shield was used as the test sample. Circled points are experimental data; the curve was plotted using equations from Merewether and Ezell [60]. The measurement frequency was 100 MHz.

The plot shown in Figure 7.17b was made by extending the length of the cable segment used in Figure 7.17a to 140 feet. As before, the outer braid shield of the triaxial cable was used to apply the external stimulus, and the inner braid shield was used as the test sample, and circled points are experimental data. The curve was plotted using equations from Merewether and Ezell [60]. The measurement frequency was 100 MHz.

It is clear the data illustrated in Figure 7.17 do not resemble the straight-line behaviour implied by (7.28). According to the authors [62], this is because the straight-line behaviour is an artefact of the setup and can only be achieved by meeting several conditions, listed here:

1. The drive (external stimulus) circuit must be terminated in its characteristic impedance.
2. The velocities of propagation in both the external and internal circuits must be the same.
3. The attenuation of the external and internal circuits must be equal to zero.

These are rather special conditions that are rarely, if ever, met in "real" life. Based on these findings, (7.28) represents a "worst-case" condition, with the downside that (7.28) will likely yield a very conservative design, driving penalties in cost, weight, and volume. Note the quarter-wavelength at 100 MHz is roughly 2.5 feet, suggesting that (7.28) fails to accurately predict SE behaviour past one-half wavelength.

7.5.3 Brief discussion of cable surface transfer admittance

The normal component of any electric field will easily be able to penetrate directly through shield apertures. This admittance is a capacitive characteristic, and unlike the transfer impedance, transfer admittance is affected by the environment within which the cable is situated and is not simply the reciprocal of the transfer impedance. Surface cable transfer admittance is mathematically defined as

$$Y_T \triangleq -\frac{1}{V_0}\frac{\partial I}{\partial z}\bigg|_{V=0} \tag{7.29}$$

where

$\partial I/\partial z =$ the RMS current per unit length flowing into the internal conductor(s)
$V_0 =$ the voltage between the inner conductor(s) and the external common reference
$V =$ the voltage applied to the inner conductor(s)

The transfer admittance yields the short-circuit current induced in the internal conductor(s) with the internal conductor(s) shorted to the shield. Imperfect shields, such as braided shields, have one or more apertures through which the external environment is able to capacitively couple to the internal conductor(s). Solid shields also have a transfer admittance, but it is negligible compared to imperfect shields. The determination of this parameter requires the characteristics of the external shield circuit to be considered, as well as the internal circuit. The transfer admittance of an aperture in a shield having a concentric return was derived by Kaden [64]. This derivation developed an electric dipole moment to represent electrically small apertures and the charge induced on internal conductor (s) by this dipole moment. The analysis yielded a mutual capacitance between the aperture and the internal conductor(s), typically represented by

$$Y_T = j\omega C_{12} \tag{7.30}$$

Note this discussion should be ringing some bells regarding earlier discussion of crosstalk.

The International Electrotechnical Commission (IEC) introduced the capacitive coupling impedance parameter, Z_f, a fictitious transfer impedance producing the same effects as Y_T. The parameter appears in

$$Y_T = j\omega C_{12} = \frac{Z_f}{Z_{c1} Z_{c2}} \tag{7.31}$$

where Z_{c1} and Z_{c2} are the characteristic impedances of the internal circuit and the external circuit, respectively.

The parameter may also be expressed as

$$Z_f = \frac{\omega \nu \mu p \sqrt{\varepsilon_{r0}/\varepsilon_{ri}}}{(\pi D_m)^2} \tag{7.32}$$

where

ν = the number of openings per metre in the shield
p = the electric polarizability of the braid apertures
D_m = the mean diameter of the shield
ε_{r0} = the relative dielectric of the external cable insulation
ε_{ri} = the relative dielectric of the internal cable insulation

The numeric value of electric polarizability of several geometric shapes is available in references [65,66]. Transfer admittance is generally not a parameter of interest if the shield circuit is low impedance (the shield is connected to the common reference at both ends, for example), or if (7.31) is small, or if the source and load impedances of the internal circuit are both small. For a shield terminated at a single end, the transfer admittance may well be of much greater importance. The interested reader, or the engineer who wishes to accumulate more detailed information about cable surface transfer admittance, is encouraged to seek out additional information regarding this topic in the literature.

7.5.4 Discussion of typical cable surface transfer impedance behaviour

Typical values of surface transfer impedance for various cable types are shown in Figure 7.18 [59,67], followed by some brief commentary regarding each of the curves presented.

The first curve of interest is that for the single braid cable design, much like a common coaxial cable such as M17/28C-RG058. Specific details of the braid design are lacking, but the optical coverage for the measured cable was identified

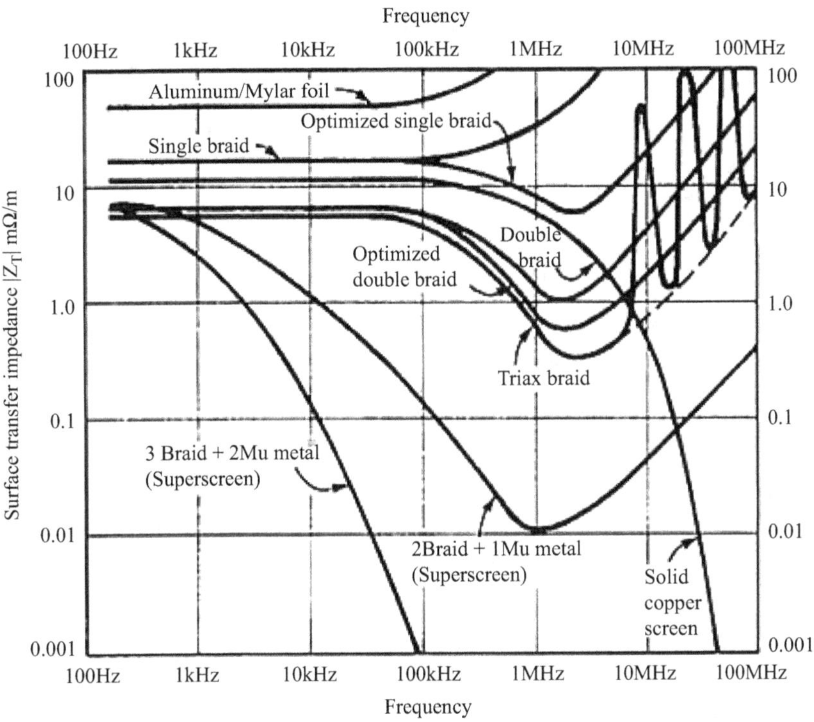

Figure 7.18 Comparison of transfer impedance of various cable shield designs

as being approximately 94%. Below roughly 100 kHz, the curve is flat and is dominated by diffusion through the shield material. Above roughly 100 kHz, the curve exhibits an increasing trend, as expected, now dominated by the inductive portions of the transfer impedance.

The second curve of interest is that for a cable having a double layer of aluminium and Mylar foil wrapped about the cable pair. Foil shields may be applied longitudinally, a so-called cigarette wrap, or they may be applied spirally, a so-called tape wrap. Foils used for this purpose are typically quite thin, bringing two separate concerns to the table. The thinness of the conductive foil layer will exhibit a comparatively high DC resistance, and the foil will tend to provide low mechanical integrity. As a means of combatting the low DC resistance, an uninsulated "drain" is typically run along the length of the cable in intimate contact with the surface of the conductive aluminium. Regardless of the type of wrap, this style of shield has a seam that generally spirals around the length of the cable. The presence of this seam, or gap, between the inner layer and outer layer of the wrapped foil has been found to have a profoundly negative impact on the transfer impedance, introducing as much as a 70 dB change [refer to (7.23) in this chapter]. Various schemes have been used to mitigate the impact of the seam, as illustrated in Figure 7.19.

Figure 7.19 Various common foil cable shield wrap styles

The shaded region surrounding the two conductors represents the conductive side of the aluminium wrap. The large black dot represents the "drain" wire. The configuration on the far right is the preferred design. The low-frequency transfer impedance of this cable shield design is much higher than that of the braid, reflecting the much higher DC resistance/diffusion behaviour of the foil. The onset and behaviour of the upward trend in the transfer impedance of this shield design occur because of the manner in which the foil is wrapped, which allows significant coupling through the seam. Although it is not known for certain what the cable shield design is for this cable, it is assumed the wrap in this case is similar to either the first or second of the two styles shown in the middle of Figure 7.17, the second style being considered a "worst case" approach. In any event, the performance of the cable design for which the data are shown is approximately 9 times worse than that of the braid design. Note that if the design shown on the right-hand side of Figure 7.18 is used, the high-frequency performance of the cable may be expected to be much better than the data illustrated in Figure 7.17 imply.

The third curve of interest is that labelled as "Solid Copper Screen". This cable design had a semi-rigid copper sheath with 100% optical coverage, and its behaviour reflects the dominance of skin depth effects and the process of diffusion, particularly at the point at which the skin depth becomes less than the sheath thickness. This is the basis for a solid shield as originally investigated by Schelkunoff.

The next curves of interest include those labelled as "Double Braid", "Optimized Single Braid", and "Optimized Double Braid". The term "optimized" refers essentially to the use of the weaving approach to intentionally minimize any openings in the braid structure. In the case of double braids, the two shield layers were not insulated from each other, resulting in simply thicker shields having smaller, more tortuous apertures and consequently lower DC resistance. These curves illustrate the improvement of transfer impedance behaviour, approaching that of a solid shield, as a function of shield thickness and shield weave "tightness".

Next up is the curve labelled "Triax Braid". This cable also has a double braid-layer construction, but in this case the two layers are separated by a dielectric, forming a second transmission line, the first being the transmission line formed between the inner shield and the enclosed conductor(s). The resonances that appear as frequency approaches 10 MHz were reported as artefacts of the measurement setup; the correct (or at least the expected) behaviour is shown by the dashed line. That said, it is important to note that shields in this configuration require termination into a characteristic impedance. Failing this, the risk of experiencing this sort of resonant behaviour is heightened. The double braid configuration previously discussed does not exhibit this behaviour since the two shields in that configuration were substantially in ohmic contact over the length of the cable. Note the arrow pointing to "Double Braid" does not clearly point at the "Double Braid" curve and should be flipped over to point at the curve immediately beneath the arrow.

The remaining two cables, labelled as "Superscreen" cables, had 2 or 3 braid shield layers with 1 or 2 interstitial layers of so-called "mu" metal. "mu" metal is a magnetically soft ferromagnetic material that exhibits an increasingly higher permeability as frequency falls below 100 kHz. The notable downward trend of the transfer impedance of these "Superscreened" shield configurations results from the combination of the shield configuration, enhanced by the very high permeability magnitudes of the "mu" metals. The term "Superscreen" refers to a cable design that employs multiple shield layers, often including solid components such as foil, or in the case of the remaining two cables, "mu" metal.

Note the data illustrated in Figure 7.17 are, with the sole exception of the aluminium and Mylar foil-wrapped configuration, for coaxial-style cable configurations. Nonetheless, these data are strongly representative of the behaviour of all manner of cable configurations that employ shielding.

7.6 Options for termination of shields left open at one end

Several schemes have been put forward in the past to somehow configure a shield circuit that is only connected to the common reference at one end so that it works well at both lower and higher frequencies while still maintaining the apparent elimination of a "ground loop". These schemes suggest the realization that a cable shield terminated at a single end is relatively useless above the resonant frequency of the shield circuit.

7.6.1 Use a series capacitor

An often-heard suggestion is to intentionally incorporate a capacitor in series with the shield circuit, based on the concept that the shield is open at one end at low frequencies and exhibits a decreasing impedance to the common reference as frequency increases. This approach can provide an effective and simple solution assuming external and internal frequencies remain low enough, but the applications wherein this is true have significant constraints. If this approach is selected, the aspects of the capacitor selected for this function as discussed in the following material should be carefully considered:

> All capacitors have both an equivalent series resistance (ESR) and an Equivalent Series Inductance (ESL). In a simplified model, the combination of the capacitance and the ESL forms a series resonant circuit that

exhibits a minimum series impedance at the primary self-resonant frequency (SRF). Multiple equivalent circuit models have been introduced by manufacturers and researchers. Some examples of these models are illustrated in Figure 7.20. Likely the version most well-known is that identified as (b) in the Figure.

1. ESR and ESL values are not well controlled and will vary from one component to another, as well as a function of temperature. This variation will cause the capacitance and the SRF to shift as a function of temperature. Figure 7.21 illustrates typical variation of capacitance as a function of temperature.
2. Below the SRF, the impedance of the capacitor falls by 20 dB per decade of frequency. Above the SRF, capacitors tend to behave as inductors and the impedance increases by 20 dB per decade. The magnitude of the series impedance at resonance is dominated by the resistance of the ESR. Figure 7.22 illustrates typical variation of capacitor impedance and phase as a function of frequency.

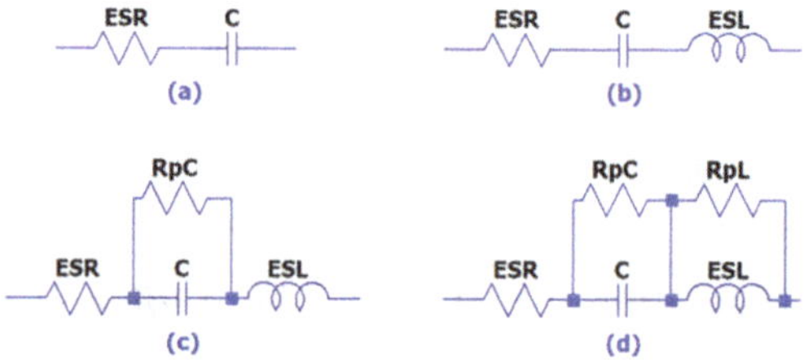

Figure 7.20 Common capacitor models

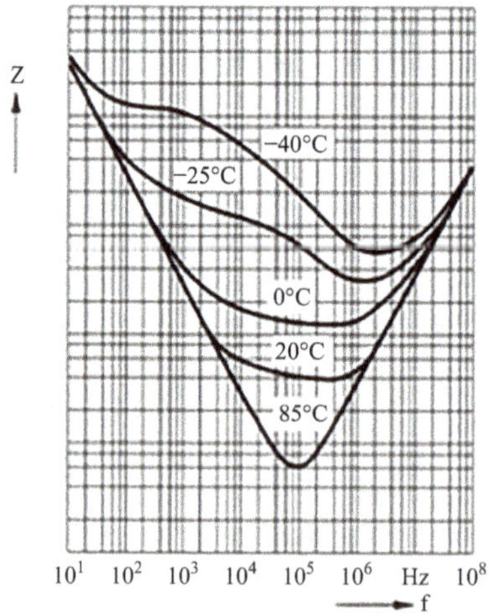

Figure 7.21 Typical capacitor impedance as a function of temperature and frequency [68]

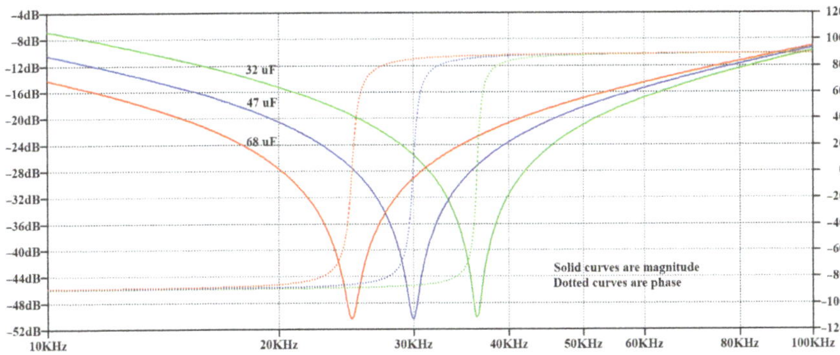

Figure 7.22 Plots of typical capacitor impedance and phase versus frequency

3. The larger the capacitance, the larger the parasitic inductance and the lower its resonant frequency.
4. The phase of the capacitor's impedance will invert at the resonant frequency.
5. Above the SRF, interference noise voltages will begin to appear on the shield connection to the common reference via the capacitor and act to negate any shielding action. To avoid this behaviour, any capacitor selected for the purpose described should exhibit an SRF well above the highest expected frequency in any possible interference signals.

Clearly this sort of solution introduces the potential for uncontrollable variations over manufacturing characteristics, operational temperature range, application operational frequency range, and the nearly inevitable exposure to an uncontrolled RFI environment. Can it work once? Sure. Can the impedance performance be reasonably expected to stay within a small and predictable range thousands of times in a manufacturing environment? Most likely not. Is it possible to levy constraints on when the performance does remain within a predictable range? Sure, if the customer is amenable to the constraints to be levied.

7.6.2 Use a series resistor in parallel with a capacitor

Over the past quarter century, several vendors and groups have published recommendations for how a shielded cable connected to a universal serial bus (USB) device should be terminated.

One vendor's recommendations include: "Connect the SHIELD connection to GND through a resistor. This helps *isolate* it and reduces EMI and RFI emissions. Keep this resistor close to the USB connector. Some experimentation may be necessary to obtain the correct value" [69]. A different vendor's recommendations include: "Tying the shield directly to ground would create a direct path from the ground plane to the shield, turning the USB cable into an antenna. To limit the USB cable antenna effect, it is recommended to connect the shield and ground through an RC filter" (see Figure 7.23) [70].

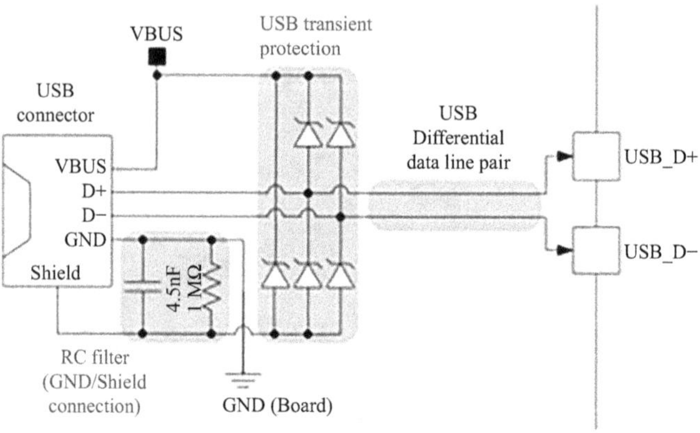

Figure 7.23 Vendor recommended shield and "ground" plane RC filter connection

Yet another vendor's recommendations include: "Full speed devices use a shielded cable which requires that the connector shell be *tied to the ground plane*. It is important to note that a ground plane does not behave like an equipotential surface at high frequencies. The location of the connector shell's termination to the ground plane is critical. The connection needs to be made to the quietest area of the ground plane to prevent noise from the ground plane from coupling to the shield" [71]. How does one simply and easily determine the "quietest area" of a printed circuit board ground plane?

Contrary to the foregoing, we have the USB Type-C Cable and Connector Specification, Release 2.0, published in August 2019 by the USB 3.0 Promoter Group: Apple Inc., Hewlett-Packard Inc., Intel Corporation, Microsoft Corporation, Renesas, STMicroelectronics, and Texas Instruments [72]. This document contains several requirements for USB cable shield termination, to wit:

Section 3.2.1 (6):

The USB Type-C receptacle shall provide an EMC ground return path through one of the following options:

1. a system of specific points of contact on the receptacle outer shell (e.g., spring fingers or spring fingers and formed solid bumps),
2. internal EMC pads, or
3. a combination of both points of contact on the receptacle outer shell and internal EMC pads.

Section 3.2.1 (11):

The receptacle shell shall be connected to the PCB ground plane.

Section 3.10.1:

The cable's external braid should be physically connected to the plug's metal shell as close to 360° as possible to control EMC. Without appropriate shielding termination, even a perfect cable with zero intra-pair skew may not meet EMC requirements. Copper tape may be needed to shield off the braid termination area.

Note the recommendation that "Shield and GND grounds shall be connected within the USB Type-C and USB 2.0 Standard-A plugs on both ends of the cable assembly" appears no less than 8 separate times in the document.

In an earlier USB specification, Universal Serial Bus Specification Revision 2.0, dtd 27 Apr 2000 (the signatories included Compaq, Hewlett-Packard, Intel, Lucent, Microsoft, NEC, and Philips) [73], a wrapped aluminized Mylar with ¼ turn overlap was cited, inside a 65% coverage braided shield. Note that although the USB 2.0 Specification raised the data transfer rate to 480 Mbps, bus limitations reduced this to 280 Mbps. Contrast this with the highest USB 3 speed of 20 Gbps. This document also contains several requirements for USB cable shield termination, to wit:

Section 6.6:

A high-/full-speed cable consists of one 28 to 20 AWG non-twisted power pair and one 28 AWG twisted data pair with an aluminium metallized polyester inner shield, 28 AWG stranded tinned copper drain wire, > 65% tinned copper wire interwoven (braided) outer shield, and PVC outer jacket. Low-speed cable consists of one 28 to 20 AWG non-twisted power pair and one 28 AWG data pair (a twist is recommended) with an aluminium metallized polyester inner shield, 28 AWG stranded tinned copper drain wire, and PVC outer jacket. A > 65% tinned copper wire interwoven (braided) outer shield is recommended.

Section 6.8:

The shield must be terminated to the connector plug for completed assemblies. The shield and chassis are bonded together. The user selected a grounding scheme for USB devices, and cables must be consistent with accepted industry practices and regulatory agency standards for safety and EMI/ESD/RFI.

Unfortunately, this guidance appears to fall short of specifying how the shield is to be terminated, although the implication is that it is permitted to use the drain wire for any cable shield terminations.

All of the foregoing seems somewhat contradictory. Don't connect the shield to the ground plane. Connect the shield to the ground plane. The guidance seems to be unclear on whether the shield should be connected to ground at all ends or only one. Nor does it provide any guidance as to which end or ends such a connection should be made, or what sort of termination, 360° into a backshell, or use a 28 AWG drain wire to form a pigtail connection. No wonder confusion reigns over this shielding topic.

Kimmel and Gerke nailed it back in March of 2010:

> The common USB port has four wires, two for differential signal and two for voltage/return. These wires are surrounded by the shield, which is terminated to the connector shell at each end. The USB cable shield is not bulletproof, but it seems to work pretty well when implemented per spec. But we are

> finding a lot of cases where the cable shield is not properly terminated, or not terminated at all. In some cases, there is simply nowhere to terminate the shield. In other cases, the designer is under the impression that single-point grounds are appropriate. In still other cases, the designer does not understand what constitutes an acceptable ground connection. [74]

Despite comments such as these from the EMC community and the guidance from USB specifications dating back some 25 years, more than one vendor has recommended a parallel RC circuit, consisting typically of a 1.0-MΩ resistor and a 4.7-nF capacitor, and another recommending a 1.0-MΩ resistor and a 1.0-nF capacitor, to be interconnected between the shield, if any, and the "ground plane" of the circuit board.

The reasoning behind these recommendations seems to be the capacitor passes AC while blocking DC, and the resistor provides a bleed path for accumulated electrical charge on the electronics. The capacitor thus directs high-frequency interference generated within the circuit into the chassis ground (strongly suggesting there actually is a conductive enclosure), ostensibly to prevent it from creating antenna radiation, and suppresses transient common-mode voltage differences between high-frequency interference sources and the circuit. The resistor serves to discharge accumulated electrostatic charge, thereby safeguarding the circuit from ESD damage, particularly during ESD testing. If only a capacitor connects the signal ground and chassis ground, the signal ground remains floating and can gradually accumulate electrostatic charge.

One wonders why internally generated high-frequency signals from USB circuit boards do not simply radiate from the circuit board itself or inside the enclosure, instead preferring a conductive path through a capacitor to avoid any current flow on the cable shield. Equally perplexing is why properly designed and functional electronics mounted inside a surrounding metallic enclosure would not simply connect the circuit ground to the inside of the enclosure at a single point, resolving the need for either the capacitor or the resistor. As for the coupling of ESD effects, properly designed and terminated shielding combined with an electromagnetically sealed enclosure and appropriate filtering and diode blocking will mitigate external ESD far more effectively than a single large resistance. In fact, a number of vendors offer precisely such filtering and diode blocking circuits in conveniently sized and rated packages that do not require large spaces on the circuit board.

7.6.3 Use a pigtail

A pigtail shield termination always introduces an undesirable inductance into the shield circuit. As frequency increases, an increasingly higher amount of voltage is dropped across the pigtail(s), as illustrated in Figure 7.24.

Figure 7.24 is identical to Figure 77, taken from [75], and shows the increasing losses associated with the use of various lengths of pigtail terminations. Clearly the longer the pigtail, the greater the voltage is at the lower the frequency. Resonant behaviour for this 0.6 m cable appears to commence around 100 MHz. The wavelength at 100 MHz is 3 m, so a quarter-wavelength is 0.75 m. At 0.6 m, the

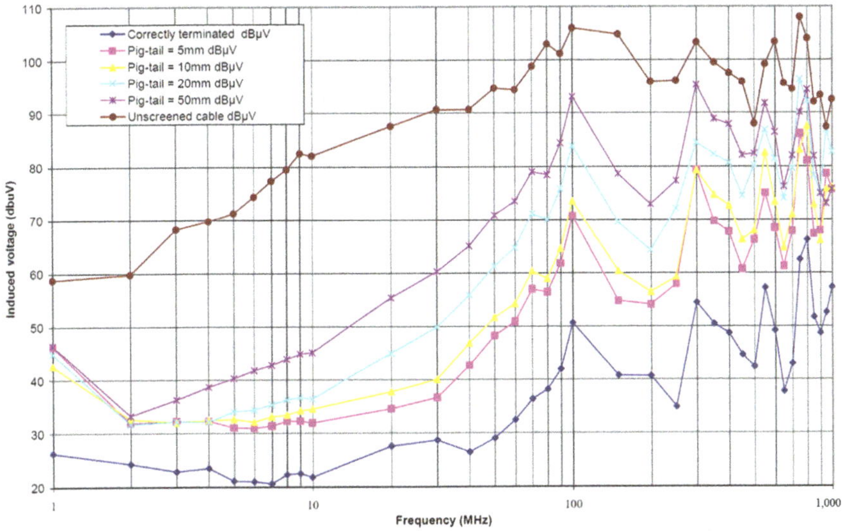

Figure 7.24 Induced voltage on a 0-6 m cable from a 1-V/m electric field for varying lengths of pigtail

cable length is near this quarter-wavelength magnitude, explaining the onset of resonant behaviour at this frequency.

Note the pigtail-terminated shielding behaviour at the lower end of the frequency range is 30+ dB below the unterminated shield. Below 2 MHz, the voltage is dominated by the decreasing resistance of the pigtail. Above 2 MHz, inductance begins to dominate the impedance of the pigtail. The margin at the lower end of the frequency range quickly deteriorates as frequency increases. Above the full-wave resonance at 500 MHz, introduction of the pigtail termination yields induced voltages on the enclosed conductor(s) comparable to those observed with no shield at all. The performance of the "correctly terminated" (i.e., connected to the common reference at both ends using a 360° peripheral termination) shield is always 10 dB or more below that of the shortest pigtail.

These data strongly suggest that use of a pigtail for the termination of a cable shield may be entirely acceptable, as long as the length of the cable is less than one quarter of the operational wavelength of both the shielded circuit and the operational external environment, and the pigtail is as short as possible for the configuration. Use of pigtail termination is not likely to be a good decision for any application much above 50 MHz.

7.7 General guidelines for the selection of cable shielding

As might be obvious from Figure 7.18, cable shields are available in several types. The most common types include metallic braid, foil (generally some form of metallized Mylar), spiral (or served), tape wound, and/or some other combination of two or more separate types (so-called Superscreen (alternating layers of braid and foil or mu-metal) is part of this category). Each type of shield brings both good and bad to the table, so as with most engineering challenges, it boils down to a trade-off. The following material is taken from [76].

7.7.1 Metallic braid shields

Metallic braid shields are usually formed from bare, tinned, or silver-plated copper strands. Steel and other metals can be used for physical protection. In a manufacturing environment, they are typically woven over a conductor or conductors in the same manner as a textile braid. It is also possible to pull a larger braid over a smaller diameter cable, but this is generally not feasible at the manufacturing level. The SE of metallic braid shields is generally proportional to the amount of braid coverage over the wire, measured in percent optical coverage, with 75–85% coverage being standard. Military standards allow for optical coverage equal to or greater than 90%; 100% optical coverage with a braided shield is technically unobtainable. Braided shields provide mechanical strength as well as flexibility and durability. Greater braid optical coverage requires more time and cost to produce. Braided shields are most effective at low frequencies.

Key factors to consider when using metallic braid shields include:

Greater flex life and flexibility
Increased mechanical strength
Offers physical protection
Most effective at lower frequencies
Braids can be combined with foil for maximum performance
Typically 40% to 96% optical coverage
Can be difficult to terminate, but appropriate backshell designs are readily available
Can be slow and costly to produce
Typically more costly than foil

7.7.2 Foil shields

Foil shields can provide 100% shield coverage, depending on the seam design, and typically comprise an aluminium and Mylar composite tape that may be configured in a "cigarette" wrap, or it may be tape-wrapped spirally around the enclosed conductor(s). Foil shields are typically lightweight, inexpensive, and easy to apply in a manufacturing environment. Foil shields are normally terminated with a drain wire in continuous longitudinal contact with the electrically conductive part of the shield. The drain wire provides easy termination but forces use of a pigtail connection. Foil shields, especially in multi-pair constructions, can be colour-coded. Foil provides good flexibility but typically exhibits limited flex life.

Key factors to consider when using foil shields include:

Thin and lightweight
Good flexibility, mediocre flex life
Can reach 100% coverage
Easy to produce
Less costly than braids
Drain wire is easy to terminate, but can restrict the operational frequency range
Foil shields can be combined with braid shields for maximum performance
Little to no mechanical strength
Little physical protection

7.7.3 Spiral (or served) shields

Spiral (or served) shields are usually formed from bare, tinned, or silver-plated copper strands. Steel and other metals can be used for physical protection. The main advantage to a spiral or serve shield is that they are more flexible than braid shields and provide for easy termination. These shields are constructed by spirally wrapping the shield material in one direction around the conductors. A reverse spiral shield is formed by wrapping shielding material in both directions on separate layers with no interweaving of wires.

Spiral or served shields can obtain shield coverage of 95%. This type of shield tends to behave like a solenoid and generally exhibits a large mutual coupling, including a large transfer admittance, between the shielding layer(s) and the enclosed conductor(s), caused by the continuous longitudinal gaps between each successive wrap of wire. For these reasons, this type of shield is most often recommended only for low-frequency audio applications.

Key factors to consider when using spiral (served) shields include:

May behave like a solenoid (longitudinal magnetic field on cable axis)
Inexpensive to produce
Good flexibility
Easy to terminate
Can achieve up to 95% optical coverage
Can be delicate (generally recommended for sound stage usage only)
Limited frequency coverage – audio use only

7.7.4 Tape shields

Tape shields are formed from a variety of materials; the most common are various widths of thin strips of copper, aluminium, and bronze. Tape shields are usually spirally wrapped around the conductors that are being shielded. A drain wire may be used to facilitate termination but introduces the inductance associated with the pigtail thus formed. Copper and aluminium are usually combined with a backing such as Mylar or polyester, much like foil shields. Tape shields can reach 100% coverage. Properties are similar to foil shields.

Key factors to consider when using tape shields include:

May exhibit solenoidal-like behaviour similar to that of a spiral (served) shield
Performance generally degrades as frequency increases
Thin and lightweight
Good flexibility, mediocre flex life
Can reach 100% coverage
Easy to produce
Less costly than braids
Easy to terminate (may require a drain wire)
Tape shields can be combined with braid shields for maximum performance
Low mechanical strength
Little physical protection

7.7.5 Combination shields

Combination Shields are formed of two or more shields combined in the same cable. The most common are a braid shield over a foil shield or a braid shield over a braid shield. Multi-layer "Superscreen" shields fall into this category.

Key factors to consider when using combination shields include:

Maximum shield effectiveness
Good physical protection
Easy to terminate
Foil layers with enhanced optical coverage
Moderate flexibility, with good flex life
Low DC resistance
Good mechanical strength if metal braids are used
Can be costly to produce

References

[1] M.D. Heerema, *HP11949B Designing for Electromagnetic Compatibility*, Hewlett Packard, 1997.
[2] J.P. Mills, *Electromagnetic Interference Reduction in Electronic Systems*, PTR Prentice Hall, Englewood Cliffs, NJ, 1993.
[3] H. Ott, *Electromagnetic Compatibility Engineering*, John Wiley & Sons, Hoboken, NJ, 2009.
[4] C.R. Paul, R.C. Scully, and M.A. Steffka, *Introduction to Electromagnetic Compatibility*, 3rd ed., John Wiley & Sons, Hoboken, NJ, 2023.
[5] A.B. Tsaliovich, *Electromagnetic Shielding Handbook for Wired and Wireless EMC Applications*, Kluwer Academic Publishers, Norwell, MA, 1999.
[6] E.F. Vance, *Coupling to Shielded Cables*, Robert E. Krieger Publishing Co, Malabar, FL, 1987.
[7] J.J. Goedbloed, *Electromagnetic Compatibility*, Prentice Hall International (UK) Ltd, Hemel Hempstead, Hertfordshire, 1992.

[8] M. Coenen and J.J. Goedbloed, *Electromagnetic Compatibility*, MYbusinessmedia BV-Deventer, 2010.

[9] E.A. Laport, *Radio Antenna Engineering*, McGraw-Hill, New York, 1952.

[10] *Interference Prevention Techniques Handbook*, Boeing Airplane Company, Document No. D2-7482, *Rev.*, Nov 1961.

[11] D.R.J. White and M. Mardiguian, *Electromagnetic Shielding*, Vol 3 of A Handbook Series on Electromagnetic Interference and Compatibility, Interference Control Technologies, Inc., Gainesville, VA, 1988, p 1.12.

[12] S.A. Schelkunoff, "The Electromagnetic Theory of Coaxial Transmission Lines and Cylindrical Shells," *The Bell System Technical Journal*, vol. 13, no. 4, pp. 532–579, 1934.

[13] E.F. Vance, "Shielding Effectiveness of Braided Shields," Note 172, EMP Interaction Notes, Air Force Weapons *Laboratory, Kirtland AFB, NM*, Apr 1974.

[14] E.F. Vance, "Shielding Effectiveness of Braided Shields," *IEEE Transactions on Electromagnetic Compatibility*, vol. EMC-17, pp. 71–77, 1975.

[15] T. Waldron and K. Armstrong, "Bonding Cable Shields at Both Ends to Reduce Noise," EMC Standards, 9 Bracken View, Brocton, Stafford ST17 0TF, UK. https://www.emcstandards.co.uk/files/bonding_cable_shields_at_both_ends_to_reduce_noise_for_emcj_may_02_pdf_version_4_may_02.pdf

[16] J. Freeman, "Techniques to Enhance Op Amp Signal Integrity in Low-Level Sensor Applications," December 2008 (Four-part series originally online at https://www.planetanalog.com/ … figures 7.9 through 7.12 and some of the related information contained in this section are taken from author's personal notes and no longer appear to be available online]

[17] A. Martin, "Things You May Not Have Heard about Shielding," *PowerPoint Presentation, Alliance for Telecommunications Industry (ATIS) Protection Engineering Group (PEG) Meeting*, 25–27 Mar, 2014, Littleton, CO.

[18] R. Tiedemann, "Current Flow in Coaxial Braided Cable Shields," *IEEE Transactions on Electromagnetic Compatibility*, vol. 45, pp. 71–77, 1975. DOI:10.1109/TEMC.2003.815562.

[19] S. Sali, "An Improved Model for the Transfer Impedance Calculations of Braided Coaxial Cables," *IEEE Transactions on Electromagnetic Compatibility*, vol. 33, no. 2, pp. 139–143, 1991. DOI:10.1109/15.78351.

[20] H. Schippers, J. Verpoorte, and R. Otin, "Electromagnetic Analysis of Metal Braids," 2011 *IEEE International Symposium on Electromagnetic Compatibility (EMC Europe 2011) York, UK, Sep 26–30, 2011.* https//ieeexplore.ieee.org/document/6078613

[21] J. Verpoorte, H. Schippers, and J.H.G.J. Lansink Rotgerink, "Advanced Models for the Transfer Impedance of Metal Braids in Cable Harnesses," *2018 IEEE International Symposium on Electromagnetic Compatibility*, May 14–18, 2018, Suntec City, Singapore. DOI:10.1109/ISEMC.2018.8393764.

[22] R.J. Peel, "Simple Relations Between Shielding Effectiveness and Transfer Impedance/Admittance for Cables," *1988 IEEE International Symposium on Electromagnetic Compatibility*, Aug 2–4, 1988, Seattle, WA. DOI:10.1109/ISEMC.1988.1401.

[23] R.J. Peel, "Relationship of Shielding Effectiveness to Transfer Impedance/Admittance for Cable Emission and Susceptibility," *1984 IEEE International Symposium on Electromagnetic Compatibility*, Apr 24–26, 1984, San Antonio, TX. DOI:10.1109/ISEMC.1984.7570981.

[24] R.J. Peel, "Relationship of Shielding Effectiveness to Transfer Impedance/Admittance for Shielded Cables," *1982 IEEE International Symposium on Electromagnetic Compatibility*, Sep. 8–10, 1982, Santa Clara, CA. DOI:10.1109/ISEMC.1982.7567764.

[25] A.R. Martin and S.E. Emert, "The Shielding Effectiveness of Long Cables, II: LT and GTR," *IEEE Transactions on Electromagnetic Compatibility*, vol. EMC-22, no. 4, pp. 269–275, 1980. DOI:10.1109/TEMC.1980.303866.

[26] A.R. Martin and M. Mendenhall, "A Fast, Accurate, and Sensitive Method for Measuring Surface Transfer Impedance," *IEEE Transactions on Electromagnetic Compatibility*, vol. EMC-26, no. 2, pp. 66–70, 1984. DOI:10.1109/TEMC.1984.304188.

[27] A. Frobel, "Cable Shielding to Minimize Electromagnetic Interference." https://www.eeeic.org/proc/papers/55.pdf

[28] C. Brainard and K. Smith, "Coaxial Cable-The Neglected Link," *QST Magazine*, Apr 1981, pp. 28–31. https://www.arrl.org/files/file/Technology/tis/info/pdf/8104028.pdf

[29] A.R. Coates, A. Gavrilakis, M.M. Al-Asadi, A.P. Duffy, K.G. Hodge, A.J. Willis, *et al.*, "*EMC Behaviour of Cable Screens*." https://www.researchgate.net/profile/Alistair-Duffy/publication/253152574_EMC_behaviour_of_cable_screens/links/00b7d529dad20d08ae000000/EMC-behaviour-of-cable-screens.pdf?origin=scientificContributions

[30] J.W.E. Jones, "Achieving Compatibility in Inter-Unit Wiring," *6th Symposium and Technical Exhibition on Electromagnetic Compatibility*, Zurich, Switzerland, 1985, pp. 139–144. DOI:10.23919/EMC.1985.10798863.

[31] E.F. Vance, W. Graf, and J.E. Nanevicz, "Unification of Electromagnetic Specifications and Standards, Part 1-Evaluation of Existing Practices." https://apps.dtic.mil/sti/tr/pdf/ADA121700.pdf

[32] W. Graf, J.M. Hamm, and E.F. Vance, "Unification of Electromagnetic Specifications and Standards, Part II-Recommendations for Revisions of Existing Practices." https://apps.dtic.mil/sti/tr/pdf/ADA144175.pdf

[33] Engineering Staff, "The Role of Cables and Connectors in Control of EMI," *EMC Technology*, vol. 1, no. 3, 1982, pp. 16–26.

[34] M. Mardiguian, "Transfer Impedances of Balanced Shielded Cables," *EMC Technology*, vol. 1, no. 3, pp. 54–58.

[35] E.F. Vance, "Cable Grounding for the Control of EMI," *EMC Technology*, vol. 2, no. 1, pp. 54–58, 1983.

[36] B. Demoulin and P. Degauque, "Effect of Cable Grounding on Shielding Performance," *EMC Technology*, vol. 3, no. 4, pp. 65–73, 1984.

[37] MIL-STD-461, Revision A, Aug 1968.

[38] MIL-STD-461, Revision B, Apr 1980.

[39] MIL-STD-461, Revision G, Dec 2015.

[40] R.D. Leach, "Electronic Systems Failures and Anomalies Attributed to Electromagnetic Interference," *NASA Contractor Report 1374*, Jul. 1995. https://cecas.clemson.edu/cvel/pdf/nasa-rp1374.pdf

[41] R.M. Lawton, "System Guidelines for EMC Safety-Critical Circuits: Design, Selection, and Margin Demonstration," *NASA Contractor Report 4759*, Oct 1996. https://ntrs.nasa.gov/api/citations/19970005154/downloads/19970005154.pdf

[42] M. Faraday, "On Static Electrical Induction Action," *London, Edinburgh, and Dublin Philosophical Magazine and Journal of Science*, series 3, vol. 22, no. 144, pp 200–204, 1843.

[43] J.C. Maxwell, *An Elementary Treatise on Electricity*, chapter 2, Oxford University Press, Oxford, England, pp. 16–21.

[44] J.C. Maxwell, *A Treatise on Electricity and Magnetism, Vol. 1*, Oxford at the Clarendon Press, London, UK, 1873, pp. 32–38.

[45] *Electromagnetic Compatibility Design Guide for Avionics and Related Ground Support Equipment, NAVAIR AD1115*, 3rd ed. 1998.

[46] *Design Principles and Practices for Controlling Hazards of Electromagnetic Radiation to Ordnance, NAVSEA OD 30393*, 2nd ed., 2001.

[47] Bell Telephone Laboratories, *Physical Design of Electronic Systems*, Vol. 1 of 4, Prentice-Hall Englewood Cliffs, NJ, 1970.

[48] R.T. Milton, *Design Handbook for Electromagnetic Compatibility*.

[49] H.W. Ott, "Ground-A Path for Current Flow," *1979 IEEE International Symposium on Electromagnetic Compatibility*, Oct 9–11, 1979, San Diego, CA. DOI:10.1109/ISEMC.1979.7568808.

[50] E.B. Joffe and K. Lock, *Grounds for Grounding*, John Wiley & Sons, Hoboken, NJ, 2010, section 7.3.6.1.

[51] S.A. Schelkunoff, "The Electromagnetic Theory of Coaxial Transmission Lines and Cylindrical Shields," *The Bell System Technical Journal*, vol. 13, no. 4, pp. 532–579, 1934. DOI:10.1002/j.1538-7305.1934.tb00679.x.

[52] W. Xiaoling, L. Chao, D. Hao, and W. Lixin, "An Improved Model for the Transfer Impedance Calculations of Braided Coaxial Cables," *Proceedings of The 7th International Power Electronics and Motion Control Conference*, Harbin, China, Jun. 2–5, 2012. DOI:10.1109/IPEMC.2012.6258960.

[53] T. Kley, "Optimized Single-Braided Cable Shields," *IEEE Transactions on Electromagnetic Compatibility*, vol. 35, no. 1, pp. 1–9, 1993. DOI:10.1109/15.249390.

[54] E.F. Vance, "Interaction Note 172 Shielding Effectiveness of Braided Wire Shields," *Stanford Research Institute*, Menlo Park, CA Apr 1974. https://summa.unm.edu/notes/In/0172.pdf

[55] M. Tyni, "The Transfer Impedance of Coaxial Cables with Braided Outer Conductors," *FV. Nauk, Inst. Telekomum Akust. Politech Wroclaw, Ser. Konfi*, 1975, pp. 410–419. DOI:10.23919/EMC.1976.10826424.

[56] H. Schippers and J. Verpoorte, "Hole Inductance in Braided Cable Shields," *2015 IEEE International Symposium on Electromagnetic Compatibility*, Aug 16–22, 2015, Dresden, Germany. DOI:10.1109/ISEMC.2015.7256132.

[57] A. Mushtaq, S. Frei, K. Siebert, and J. Barenfanger, "Analysis of Shielding Effectiveness of HV Cable and Connector Systems used for Electric Vehicles," *2013 International Symposium on Electromagnetic Compatibility (EMC Europe 2013)* Brugge, Belgium, Sep. 2–6, 2013. https//ieeexplore.ieee.org/document/6653235

[58] F.M. Tesche, M.V. Ianoz, and T. Karlsson, *EMC Analysis Methods and Computational Models*, John Wiley & Sons, Hoboken, NJ, 1997.

[59] A.R. Martin, "An Introduction to Surface Transfer Impedance," *EMC Technology*, vol. 1, no. 3, pp. 44–52, 1982.

[60] D.E. Merewether and T.F. Ezell, "The Effect of Mutual Inductance and Mutual Capacitance on the Transient Response of Braided-Shield Coaxial Cables," *IEEE Transactions on Electromagnetic Compatibility*, vol. EMC-18, no. 1, pp. 15–20, 1976. DOI:10.1109/TEMC.1976.303450.

[61] L.O. Hoeft and J.S. Hofstra, "Measured Electromagnetic Shielding Performance of Commonly Used Cables and Connectors," *IEEE Transactions on Electromagnetic Compatibility*, vol. 30, no. 3, pp. 260–275, 1988. DOI:10.1109/15.3304.

[62] A.R. Martin and S.E. Emert, "Shielding Effectiveness of Long Cables," *1979 IEEE International Symposium on Electromagnetic Compatibility*, Oct 9–11, 1979, San Diego, CA. DOI:10.1109/ISEMC.1979.7568782.

[63] R.J. Oakley, "Surface Transfer Impedance Measurements – A Practical Aid to Communication Cable Shielding Design," *The 18th International Wire & Cable Symposium*, Atlantic City, New Jersey, Dec 1969. https://apps.dtic.mil/sti/tr/pdf/AD0698720.pdf

[64] H. Kaden, *Wirbelstrome und Schirmung in der Nachrichtentechnik*, Springer-Verlag, Berlin, Germany, 1959.

[65] E. Arvas and R.F. Harrington, "Computation of the Magnetic Polarizability of Conducting Disks and the Electric Poloarizability of Apertures," *IEEE Transactions on Antennas and Propagation*, vol. AP-31, no. 5, pp. 719–725, 1983. DOI:10.1109/TAP.1983.1143122.

[66] F.D. Meulenaere and J.V. Bladel, "Polarizability of Some Small Apertures," *IEEE Transactions on Antennas and Propagation*, vol. AP-25, no. 2, pp. 198–205, 1977. DOI:10.1109/TAP.1977.1141568.

[67] M. Green, "*Optimised and Superscreened Cables*," Report, Wire and Cable Division, Raychem LTD, Swindon, UK, 1977.

[68] Aluminum Electrolytic Capacitors General Technical Information, TDK August 2022, available at url: https://www.tdk-electronics.tdk.com/download/530704/e724fb43668a157bc547c65b0cff75f8/pdf-generaltechnicalinformation.pdf

[69] Cypress Guide to a Successful EZ-USB® FX2LP™ Hardware Design AN15456. https://www.infineon.com/dgdl/Infineon-AN15456_Guide_to_a_Successful_EZ-USB_FX2LP_Hardware_Design-ApplicationNotes-v08_00-EN.pdf?fileId=8ac78c8c7cdc391c017d07356bb458bf&utm_source=cypress&utm_medium=referral&utm_campaign=202110_globe_en_all_integration-files&redirId=File_1_3_942

[70] Microchip IEEE 802.15.4 Sub-GHz System in Package Datasheet. https://onlinedocs.microchip.com/oxy/GUID-4B3D771E-5647-4191-AD45-C897B0D23071-en-US-3/GUID-35928B7E-EC10-4750-B54B-2FB7CAAF1128.html
[71] Intel EMI Design Guidelines for USB Components (difficult to find; copy available upon request).
[72] USB Type-C Cable and Connector Specification, Release 2.0, August 2019
[73] Universal Serial Bus Specification Revision 2.0, Apr. 27, 2000.
[74] News Report (ENR), Mar.-Apr. 2010, in the section entitled EMC Notebook, EMI and USB, W.D. Kimmel and D.D. Gerke.
[75] Ministry of Defence, Defence Standard 59-411, Part 5, Issue 2, Mar. 31, 2014, p. 119.
[76] Popular Wire & Cable Shielding Types. https://www.iewc.com/resources/technical-guide/popular-shielding-types#:~:text=Metallic%20Braid%20Shields%20are%20usually,manner%20as%20a%20textile%20braid

Index

www.ingramcontent.com/pod-product-compliance
Lightning Source LLC
LaVergne TN
LVHW011849200926
848484LV00002B/429